Distributed Intelligent Circuits and Systems

World Scientific Series on Future Computing Paradigms and Applications

Series Editor: Brij B Gupta (*Asia University, Taichung, Taiwan*)

Published:

Vol. 3 *Distributed Intelligent Circuits and Systems*
by Balwinder Raj, Brij B Gupta, Shalendra Singh and Monirujjaman Khan

Vol. 2 *SDN and NFV: A New Dimension to Virtualization*
by Brij B Gupta, Amrita Dahiya and Elhadj Benkhelifa

Vol. 1 *Big Data Management and Analytic*
by Brij B Gupta and Mamta

World Scientific Series on
Future Computing Paradigms
and Applications – **Vol. 3**

Distributed Intelligent Circuits and Systems

Editors

Balwinder Raj
Dr B R Ambedkar National Institute of Technology, India

Brij B. Gupta
Asia University, Taiwan

Shalendra Singh
Pranveer Singh Institute of Technology, India

Monirujjaman Khan
North South University, Bangladesh

World Scientific

NEW JERSEY • LONDON • SINGAPORE • BEIJING • SHANGHAI • HONG KONG • TAIPEI • CHENNAI • TOKYO

Published by

World Scientific Publishing Co. Pte. Ltd.
5 Toh Tuck Link, Singapore 596224
USA office: 27 Warren Street, Suite 401-402, Hackensack, NJ 07601
UK office: 57 Shelton Street, Covent Garden, London WC2H 9HE

Library of Congress Cataloging-in-Publication Data
Names: Raj, Balwinder, 1980– editor. | Gupta, Brij, 1982– editor. |
Singh, Shalendra, editor. | Khan, Monirujjaman, editor.
Title: Distributed intelligent circuits and systems / [edited by] Balwinder Raj, Dr B R Ambedkar National Institute of Technology, India, Brij B Gupta, Asia University, Taiwan, Shalendra Singh, Pranveer Singh Institute of Technology, India, Monirujjaman Khan, North South University, Bangladesh.
Description: New Jersey : World Scientific, [2024] | Series: World Scientific series on future computing paradigms and applications ; vol. 3 | Includes bibliographical references and index.
Identifiers: LCCN 2023039687 | ISBN 9789811279522 (hardcover) |
ISBN 9789811279539 (ebook for institutions) | ISBN 9789811279546 (ebook for individuals)
Subjects: LCSH: Autonomous distributed systems. | Distributed artificial intelligence--Industrial applications. | Wireless sensor networks. | Artificial intelligence. | Machine learning. | Robotics.
Classification: LCC TA168 .D57 2024 | DDC 621.39--dc23/eng/20231226
LC record available at https://lccn.loc.gov/2023039687

British Library Cataloguing-in-Publication Data
A catalogue record for this book is available from the British Library.

For any available supplementary material, please visit
https://www.worldscientific.com/worldscibooks/10.1142/13505#t=suppl

Desk Editors: Sanjay Varadharajan/Amanda Yun

Typeset by Stallion Press
Email: enquiries@stallionpress.com

https://doi.org/10.1142/9789811279539_fmatter

Preface

The current trend in innovation is exploring the potential of natural and artificial intelligence for emerging applications in the form of hybrid distributed intelligent systems. The innovations directed toward developing new intelligent systems build upon existing algorithms as a base and employ smart technologies. The main objective of this book is to provide a complete insight into the recent advances in distributed intelligent circuits, systems, and their applications. This evolving era is witnessing the interfacing of artificial intelligence with machines and incorporating the intelligence factor into the decision-making capability of smart machines. Distributed intelligence would be the key enabler of innovation in machine-to-machine communication. The motive of this research-focused book is to provide relevant theoretical frameworks that include basic models, algorithms, circuit design, and the latest developments in the experimental aspects of the field of distributed intelligence systems for various industrial applications.

There are numerous challenges in the development of models for distributed intelligence systems for environmental monitoring, but these can be mitigated with artificial intelligence, machine learning, and deep learning. This book will help identify the areas where we encounter numerous challenges and apply diverse solutions for the development of advanced intelligent systems in environmental monitoring.

Chapter 1: The working approach is changing with the digitalization and advancement of technical tools, along with increased security risks. One of the most common and biggest threats that have been observed is attacks on communicational applications and networks. This threat is caused by denial of service (DoS) and distributed denial of service (DDoS). The behavior of the attacks performed through DoS and DDoS could be understood with an analysis of each step and attack process. A piece of brief information related to the attack through DoS or DDoS helps in understanding its effects, such as the development of countermeasures for the migration. Different types of networks, such as software-defined networks (SDN), virtual networks, and traditional networks, are classified in the model of DoS or DDoS. This chapter describes an analytical view of the different types of DoS or DDoS for network architecture. Mathematical modeling of the attacks through DoS or DDoS on different models, such as epidemic, hierarchical, analytical, and traffic, is described in this chapter.

Chapter 2: This chapter provides an overview of the message-oriented architectural pattern. The core components of this type of system are explained, and an overview of common messaging protocols is provided. Two binary messaging protocols (AMQP and MQTT) and a text-based (STOMP) one are analyzed. The exchange of information in these protocols, the message format they expect, and the message exchange order are described. The core components of the message-oriented architectural pattern are the message and the message broker. Message brokers provide implementations of messaging protocols. Five commonly used message brokers (Google Cloud Pub/Sub, Azure Service Bus, RabbitMQ, NATS, and Apache Kafka) and their components are introduced and explained, and common use cases for each broker are listed. A message broker contains a topic and/or queue as a means of exchanging messages between message producers (publishers) and message consumers. The way messages are exchanged between these components is described. An example of a message-oriented architecture is implemented. The central component of the architecture is a Spring Boot application which communicates with the Apache Kafka broker and additional external components. Finally, two use cases of such a component are explained: a chat application in the React framework and a high-throughput message producer in the Golang programming language.

Chapter 3: Computer-aided diagnostic (CAD) systems are gaining popularity because they enhance the analysis and comprehension of clinical images, as opposed to the preliminary identification of different illnesses by physicians and medical experts. Likewise, bone fractures are a frequent occurrence owing to strain, accidents, and osteoporosis. In addition, bone is a solid structure that supports the whole body. Therefore, bone fractures are regarded as one of the most pressing issues in the later stages of life. The detection of bone fractures using computer vision is gaining importance in CAD systems since it has the potential to minimize the doctor's burden by eliminating simple cases. In this chapter, the identification of tibia fracture types is analyzed utilizing several techniques of image processing. The purposes of this investigation are to determine whether or not there is a fracture in the lower leg bone from the X-ray scans and to characterize the kind of fracture that is present (tibia). There are three primary stages involved in the process of developing a system to detect tibia fractures. In order to classify the different kinds of fractures and determine the locations of fractures, preprocessing, feature extraction, and classification are used. During preprocessing, the sharpening technique known as unsharp masking is used on the image in order to enhance it and draw attention to the image's edges. After the image has been sharpened, the Harris corner detection approach is used on it to generate corner feature points for further feature extraction. Then, in order to detect fractures or non-fractures and, if detected, classify the fracture, two different classification techniques are used. A simple decision tree is employed for the classification of fractures, while the *K*-nearest neighbor algorithm is used for the classification of the fracture type. In this work, many forms of fractures, including normal, transverse, oblique, and comminute, are described. In addition, fracture locations may be determined by examining the Harris corner points that have been formed. Last but not least, the outputs of the system are evaluated utilizing two different performance assessment approaches. The first component is a performance evaluation that may be conducted in fracture or non-fracture (normal) conditions, and it can provide one of four possible results: TP, TN, FP, or FN. The second goal is to use the Kappa approach in order to conduct an analysis of the accuracy of each fracture type under various error settings. The MATLAB programming environment was used throughout the building of the system, which offers a vast array of image processing capabilities. The technique has an 82% classification accuracy for fracture types. This image processing approach is extraordinarily useful for several applications, including biomedical, security, satellite imaging, personal imaging, and medicine. Fracture detection employs image processing techniques such as picture augmentation, feature segmentation, and feature excitation.

This study employs an astute edge detection technique for segmentation. Using a shrewd technique, accurate information can be extracted from the bone image. This research's primary objective is to identify bone fractures using image processing in MATLAB. Preprocessing, segmentation, and fracture identification are the phases that comprise the proposed system. In the feature excitation stage, the Hough transform method is used for the line detection of inline images. The system's primary function is the extraction of features. Multiple experiments demonstrate that the suggested approach is very precise and effective.

Chapter 4: In light of the rapid urbanization throughout the world, it is crucial to combine intelligent and environmentally friendly city design in terms of the carriage of passengers and commodities. This will play an important role in shaping how future civilizations handle transportation for decades to come. Intelligent transportation is a concept which is developed over time, contemporaneous with the increase in interest in the so-called "Intelligent Cities," in line with the United Nations' emphasis on the sustainability elements of development over the long term. However, other factors, such as promoting active and inclusive transportation, supporting the use of alternative ecologically friendly fuels, and engaging with individuals, must also be taken into account. Some of the crucial components of smart mobility include a decline in traffic congestion and new route optimizations with a reduced ecological footprint. Internet-of-Everything, artificial intelligence, blockchain, and Big Data technologies can become key points of initiation as well as core supports that encourage the growth of innovation, which will revolutionize the present conceptual framework for cities and their inhabitants. These technologies will also play a key role in the advancement of revolutionary new ideas that will alter every aspect of our lives. The development of current urban areas into intelligent cities will bring about a slew of modifications over the next several decades at least. Some of the impacts on services and applications include optimization of logistics, heavy traffic improvement, and driverless vehicles, and these changes will incorporate a variety of shifts. This study offers a comprehensive analysis of the most emerging advancements and solutions that have been offered within the realm of intelligent connectivity and the supporting advancements in science that are necessary for its success. The purpose of this chapter is to convey a breakdown of how intelligent mobility contributes to the emergence of smart cities by defining specific aspects and primary merits of using intelligent transportation within the context of a smart city ecosystem. This chapter discusses a variety of other potential advantages and constraints associated with smart mobility. In addition, the primary services and applications that are anticipated to emerge in the next few years in the realm of smart connectivity are probed, along with future potential trends and the scope of the industry.

Chapter 5: The evolution of communication technologies with high-frequency radio-frequency (RF) devices increased the demand for compact and efficient designs. Micro-electromechanical systems (MEMS) technology revolutionized microwave and RF applications because of its ability to be engineered into miniaturized devices that are highly linear and power efficient. It is more challenging to perform numerical analysis and optimization of such complex MEMS devices. Electromagnetic (EM) simulation-based optimization software using different methodologies employing coupled domains and time domain analysis of MEMS devices requires repeated simulation, which makes it computationally expensive. The artificial neural network (ANN) model is an alternative to these conventional simulation-based design methodologies to expedite the design process. ANN models for RF and microwave modeling are known to be effective, precise, and flexible. ANN is capable of producing accurate results with less computational time than sophisticated EM models. An overview of various RF MEMS components and an introduction to ANN are provided in this chapter. In addition, this chapter presents the concept of modeling an RF MEMS shunt switch using ANN as a case study.

Chapter 6: Coronavirus disease 2019 (COVID-19), which causes severe acute respiratory syndrome (SARS-CoV-2), is a persistent global epidemic. The coronavirus was discovered for the first time on March 11, 2020, following an outbreak in the Chinese city of Wuhan in December 2019. This virus causes respiratory issues, fever, coughing, and shortness of breath. In the worst circumstances, infection might result in mortality due to pneumonia, severe acute respiratory syndrome, and other conditions. The second COVID-19 wave in India showed how ill-prepared the country's healthcare systems were to deal with the sudden rise in the number of daily positive cases. Infected patients were left gasping for air during the second wave of COVID-19 as a result of the oxygen and bed shortage issues, particularly in India. This chapter proposes a wireless, Internet of Things (IoT)-based healthcare system featuring an automatically operated Ambu bag that might be utilized to save patients' lives. The system can use NodeMCU with IoT technology to continually monitor a patient's oxygen saturation, pulse rate, and body and room temperatures using a DS18B20 temperature sensor from any location in the world.

Chapter 7: The management of increased demand for the utilization of available road infrastructure thanks to urbanization has attracted the interest of researchers in the domain of intelligent transportation systems (ITS). This helps the commuter travel in congestion-free, accident-free, and efficient road scenarios. Vehicular *ad hoc* networks (VANETs) form a vital element of the ITS. VANET is defined as a subcategory of mobile *ad hoc* networks (MANETs) with some predefined distinct properties. Unlike MANET, VANET has no constraint on the node's battery life, but there is a bandwidth constraint in this scenario. VANET supports two types of communication: either among vehicles or among vehicles and infrastructure. The key objectives of VANETs are to modify public safety and transportation efficiency in the highly dynamic and mobile scenario. This chapter provides detailed information regarding the VANETs and their characteristics, network layer-wise communication protocols, shortcomings, features, and applications, and finally the challenges and future scope.

Chapter 8: Generally, the current application areas in communication are completely dependent on the establishment and operations of (camera-based/non-camera-based) wireless sensor networks for cooperative communication (CC). CC, being an ideal sub-part of the upcoming 6G telecom infrastructure, ensures that optimum operations can be analyzed and evaluated based on the lifeline of the model of a network. It should not be very power-consuming (i.e., it must be energy efficient with better throughput), and its reliability for the transmission of data packets should be optimal. The majority of researchers working in sensor-based fields have shifted their attention to the application domains of 6G (preferably in the CC domain) with the aim of optimizing the energy, power, and life span of sensor-based networks while also taking reliability into consideration (camera-based or non-camera-based). In light of this, this chapter gives a broader analytical review of CC based on recent research, followed by a discussion on research gaps and performance issues in current wireless networks leading up to CC. Finally, the implementation of several clustering schemas for data transmission in MATLAB is described, and their performances are compared in terms of energy efficiency.

Chapter 9: Although research in autonomous vehicles had started in the early 20th century, due to a lack of efficient and highly powerful computational devices, they were only controlled by radio wave technologies, widely known as remotes, despite rapid developments in the related technologies. Then, in the 1950s, research and development activities picked up pace with the development of computers. With the introduction and advancement of artificial intelligence (AI) and computing hardware technologies, autonomous vehicles have become a primary field of research and development for researchers seeking to address various day-to-day problems faced by drivers, traffic controllers, and human transportation as a whole, such as traffic jams, road accidents, expensive infrastructure maintenance, and the engagement of human resources (drivers) in inefficient and repetitive work. Considering these factors, automobile and tech giants have identified a new market opportunity with the potential for huge profits. This has resulted in rapid research and development as companies strive to be the first to capture the market. As we progressed with new developments, various technical, social, and administrative problems and challenges started arising which needs to be resolved. This chapter contains a review and analysis of various studies carried out in this field and proposes alternative approaches wherever needed. The main technical challenge is to design and develop a superfast and energy-efficient AI model which must be capable of performing millions of calculations in microseconds, which would require a very powerful GPU and a highly efficient system architecture design. As infrastructure is designed and developed solely for human drivers everywhere, it becomes a challenge for the physical testing and deployment of autonomous vehicles. As such, there are either no clear administrative rules and regulations or no rules at all for such vehicles. Even if all these technical and administrative challenges of autonomous vehicles are overcome, social challenges may remain since the extent to which customers/passengers accept and adopt them will usher in completely new changes to conventional transportation.

Chapter 10: In the current scenario, technology scaling is being implemented in digital integrated circuits to improve their power, speed, size, and cost-effectiveness. On the other hand, scaling induces reductions in supply voltage and intrinsic transistor gain, because of which analog circuit designers face various design challenges. This paved the way for using a time domain approach to analog designs, which is based on the representation of signals in time domain parameters, such as time period, transition time, and time delays. These signals are processed using time domain signal processing implemented by time-mode circuits, such as voltage-to-time converters, time amplifiers, time registers, and time-to-digital converters (TDCs). This chapter incorporates various reported time-mode circuits, but emphasis is placed on TDCs. It covers the details of a wide range of TDC applications and performance metrics. Further, a classified review of various time interval measurement techniques used to design TDCs is given. The key design concepts of each technique are comprehensively discussed, along with their merits, demerits, salient features, and applications. It concludes with a timeline of the evolution of various TDC techniques, right from the 1960s until today, using a flow diagram, and provides a comprehensive bibliography.

Chapter 11: The cornerstone of the Indian economy is the agricultural sector. A large percentage of individuals around the world — more than 40% — have agriculture as their primary line of employment. The advancement of autonomous vehicles in agriculture has gained considerable spotlight in recent years. Soil quality is unaltered by robotization, which also significantly reduces manpower expenses and boosts overall productivity. The machinery described in this project is proficient at performing several farming operations at once. An environmentally safe source of energy is utilized by this equipment: Energy from the sun's radiation is accumulated and transformed into electrical energy, and a direct current motor then transforms this electrical power into mechanical power to rotate an anchor during a digging operation. For irrigation operations and seed sowing, respectively, we deploy seed hoppers and water reservoirs. The machine can change its course and maneuver over varied ground contours, excavate, sow seeds, and water the ground. It also decreases the expense of planting seeds and the manpower required.

https://doi.org/10.1142/9789811279539_fmatter

About the Editors

Balwinder Raj is an Associate Professor at the Dr. B. R. Ambedkar National Institute of Technology Jalandhar, India. He has more than 15 years of teaching and research experience. Dr. Raj has authored/co-authored 5 books, 12 book chapters, and more than 100 research papers in peer-reviewed international and national journals and conferences. He became a Member of the Institute of Electrical and Electronics Engineers (MIEEE) in 2006. His research areas of interest include classical/non-classical nanoscale semiconductor device modeling, nanoelectronics and their applications in hardware security, FPGA implementation, and the design of sensors and circuits, FinFET-based memory, low-power VLSI, and digital/analog VLSI.

Brij B. Gupta is the Director of the International Center for AI and Cyber Security Research and Innovations and a Distinguished Professor at the Department of Computer Science and Information Engineering (CSIE), Asia University, Taiwan. In more than 17 years of his professional experience, he has published over 500 papers in journals/conferences, including 35 books and 12 Patents with over 25,000 citations. He has received numerous national and international awards, including the Canadian Commonwealth Scholarship (2009), the Faculty Research

Fellowship Award (2017), the Visvesvaraya Young Faculty Research Fellowship Award from the Indian Ministry of Electronics and Information Technology, Government of India, the IEEE GCCE outstanding and Women in Engineering (WIE) paper awards, and National Institute of Technology Kurukshetra, India's Best Faculty Award (2018 and 2019), respectively. Prof. Gupta was selected for Clarivate's Web of Science Highly Cited Researchers in Computer Science (top 0.1% researchers in the world) consecutively in 2022 and 2023. He was also listed in Stanford University's ranking of the world's top 2% of scientists in 2020, 2021, 2022, and 2023.

Dr. Gupta is also a Visiting/Adjunct Professor with several universities worldwide and an IEEE Senior Member (2017). He was selected as the 2021 Distinguished Lecturer in IEEE Consumer Technology Society (CTSoc). Dr. Gupta is also serving as a Member-at-Large of IEEE Consumer Technology Society's Board of Governors (2022–2024). He is also the Editor-in-Chief of the *International Journal on Semantic Web and Information Systems, International Journal of Software Science and Computational Intelligence, Sustainable Technology and Entrepreneurship,* and *International Journal of Cloud Applications and Computing*, the Lead Editor of a Book Series with CRC and IET Press, and the Associate/Guest Editor of various journals and transactions. He also served as a member of the technical program committee of more than 150 international conferences. His research interests include information security, cyber physical systems, cloud computing, blockchain technologies, intrusion detection, AI, social media, and networking.

Mohammad Monirujjaman Khan is an Associate Professor in the Department of Electrical and Computer Engineering at North South University, Dhaka, Bangladesh. He received his BEng degree in electrical and electronic engineering from Queen Mary University of London (QMUL), UK, with first-class honors. Dr. Khan received the Draper's Company Undergraduate Prize in 2008 from QMUL for outstanding academic merit. He also completed his Ph.D. in electrical and electronic engineering at QMUL. His Ph.D. was funded by a full QMUL scholarship. After completing his Ph.D., Dr. Khan worked as a postdoctoral research assistant at the School of Electronic Engineering and Computer Science at QMUL under the IMPACT QM KTA Scheme 1 project funded by the Engineering and Physical Science Research Council (EPSRC).

Shailendra Singh is an Assistant Professor in the Department of Electronics and Communication Engineering at the Pranveer Singh Institute of Technology, Kanpur, India. He received his BE degree in electronics and communication engineering and his M.Tech. in microelectronics from Panjab University, Chandigarh, India, in 2011 and 2013, respectively. He worked as an Assistant Professor in the Department of Electronics and Communication Engineering at the Dr. B. R. Ambedkar National Institute of Technology, Jalandhar, India, from 2015 to 2017. He then successfully completed his doctoral research on vertical TFET devices for low-power applications within a period of 3.5 years. After completing his Ph.D., Dr. Shailendra Singh worked as a doctoral researcher from 2017 to 2020 at the Nanoelectronics Research Lab at National Institution of Technology Jalandhar, focusing on nanoscale device development. During this period, he published over 17 research papers in journals, eight of which were SCI-indexed. He has also authored four book chapters and several chapters/papers in the proceedings of national and international conferences. His research interests include the modeling and simulation of semiconductor devices, especially vertical TFETs for low-power applications.

https://doi.org/10.1142/9789811279539_fmatter

Acknowledgments

This book is based on research carried out by several authors. Many people have greatly contributed to this book on distributed intelligent circuits and systems. We are grateful to a number of friends and colleagues for encouraging us to write this book. We, as the editors, would like to acknowledge all of them for their valuable support and generous ideas in improving the quality of this book. With our feelings of gratitude, we would like to introduce them in turn. The first mention is to the authors and reviewers of all the chapters of this book. Without their outstanding expertise, constructive reviews, and devoted effort, this comprehensive book would be without substance. The second mention is to the staff of World Scientific Publishing for their constant encouragement, continuous assistance, and untiring support. Without their technical support, this book would not have been completed. The third mention is to the editors' families for being the source of continuous love, unconditional support, and prayers not only for this work but throughout our lives. Last but far from least, we express our heartfelt thanks to the Almighty for bestowing us the courage to face the complexities of life and see this work through to completion.

https://doi.org/10.1142/9789811279539_fmatter

Contents

https://doi.org/10.1142/9789811279539_0001

Chapter 1

An Analytical View of DoS or DDoS Attacks for Network Architecture

Sahilpreet Singh[*,†,§], Arjan Singh[‡,¶], and Vishal Goyal[†,‖]

[†]*Department of Computer Science, Punjabi University Patiala, Patiala, India*

[‡]*Department of Mathematics, Punjabi University Patiala, Patiala, India*

[§]*ersahilpreetsingh@gmail.com*

[¶]*arjanpu@gmail.com*

[‖]*vishal.pup@gmail.com*

Abstract

The working approach is changing with the digitalization and advancement of technical tools, along with increased security risks. One of the most common and biggest threats that have been observed is attacks on communicational applications and networks. This threat is caused by denial of service (DoS) or distributed denial of service (DDoS). The behavior of the attacks performed through DoS and DDoS could be

[*]Corresponding author.

understood with an analysis of each step and attack process. A piece of brief information related to the attack through DoS or DDoS helps in understanding its effects, such as the development of countermeasures for the migration. Different types of networks, such as software-defined networks (SDN), virtual networks, and traditional networks, are classified in the model of DoS or DDoS. This chapter describes an analytical view of the different types of DoS or DDoS for network architecture. Mathematical modeling of the attacks through DoS or DDoS on different models, such as epidemic, hierarchical, analytical, and traffic, is described in the chapter.

Keywords: DoS attacks, SDN, traditional network, machine learning, mathematical models

1.1. Introduction

A great deal of effort is required for the development and management of traditional networks since the plane required for forwarding and control lies on the same device. While performing modifications on the network administrators, it is most important to take care of large-scale networks. According to Sultana *et al.* [18], a deflection in the configuration of the network could lead to an increase in complexity as the connected device changes its configuration accordingly. The changes caused by raising various issues, including human errors, malfunctioning of networks, incompatibility among devices, and service outages, increase with the defect in configuration. At present, the emerging networking approach is software-defined networking (SDN), which is an advanced technology in which the control plane is separated from the routers or switches. The effectiveness of SDN could be understood by its approach that makes the network more flexible and easier to manage. The main concern of the current chapter is to analyze the development of an intrusion detection system (IDS) for the cloud of things through the application of machine learning (ML). The security issues occurring in the network are caused by denial of service (DoS) or distributed denial of service (DDoS) [1–5].

Over the years, different types of challenges, including security issues and connectivity, have been faced by communication networks. Security levels provided to the communication networks could be understood by categorizing them based on availability, confidentiality, and integrity. The security threats caused by the presence of DoS or

DDoS could be one of the most critical on the internet. The magnitude of security attacks, such as DoS or DDoS, was observed to be around 1.65 Tbps in 2018 [7]. Efficient techniques for migration are developed with mitigation techniques. The securities of traditional networks are affected by DoS, and the attacks could be classified into several categories, which help in understanding their dynamics. The attacks occurring on SDN through DDoS could be analyzed through the division into parts of the affected plane, such as the control, application, and data planes. The different types of DoS or DDoS are well explained across the architecture of networks in this chapter. It also provides detailed information related to the security issues of communication networks in different sections [6–10].

1.2. Classification of Networks Architectures for DoS or DDoS

DoS or DDoS attacks can affect the strength or security of the network and can be better understood by classifying them. The classification of DoS or DDoS could be done across the architecture of networks [11–16]. DoS or DDoS is divided into three categories: traditional network, virtual network, and SDN.

1.2.1. *Attacks of DoS or DDoS on Traditional Architectures*

Switch devices or separate routing are present within traditional networks and are effective in forwarding and controlling. Khan and Kim [30] stated that the architecture of traditional networks has different planes, including data and control. The security of traditional networks is primarily affected by DoS attacks that are embedded in devices of a single network [17–24]. A variety of brands were applied for routing devices, with interfaces and protocols provided for each device. The configuration languages for each device are unique in traditional networks. The architecture of traditional networks has the strength to provide information related to skills and high-level knowledge. The cost of security arises for this network due to its multiple features, such as complex and heterogeneous networks. The security attacks placed through DDoS on traditional networks affect the quality and services of connectivity (see Figure 1.1). Different vectors of attack could be applied to the traditional networks [25–32].

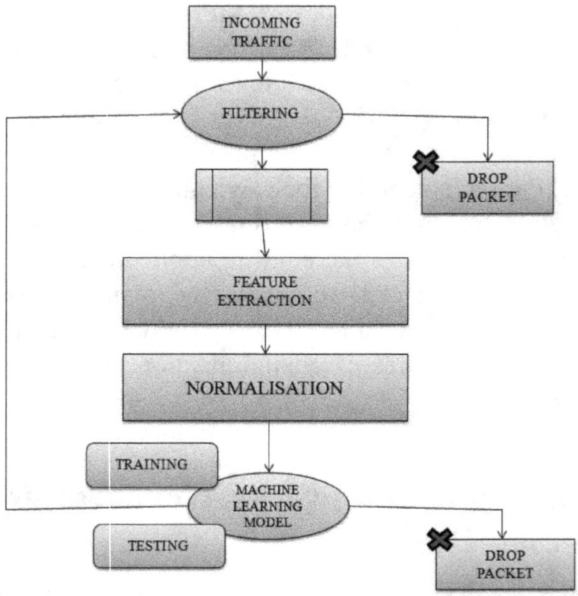

Figure 1.1. An ML-Based Framework for Detection of DOS Attacks on Network

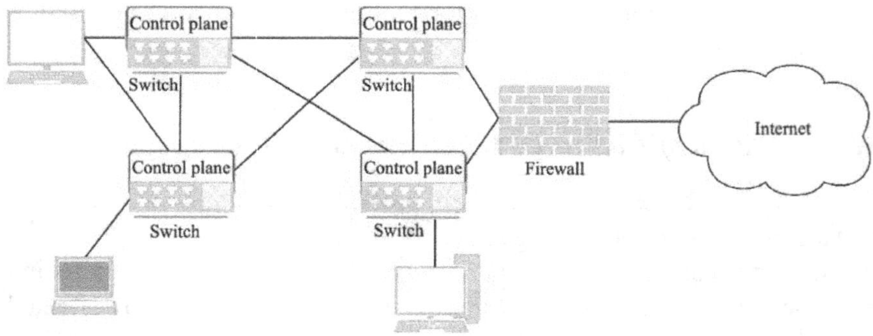

Figure 1.2. Architecture of Traditional Networks

Source: Miles *et al.* [11].

Figure 1.2 shows the architecture of traditional networks, in which different components, such as switches, control planes, and firewalls, are connected to the internet. The firewall present between the connection to the internet and the control plane helps in avoiding flooding or security attacks. Figure 1.3 shows the classification of DDoS attacks on traditional networks [33–39]. The security attacks on the networks, DoS or DDoS,

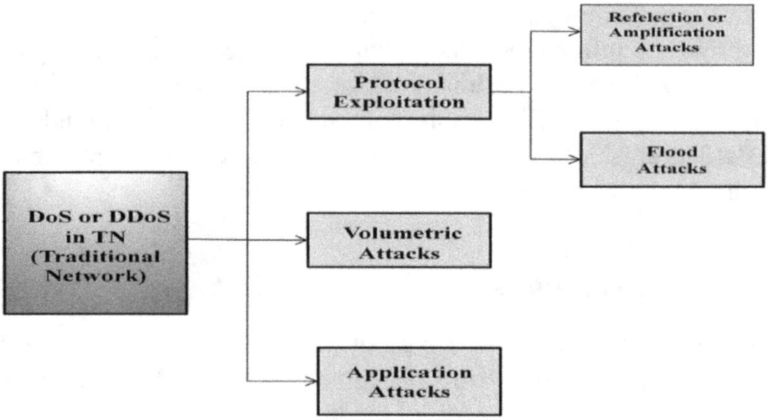

Figure 1.3. Classification of DDoS Attacks on the Traditional Networks

could be classified into three different parts: protocol exploitation, volumetric attacks, and application attacks. These attacks work as multiple vectors that could be launched one at a time (Asharf *et al.* [23]). Damage to the communication networks is targeted through multi-vectors.

1.2.2. *Protocol Exploitation*

The security threat faced by traditional attacks could be affected by protocol exploitation, which is also known as state exhaustion attacks (Khare *et al.* [4]). The state of connection could be completed with the building of network devices. The network devices are built with the protocols of networks and vulnerabilities that consume the stated tables.

1.2.3. *Volumetric Attacks*

The depletion of required network or hardware resources focuses on flooding with requests, along with traffic and the entire focus of attacks on the bandwidth are required. This type of attack could be identified as having reflection or amplification and flooding. As per Khater *et al.* [17], attacks that are in the form of reflection are effective against vulnerabilities of exploiting the spoof. The prime indicator of reflection attacks is the application of forged requests to the victims by creating traffic. Exhaust bandwidth is sought through flooding or volumetric attacks, and it applies large traffic volume or processing strength [40–45]. Small requests are made through amplification attacks, which are effective in creating larger responses from targeted victims. For example, the domain name system (DNS) completes the DNS table with the DNS server through small requests. The crashing of the server through the DNS table is possible due to the amplification attacks.

1.2.4. *Application Attacks*

Application layer attacks are a type of attack that occurs in traditional attacks through DoS or DDoS. The main target of application-layer attacks is to damage the protocols of the link or server, primarily affecting the SSL and HTTP (Yang *et al.* [21]). The effectiveness of the attacks could be analyzed from the fact that they could be as vulnerable in themselves as they could be for the securely coded [46–52]. The attack is one of the most dangerous among the DoS or DDoS attacks on traditional networks, as it could generate a huge amount of traffic in a short time. The identification of application-layer attacks is a bit difficult as their nature is quite stealthy and creates legitimate traffic for the victims.

1.2.5. *Attacks of DoS or DDoS on Virtual Networks*

Figure 1.4 shows the classification of SDNs subjected to DoS or DDoS into three parts. Attacks on the SDN could be divided into specific attacks

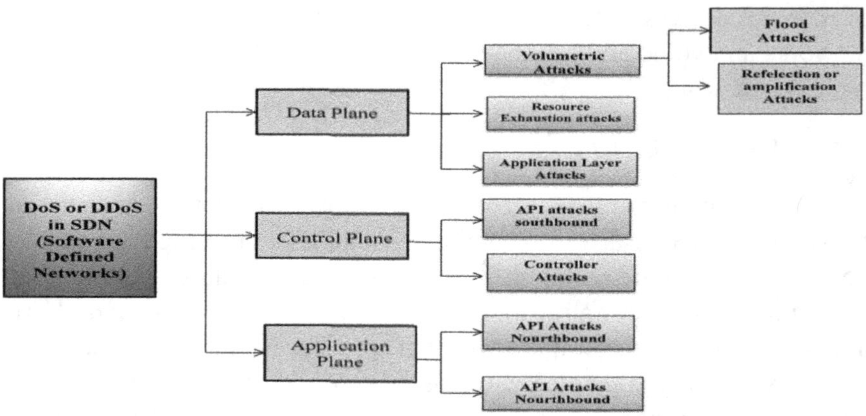

Figure 1.4. Classification of DDoS Attacks on the SDN

on the data plane, control plane, and application plane. Tabassum *et al.* [19] stated that data plane attacks are further divided into three parts: volumetric attacks, resource exhaustion attacks, and application-layer attacks [53–59]. The attacks on the SDN in the form of volumetric attacks could be divided into two different parts: flood attacks and reflection or amplification attacks. Controller attacks and API attacks southbound are two specific parts of the control plane attacks. The development of visualization of networks provides a virtual environment with resources of networking, physical computing, and storage. Network function virtualization (NFV) is effective in the development of capabilities and new features. NFV is emerging as a technology for the non-virtualized environment that effectively comprises specific vendors of hardware and software. The provision of network services is completed with the introduction of NFV and different approaches to NFV technology. The technologies of NFV have different features that could be understood through the following points:

- Decoupling of software from the hardware of the connected systems is possible with NFV, which is effective in providing dynamic operation and flexibility for deployed networks.
- The implementation of NFV conceives the functions of the network in the form of the object of software that executes the infrastructure of NFV (Khan *et al.*, 2019).
- A virtual environment performs NFV with the application of hardware devices that is effective for different purposes. The main targets of NFV are the reduction of costs, increasing the efficiency of operations, and optimizing the management of the network.

Figure 1.5 shows the architecture of NFV, which was proposed by the European Telecommunications Standards Institute (ETSI). As per Elrawy *et al.* [26], the domain of NFV could be divided into three specific categories: management and orchestration of NFV, NFVI, and virtualized network functions (VNFs). The functional blocks of NFV include the NFVI, business or operational support system, service, EM, virtualization layer, and virtual resources. Reference points could be changed while the blocks interact with each other [60–65]. Changes in the infrastructure of networking, such as modifications in management, deployment, and design, bring security challenges to NFV and SDN. The system is allowed to share its physical elements, including network resources, CPUs, and

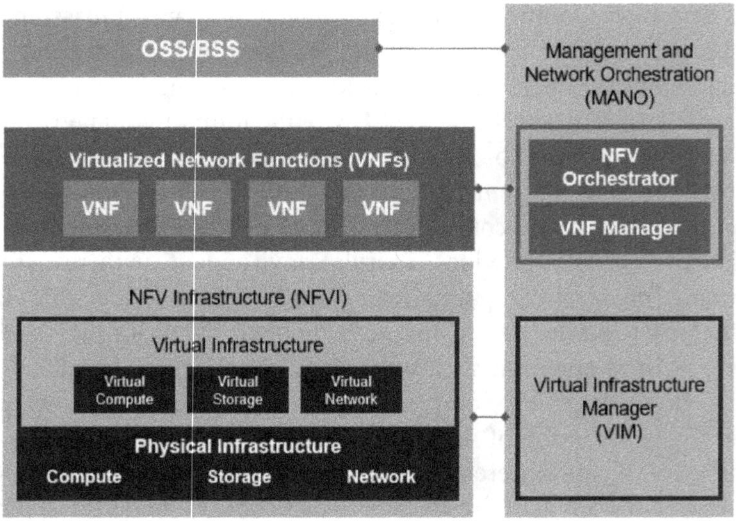

Figure 1.5. Architecture of NFV

Source: Papavassiliou [12].

memory units or storage, with the hosted virtual machines while performing virtualization. The main targets of the security attacks performed through DoS or DDoS are the hypervisor and connected virtual systems. The virtual machines could manage to address the vulnerabilities. Security attacks, such as DoS or DDoS, target the controller of SDN and deploy devices for the virtual network [66–73].

1.2.6. *SDN and NFV*

SDN gets services through the different elements of NFV, such as the VMM or hypervisor for a network controller. SDN is recognized as the base or brain of the connected system that is effective in provisioning service for VMM (Lima *et al.* [9]). NFV and SDN share specific frameworks, yet they provide support for each other. The components of a resource controller for a single infrastructure are mainly the controllers of networks and VMM. The functions and features of specific areas could be analyzed using the performance of SDN and NFV. The interaction of SDN and NFV could be identified with the proper integration of networks. Shared topics are required to be agreed upon with the concepts of NFV and SDN, as they share common information over security [74–79].

1.2.7. *Attacks of DoS or DDoS on SDN*

The categorization of the plane into three different parts, namely the data plane, control plane, and application plane, is possible with the decoupling of SDN, as shown in Figure 1.6. Northbound and southbound are effective in carrying communication among the planes of an application programming interface (API). APIs introduce the concept of programmability into SDN for DoS or DDoS attacks. Ullah and Mahmoud [20] stated that management and configuration of the network are possible with the application of open and standard programming languages. OpenFlow is recognized as one of the most common uses of APIs, with the interface of southbound. Communication is established in the SDN following the protocols of OpenFlow and using TCP between the controllers and forwarding devices. The work of the controller is performed by the control plane, and forwarding devices form the data plane in the SDN. The control plane is able to install the policies of the network within the data plane of overflow. On the basis of flow entries, networking devices could modify, drop, or forward, while matching the entry of configuration [80–87].

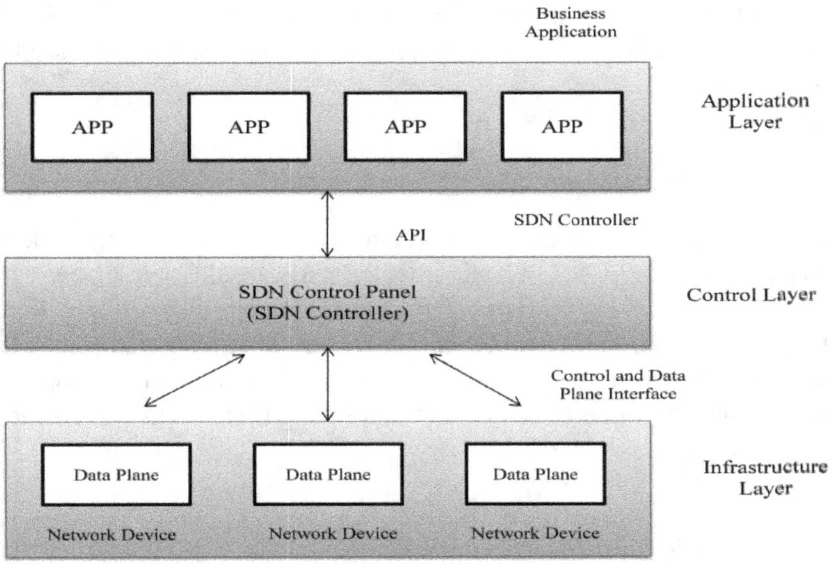

Figure 1.6. Architecture of SDN

Source: Barka *et al.* [24].

OpenFlow is one of the most common interfaces of northbound and southbound used in the API of SDN. The interfaces could be divided into different categories depending on the purpose of development that supports specific types. The selection of APIs varies with the specification of applied software in the application layer (Gonzalez-Cuautle *et al.* [27]). The application plane is crucial for multiple purposes, such as high-level interface delivery, network management, and operations. The translation of commands into the language of operation is possible through the application layer, which forwards the commands to the data plane. Various types of applications, including IDS, firewalls, mitigation tools for DDoS, and intrusion prevention systems (IPSs), could be deployed within this layer. The brain of SDN is the controller for traditional networks for the centralized architecture of DDoS attacks. The attacks on SDN through DoS or DDoS are classified into three parts: control-plane, data-plane, and application-plane attacks.

1.2.8. *Control Plane Attacks*

The management of networks is performed within the control plane that targets attacks through DDoS. The depletion of resource controllers is affected by the attacks of the control plane on SDN (Kim *et al.* [8]). The attacks of the control plane are effective in exhausting the interface bandwidth with southbound communication blocked for the control and data planes [88–91].

1.2.9. *Data Plane Attacks*

Attacks performed through the data plane mainly target routing devices, hosts, and applications. The main affected parts are the data plane, with volumetric attacks in the form of reflection or amplification and flood attacks. According to Dhaliwal *et al.* [25], targeted attacks on the application layer affects the exhaustion of resources or operations of the data plane. Attacks through the data plane have a significant effect on all traditional networks.

1.2.10. *Application Plane Attacks*

Installed applications are effective for the layer of northbound targeted attacks through DDoS for the operating SDN. The exhaustion of API is a target for application-layer attacks. Residing applications are applicable for residing for the application layer.

1.3. Attacks of DoS or DDoS and Its Mathematical Modeling

The security of networks is primarily attacked by the tools of DoS or DDoS, and for their identification, different strategies have been adopted. The limited work on modeling attacks increases the complexity of their recognition (Sharma *et al.* [66]). The strength of the defense mechanism could be evaluated with a proper understanding of the depth of the threat. The improvement in the solution for the mechanism results in proper detection and migration of specific types of attacks. In the phase of designing the networks, the testing of a model could be done for robustness with proper emulation and simulation. Most of the models are designed for traditional networks for different attacks that target either virtual networks or SDNs. With the detection of the network, the purpose of applying the methodology could be identified for its classification. The used model could either be migrated or not with the classification, as shown in Figure 1.7. The classification of used models helps in identifying the applied parameters and their gap of advantage.

1.3.1. *Attacks of DoS or DDoS on the Network and Modeling*

The presented literature provides information related to the different attacks on specific networks through DoS or DDoS. The model presented for the security attacks on traditional networks could be analyzed with mathematical modeling. The following section describes the effectiveness of mathematical modeling for traditional networks. Khraisat *et al.* [6] stated that different types of models are available for the development of traditional networks, and these models are based on traffic, hierarchy, analysis, and epidemic [91–93].

1.3.2. *Epidemic Models*

The identification of epidemic models could be analyzed as a category of analytical models. The difference in the epidemic and analytical models is both are speared due to the amount of targeted devices and its implication.

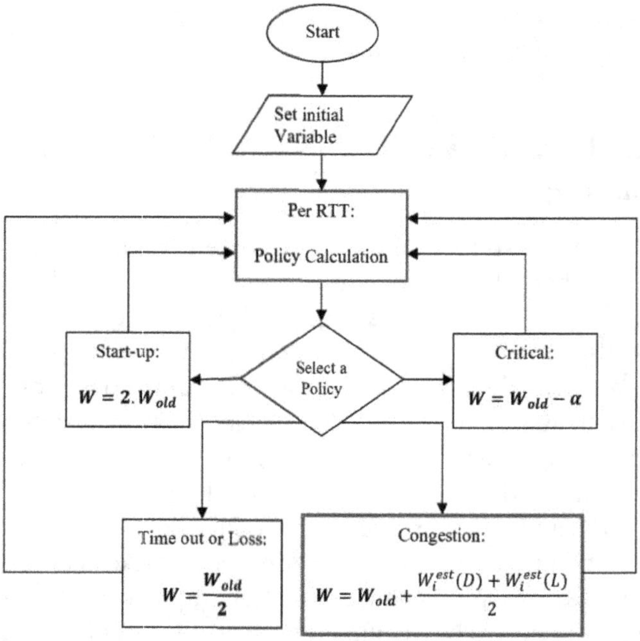

Figure 1.7. Flowchart for the Algorithm of TCP that Deals Mathematical Modelling of DoS or DDoS

Source: Jamali *et al.* [29].

The part of infected system could be termed as bots, or it could be understood as a machine that gets infected through software. A group of bots are considered as a botnet, which are used for the creation of attack traffic [5]. The entire process of security attacks could be considered the achieved elements of DDoS through essential component botnets. A proper study of mathematical modeling or calculation for the epidemic model could be possible with the propagation of botnets. The propagation of models could be done in two specific forms: SERIS and SIS. The meaning of SIS is infected and susceptible, while SERIS means susceptible, exposed, infected, recovered, and susceptible. Communication among the systems and networks in the epidemic model is possible with the presence of an epidemic framework and botnets.

1.3.3. *Analytical Models*

The parameters and conditions could be used to determine the type of model applied for the security attack. The strength of a defense party determines the interaction between the investment in security and the defining system. The exact calculation of the success probability could be recognized with the selection of the provided model that is effective in finding the value on the basis of allocated strategies and the size of botnets. The probability of resistance for the victim could be analyzed or evaluated with the application of an analytical model for specific security attacks and their consequences. The vulnerabilities of protocols could be determined with the use of this model for the attacks caused by DDoS (Salloum *et al.* [14]). Analytical models perform well in the process of forming designs for the optimization of protocols and the structure of system [94–97].

1.3.4. *Hierarchical Models*

Hierarchical models are effective in determining the effects and types of attacks caused by DoS or DDoS. The working of the model depends on the affected components and the availability of the targeted system. The model is proposed on the basic concept applied to an attack tree for the estimation of different parameters, such as advantages for the attacker, feasibility, and the likelihood of the attack. According to Iwendi *et al.* [28], the evaluation of simultaneous attacks performed through DDoS or DoS is possible with the use of hierarchical models and attack trees.

1.3.5. *Model-Based on Traffic*

A model which is based on traffic is effective in showing attack patterns along with the environment of the network. Three specific elements, namely the CPU, memory, and bandwidth, are simultaneously used to create multiple problems for the victim. Queuing theory has been used in the development of the traffic-based model that provides the exact assumptions of attacks performed through DDoS. The main concern of the queuing theory or the traffic-based model is taking account of only one targeted resource. The vulnerabilities of exploited retransmission could be mitigated with the application of TCP. To achieve the effects of security attacks on the resources, this model uses a mathematical formula for the calculation of the optimum cost. Attacks caused by DDoS could be analyzed for its effects by following the protocols of TCP that propose stabilizing mechanism of active queue management (AQM). 3G wireless networks are targeted through DDoS in the traffic-based model to recognize the vulnerable resource (Khan and Hermann [3]).

1.3.6. *Attacks of DoS or DDoS on the Network of SDNs*

Compared with traditional networks, the development of networks is possible in less time for SDNs. The architecture of the network for security attacks on the SDN for successful attempts. The representation of network is effective for the attacks that occurred due to DoS or DDoS and affect the southbound and northbound. Queuing theories are applied for the management of overflowing attacks. According to Lokman *et al.* [10], the estimation of capacity for the devices of SDN is performed with the complement of architecture. The attacks occurring through DoS or DDoS are effective in providing a change in the architecture of SDN for the controller of different communication channels. The impact of attacks results in improper mathematical calculations for the models of SDN. The centralization of architecture for the SDN is effective in presenting the effects of security attacks due to a lack of mathematical modeling.

1.3.7. *Models for the Migration and Detection*

The agents and attacks that affect the migration and detection models could be analyzed with the evaluation of linked factors. Deep analyses of the targeted models are effective in assessing the DDoS attacks with the provided the solution of migration. Mathematical modeling present for the perspective of detection and avoid the effects is possible in this model. Different approaches are used for the scope of the survey and collecting the required information. This chapter provides detailed information about the types of networks and their models, along with the effects of security effects on them. The following section presents a brief review of specific networks, such as traditional networks, SDNs, and virtual networks (Arshad *et al.* [1]).

Traditional Networks: The rate and timing of packets could be detected with the extraction of traffic created by the attackers. The detection of algorithms for the estimation of the quantity of used packets. Traditional networks are effective in using the chaos theory for the detection of implemented neural networks. The detection of DDoS attacks could be done with the mathematical modeling of recurrence qualification analysis (RQA).

Virtual Networks: Effective strategies for the presented hybrid could be done within the cloud computing environment. The level of hosts and networks in the visual networks is analyzed with the allocation of dynamic resources for the attacks caused through DDoS. The flexibility and time of delivery through the applied technologies have a great impact on the estimated models. A framework is proposed for the estimation of similar models of networks, and the defense strategy is followed for the DDoS attacks [93–97].

SDNs: The effectiveness of SDNs could be analyzed as providing protection against attacks in the form of flooding and spoofing. The emulation of attacks caused by DoS or DDoS could be effective in countering the statics of flow switches. The application of algorithms and ML models is used to classify the relevant features of genetic algorithms. The effects of attacks could be reduced with the application of mathematical modeling and different algorithms, such as genetic algorithm and entropy algorithm.

1.3.8. *Virtual Networks Attacked Through DoS or DDoS*

Virtual infrastructure could be deployed with the application of two specific networks: SDN and traditional. The attacks through the vectors could be done with DoS or DDoS, which is based on the concept of queuing theory. Cloud environment for the low rate is effective in targeting the traditional networks and its model. The technologies used for the virtual network are quite flexible for handling its virtualization and are lightweight too. The migration strategy used for the allocation of models is based on the attack's strength.

1.4. Comparative Analysis of Mathematical Models of DoS or DDoS

The classifications of the DDoS or DoS attacks on different networks were classified in the previous section. The effectiveness and features of the applied mathematical models are determined through a comparative analysis of their architectures and functions. The following models could be used for the comparative analysis of these attacks: congestion window model, epidemic model, and queuing model. An effective framework is designed for the comparative analysis of these models and the effects of attacks on the network. Different networks, including traditional and SDNs, could also be analyzed for the comparative effects of the security attacks caused through DoS or DDoS.

1.4.1. *Protocol Attack: Model of Congestion Window*

Attacks that affect the protocols of the network could be analyzed through the congestion model of the window that targets the server (single network device). The effectiveness of the network could be analyzed by targeting the three most important elements of the devices: bandwidth, the memory unit, and the CPU. Communication channels are mainly targeted by attackers that are centralized to SDN for TCP and forwarding devices (Preuveneers *et al.* [13]). The development of the present model is done to analyze the activities of the open flow in two specific congestion windows. The identified windows are effective in providing information such as the pattern of the attacks and the network environment.

DDoS attacks aim to degrade the quality of the target services or servers, or possibly make them unavailable. Different attack vectors might be employed to impact traditional networks in order to attain this purpose. The attack tree is depicted in Figure 1.7, with the appropriate attack vectors classified as follows. The target is flooded with traffic in order to exhaust network or hardware resources, with bandwidth being the primary goal. Flooding and amplification/reflection assaults are two types of volumetric attacks. Flooding attacks attempt to use a huge volume of traffic to drain bandwidth, processing capacity, or other network resources. Meanwhile, spoofing vulnerabilities are exploited through reflection attacks, in which the attacker uses faked requests to produce traffic from several devices to the target [98–101].

1.4.2. *Device Attack: Epidemic Model*

The attacks that could target the connected devices could be analyzed through the epidemic model as the attacks harm the botnet. The epidemic model is effective in targeting the model's multiple networks at a time and showing the effects of attacks caused by DDoS or DoS. The management of all forwarding devices is done with single devices that are in a centralized form. Soe *et al.* [16] stated that epidemic SIS-SERIS could be well described through the epidemic model in different computer networks. Epidemic equilibrium and free equilibrium could be analyzed properly using the epidemic model, and it also provides stability to unstable networks. The stability of the targeted system could be analyzed with the application of the epidemic model.

1.4.3. *Resource Model: Queuing Model*

The resource or queuing model is applied to the attacks that occur on the networks of a single network device, such as a server or router. The effectiveness of the queuing model could be analyzed through the centralized architecture of specific networks, such as SDN, for the attacks caused by DDoS or DoS (Khan *et al.* [2]). Resource models are used for the analysis of effects caused by security attacks on different elements of the system, such as bandwidth, CPU, and memory units. The probability exhaustion of bandwidth, memory units, and CPUs is completed with mathematical equations. Figure 1.8 represents such types of mathematical models used for the queuing model to analyze the effects of DDoS or DoS attacks.

1.4.4. *Effective Framework for Comparison and Result of Mathematical Models*

Three specific mathematical models, namely queuing, epidemic, and congestion windows, are analyzed. The queuing model targets the resources of the device and measures the attacks according to the queries [105–107]. The result obtained through the mathematical modeling, queuing model acknowledges the probability of blocked traffic. The rate of targeted

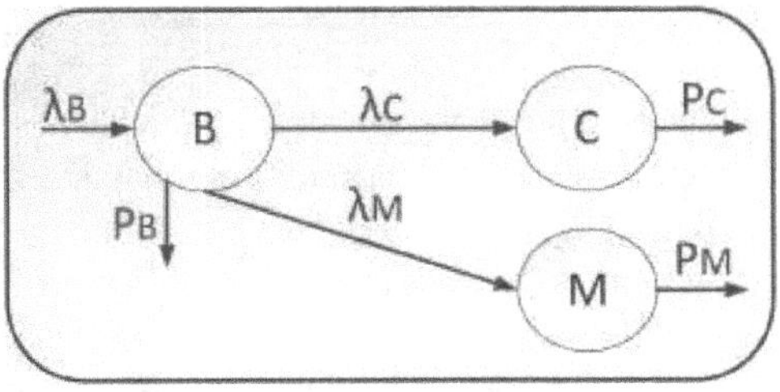

Figure 1.8. Mathematical Model for the Representation of Queuing Model

Source: Zaki *et al.* [22].

devices is obtained by applying an epidemic mathematical model. The results of applying a congestion window give the status of the probability of normalized legitimate attacks. Table 1.1 provides a comparative analysis of different models for the effects of attacks caused by DoS or DDoS [108–109].

Table 1.1. Comparison of Different Models and Effects of DoS or DDoS Attacks

Serial Number	Model Used	Target of Attack	Measuring Units for Attacks	Result	Applications
1.	Queuing	Resources of devices	Queries per second (speed of attacking and rate of services)	Probability of blocked traffic	Design for the network could be done with the use of required resources, such as software and hardware
2.	Epidemic	Network or its communication channel	Number of devices per second (rate of transition), device percentage for the networks (number of nodes recovered, exposed, recovered, and susceptible)	Specific rate of devices in percentage	Detection of antivirus, development of security, identifying network damage
3.	Congestion Window	Protocol of the connected devices or networks	Mbps (magnitude of attack burst) and bytes (buffering of bottleneck)	Probability of attacks for normalized legitimate	The size of TCP protocol could be optimized within this model for the resources of server

1.5. Discussion and Conclusion

The current chapter focused on analyzing the development of intrusion system for attacks caused by DoS or DDoS. Different mathematical models were applied for networks such as SDN or traditional networks to determine the targeted parts [102–104]. By completing the entire study, it could be observed that the classification of networks was required for acknowledging the effects of attacks caused through DoS or DDoS in a better way. Traditional networks show the maximum effects of the attacks caused through DoS or DDoS. A comparative analysis of different mathematical models, including queuing, epidemic, and congestion window models, provided information about multiple elements, such as their application and unit. Communication among the connected devices needs to be more secure with the application of OpenFlow of TCP.

References

1. Arshad, J., Azad, M. A., Amad, R., Salah, K., Alazab, M., and Iqbal, R. (2020). A review of performance, energy and privacy of intrusion detection systems for IoT. *Electronics*, *9*(4), 629.
2. Khan, M. A., Khaskheli, S. A., Kalwar, H. A., Kalwar, M. A., Marri, H. B., and Nebhwani, M. (2021). Improving the performance of reception and OPD by using multi-server queuing model in Covid-19 pandemic. *International Journal of Science and Engineering Investigations*, *10*(113), 20–29.
3. Khan, Z. A. and Herrmann, P. (2019). Recent advancements in intrusion detection systems for the internet of things. *Security and Communication Networks*, *2019*, 1–20.
4. Khare, N., Devan, P., Chowdhary, C. L., Bhattacharya, S., Singh, G., Singh, S., and Yoon, B. (2020). SMO-DNN: Spider monkey optimization and deep neural network hybrid classifier model for intrusion detection. *Electronics*, *9*(4), 692.
5. Khraisat, A. and Alazab, A. (2021). A critical review of intrusion detection systems in the internet of things: Techniques, deployment strategy, validation strategy, attacks, public datasets and challenges. *Cybersecurity*, *4*(1), 1–27.
6. Khraisat, A., Gondal, I., Vamplew, P., and Kamruzzaman, J. (2019). Survey of intrusion detection systems: Techniques, datasets and challenges. *Cybersecurity*, *2*(1), 1–22.
7. Khraisat, A., Gondal, I., Vamplew, P., Kamruzzaman, J., and Alazab, A. (2019). A novel ensemble of hybrid intrusion detection system for detecting internet of things attacks. *Electronics*, *8*(11), 1210.

8. Kim, J., Kim, J., Kim, H., Shim, M., and Choi, E. (2020). CNN-based network intrusion detection against denial-of-service attacks. *Electronics*, *9*(6), 916.
9. Lima Filho, F. S. D., Silveira, F. A., de Medeiros Brito Junior, A., Vargas-Solar, G., and Silveira, L. F. (2019). Smart detection: An online approach for DoS/DDoS attack detection using machine learning. *Security and Communication Networks*, *2019*, 1–16.
10. Lokman, S. F., Othman, A. T., and Abu-Bakar, M. H. (2019). Intrusion detection system for automotive Controller Area Network (CAN) bus system: A review. *EURASIP Journal on Wireless Communications and Networking*, *2019*(1), 1–17.
11. Miles, C., Bohrdt, A., Wu, R., Chiu, C., Xu, M., Ji, G., Greiner, M., Weinberger, K.Q., Demler, E., and Kim, E. A. (2021). Correlator convolutional neural networks as an interpretable architecture for image-like quantum matter data. *Nature Communications*, *12*(1), 1–7.
12. Papavassiliou, S. (2020). Software defined networking (SDN) and network function virtualization (NFV). *Future Internet*, *12*(1), 7.
13. Preuveneers, D., Rimmer, V., Tsingenopoulos, I., Spooren, J., Joosen, W., and Ilie-Zudor, E. (2018). Chained anomaly detection models for federated learning: An intrusion detection case study. *Applied Sciences*, *8*(12), 2663.
14. Salloum, S. A., Alshurideh, M., Elnagar, A., and Shaalan, K. (2020). Machine learning and deep learning techniques for cybersecurity: A review. In *The International Conference on Artificial Intelligence and Computer Vision* (pp. 50–57). Springer: Cham.
15. Sharma, J., Giri, C., Granmo, O. C., and Goodwin, M. (2019). Multi-layer intrusion detection system with ExtraTrees feature selection, extreme learning machine ensemble, and softmax aggregation. *EURASIP Journal on Information Security*, *2019*(1), 1–16.
16. Soe, Y. N., Feng, Y., Santosa, P. I., Hartanto, R., and Sakurai, K. (2020). Machine learning- based IoT-botnet attack detection with sequential architecture. *Sensors*, *20*(16), 4372.
17. Sudqi Khater, B., Abdul Wahab, A. W. B., Idris, M. Y. I. B., Abdulla Hussain, M., and Ahmed Ibrahim, A., 2019. A lightweight perceptron-based intrusion detection system for fog computing. *Applied sciences*, *9*(1), 178.
18. Sultana, N., Chilamkurti, N., Peng, W., and Alhadad, R. (2019). Survey on SDN based network intrusion detection system using machine learning approaches. *Peer-to-Peer Networking and Applications*, *12*(2), 493–501.
19. Tabassum, A., Erbad, A., and Guizani, M. (2019, June). A survey on recent approaches in intrusion detection system in IoTs. In *2019 15th International Wireless Communications & Mobile Computing Conference (IWCMC)* (pp. 1190–1197). IEEE.

20. Ullah, I. and Mahmoud, Q. H. (2020). A two-level flow-based anomalous activity detection system for IoT networks. *Electronics*, *9*(3), 530.
21. Yang, Y., Zheng, K., Wu, C., Niu, X., and Yang, Y. (2019). Building an effective intrusion detection system using the modified density peak clustering algorithm and deep belief networks. *Applied Sciences*, *9*(2), 238.
22. Zaki, N. H. M., Saliman, A. N., Abdullah, N. A., Hussain, N. S. A. A., and Amit, N. (2019). Comparison of queuing performance using queuing theory model and fuzzy queuing model at check-in counter in airport. *Mathematics and Statistics*, *7*(4), 17–23.
23. Asharf, J., Moustafa, N., Khurshid, H., Debie, E., Haider, W., and Wahab, A. (2020). A review of intrusion detection systems using machine and deep learning in internet of things: Challenges, solutions and future directions. *Electronics*, *9*(7), 1177.
24. Barka, E., Dahmane, S., Kerrache, C. A., Khayat, M., and Sallabi, F. (2021). STHM: A secured and trusted healthcare monitoring architecture using SDN and Blockchain. *Electronics*, *10*(15), 1787.
25. Dhaliwal, S. S., Nahid, A. A., and Abbas, R. (2018). Effective intrusion detection system using XGBoost. *Information*, *9*(7), 149.
26. Elrawy, M. F., Awad, A. I. and Hamed, H. F. (2018). Intrusion detection systems for IoT-based smart environments: A survey. *Journal of Cloud Computing*, *7*(1), 1–20.
27. Gonzalez-Cuautle, D., Hernandez-Suarez, A., Sanchez-Perez, G., Toscano-Medina, L. K., Portillo-Portillo, J., Olivares-Mercado, J., Perez-Meana, H. M., and Sandoval-Orozco, A. L. (2020). Synthetic minority oversampling technique for optimizing classification tasks in botnet and intrusion-detection-system datasets. *Applied Sciences*, *10*(3), 794.
28. Iwendi, C., Khan, S., Anajemba, J. H., Mittal, M., Alenezi, M., and Alazab, M. (2020). The use of ensemble models for multiple class and binary class classification for improving intrusion detection systems. *Sensors*, *20*(9), 2559.
29. Jamali, S., Talebi, M. M., and Fotohi, R. (2021). Congestion control in high-speed networks using the probabilistic estimation approach. *International Journal of Communication Systems*, *34*(7), e4766.
30. Khan, M. A. and Kim, J. (2020). Toward developing efficient Conv-AE-based intrusion detection system using heterogeneous dataset. *Electronics*, *9*(11), 1771.
31. Khan, M. A., Karim, M., and Kim, Y. (2019). A scalable and hybrid intrusion detection system based on the convolutional-LSTM network. *Symmetry*, *11*(4), 583.
32. Wani, A. R., Rana, Q. P., Saxena, U., and Pandey, N. (2019). Analysis and detection of DDoS attacks on cloud computing environment using machine learning techniques. In *IEEE Amity International Conference on Artificial Intelligence (AICAI)*, pp. 870–875.

33. Shafi, Q. and Basit, A. (2019). DDoS botnet prevention using blockchain in software defined Internet of Things. In *IEEE 16th International Bhurban Conference on Applied Sciences and Technology (IBCAST)*, pp. 624–628.
34. Bhardwaj, A., Sharma, A., Mangat, V., Kumar, K., and Vig, R. (2019). Experimental analysis of DDoS attacks on openstack cloud platform. In *Proceedings of 2nd International Conference on Communication, Computing and Networking*, pp. 3–13.
35. Gurulakshmi, K. and Nesarani, A. (2019). Analysis of IoT Bots against DDOS attack using Machine learning algorithm. In *IEEE 2018 2nd International Conference on Trends in Electronics and Informatics (ICOEI)*, pp. 1052–1057.
36. Dutta, I. K., Ghosh, B., and Bayoumi, M. (2019). Lightweight cryptography for internet of Insecure Things: A survey. In *IEEE 9th Annual Computing and Communication Workshop and Conference (CCWC)*, pp. 475–481.
37. Achbarou, O. and El Bouanani, S. (2019). Securing cloud computing from different attacks using intrusion detection systems. *IJIMAI*, *4*(3), 61–64.
38. Kirichek, R., Kulik, V., and Koucheryavy, A. (2019). False clouds for Internet of Things and methods of protection. In *IEEE 2016 18th International Conference on Advanced Communication Technology (ICACT)*, pp. 201–205.
39. Hassan, W. H. (2019). Current research on Internet of Things (IoT) security: A survey. *Computer Networks*, *148*, 283–294.
40. Nawir, M., Amir, A., Yaakob, N., and Lynn, O. B. (2019). Internet of Things (IoT): Taxonomy of security attacks. In *IEEE 2016 3rd International Conference on Electronic Design (ICED)*. pp. 321–326.
41. Yin, D., Zhang, L., and Yang, K. (2019). A DDoS attack detection and mitigation with software- defined Internet of Things framework. *IEEE Access*, *6*, 24694–24705.
42. Zhang, Y., Li, P., and Wang, X. (2019). Intrusion detection for IoT based on improved genetic algorithm and deep belief network. *IEEE Access*, *7*, 31711–31722.
43. Chaabouni, N., Mosbah, M., Zemmari, A., Sauvignac, C., and Faruki, P. (2019). Network intrusion detection for IoT security based on learning techniques. *IEEE Communications Surveys & Tutorials*, *21*, 2671–2701.
44. Khan, A. N., Fan, M. Y., Malik, A., and Memon, R. A. (2019). Learning from privacy preserved encrypted data on cloud through supervised and unsupervised machine learning. In *IEEE 2019 2nd International Conference on Computing, Mathematics and Engineering Technologies (iCoMET)*, pp. 1–5.
45. Bhardwaj, A., Kumar, P., Raj, B., Anand, S. (2022). Design and performance optimization of doping-less vertical nanowire TFET using gate stack technique. *Journal of Electronic Materials (JEMS)*, *41*(7), 4005–4013.

46. Singh, J. and Raj, B. (2019). Tunnel current model of asymmetric MIM structure levying various image forces to analyze the characteristics of filamentary memristor. *Applied Physics A*, *125*(3), 203.
47. Goyal, C., Ubhi, J. S., and Raj, B. (2019). Low leakage zero ground noise nanoscale full adder using source biasing technique. *Journal of Nanoelectronics and Optoelectronics*, *14*, 360–370.
48. Singh, G., Sarin, R. K., and Raj, B. (2016). A novel robust exclusive-OR function implementation in QCA nanotechnology with energy dissipation analysis. *Journal of Computational Electronics*, *15*(2), 455–465.
49. Wadhera, T., Kakkar, D., Wadhwa, G., and Raj, B. (2019). Recent advances and progress in development of the field effect transistor biosensor: A review. *Journal of Electronic Materials*, *48*(12), 7635–7646.
50. Wadhwa, G., Kamboj, P., and Raj, B. (2019). Design optimisation of junctionless TFET biosensor for high sensitivity. *Advances in Natural Sciences: Nanoscience and Nanotechnology*, *10*(7), 045001.
51. Bansal, P. and Raj, B. (2016). Memristor modeling and analysis for linear dopant drift kinetics. *Journal of Nanoengineering and Nanomanufacturing*, *6*, 1–7.
52. Singh, A., Khosla, M., and Raj, B. (2016). Circuit compatible model for electrostatic doped schottky barrier CNTFET. *Journal of Electronic Materials*, *45*(12), 4825–4835.
53. Ashima, Vaithiyanathan, D., and Raj, B. (2021). Performance analysis of charge plasma induced graded channel Si nanotube. *Journal of Engineering Research (JER)*, *14*, 146–154.
54. Tomar, Abhishek Singh, Magraiya, Vijay Kumar, and Raj, B. (2013). Scaling of access and data transistor for high performance DRAM cell design. *Quantum Matter*, *2*, 412–416.
55. Jain, N. and Raj, B. (2018). Parasitic capacitance and resistance model development and optimization of raised source/drain SOI FinFET structure for analog circuit applications. *Journal of Nanoelectronics and Optoelectronins*, *13*, 531–539.
56. Singh, Shradhya, Vishvakarma, S. K., and Raj, B. (2018). Analytical modeling of split-gate junction-less transistor for a biosensor application. *Sensing and Bio-sensing*, *18*, 31–36.
57. Gopal, M. and Raj, B. (2013). Low power 8T SRAM cell design for high stability video applications. *ITSI Transaction on Electrical and Electronics Engineering*, *1*(5), 91–97.
58. Raj, B., Mitra, J., Bihani, D. K., Rangharajan, V., Saxena, A. K., and Dasgupta, S. (2010). Analysis of noise margin, power and process variation for 32 nm FinFET based 6T SRAM cell. *Journal of Computer (JCP)*, *5*(6), 1–8.
59. Sharma, D., Mehra, R., and Raj, B. (2021). Comparative analysis of photovoltaic technologies for high efficiency solar cell design. *Superlattices and Microstructures*, *153*, 106861.

60. Kaur, P., Buttar, A. S., and Raj, B. (2021). A comprehensive analysis of nanoscale transistor based biosensor: A review. *Indian Journal of Pure and Applied Physics*, *59*, 304–318.
61. Yadav, D., Raj, B., and Raj, B. (2020). Design and simulation of low power microcontroller for IoT applications. *Journal of Sensor Letters*, *18*, 401–409.
62. Singh, S. and Raj, B. (2020). A 2-D analytical surface potential and drain current modeling of double-gate vertical t-shaped tunnel FET. *Journal of Computational Electronics*, *19*, 1154–1163.
63. Singh, J. and Raj, B. (2019). An accurate and generic window function for non-linear memristor model. *Journal of Computational Electronics*, *18*(2), 640–647.
64. Kaur, M., Gupta, N., Kumar, S., Raj, B., and Singh, A. K. (2021). Comparative RF and crosstalk analysis of carbon based nano interconnects. *IET Circuits, Devices & Systems*, *15*(6), 493–503.
65. Kandasamy, N., Ahmad, F., Ajitha, D., Raj, B., and Telagam, N. (2020). Quantum dot cellular automata based scan flip flop and boundary scan register. *IETE Journal of Research*, *66*, 535–548.
66. Sharma, S. K., Raj, B., and Khosla, M. (2019). Enhanced photosensivity of highly spectrum selective cylindrical gate $In_{1-x}Ga_xAs$ nanowire MOSFET photodetector. *Modern Physics Letter B*, *33*(12), 1950144.
67. Singh, J. and Raj, B. (2019). Design and investigation of 7T2M NVSARM with enhanced stability and temperature impact on store/restore energy. *IEEE Transactions on Very Large Scale Integration Systems*, *27*(6), 1322–1328.
68. Bhardwaj, A. K., Gupta, S., Raj, B., and Singh, A. (2019). Impact of double gate geometry on the performance of carbon nanotube field effect transistor structures for low power digital design. *Computational and Theoretical Nanoscience*, *16*, 1813–1820.
69. Jain, N. and Raj, B. (2019). Thermal stability analysis and performance exploration of asymmetrical dual-k underlap spacer (ADKUS) SOI FinFET for security and privacy applications. *Indian Journal of Pure & Applied Physics (IJPAP)*, *57*, 352–360.
70. Singh, A., Khosla, M., and Raj, B. (2019). Design and analysis of dynamically configurable electrostatic doped carbon nanotube tunnel FET. *Microelectronics Journal*, *85*, 17–24.
71. Jain, N. and Raj, B. (2019). Dual-k spacer region variation at the drain side of asymmetric SOI FinFET structure: Performance analysis towards the analog/RF design applications. *Journal of Nanoelectronics and Optoelectronics*, *14*, 349–359.
72. Singh, J., Sharma, S., Raj, B., and Khosla, M. (2018). Analysis of barrier layer thickness on performance of $In_{1-x}Ga_xAs$ based gate stack cylindrical gate nanowire MOSFET. *JNO*, *13*, 1473–1477.

73. Jain, N. and Raj, B. (2018). Analysis and performance exploration of high-k SOI FinFETs over the conventional low-k SOI FinFET toward analog/RF design. *Journal of Semiconductors (JoS)*, *39*(12), 124002-1-7.
74. Goyal, C., Ubhi, J. S., and Raj, B. (2018). A reliable leakage reduction technique for approximate full adder with reduced ground bounce noise. *Journal of Mathematical Problems in Engineering*, *2018*, 16. Article ID 3501041.
75. Anuradha, J. S., Raj, B. and Khosla, M. (2018). Design and performance analysis of nano-scale memristor-based nonvolatile SRAM. *Journal of Sensor Letter*, *16*, 798–805.
76. Wadhwa, G. and Raj, B. (2018). Parametric variation analysis of charge-plasma-based dielectric modulated JLTFET for biosensor application. *IEEE Sensor Journal*, *18*(15), 6070–6077.
77. Singh, J. and Raj, B. (2018). Comparative analysis of memristor models for memories design. *JoS*, *39*(7), 074006-1-12.
78. Yadav, D., Chouhan, S. S., Vishvakarma, S. K., and Raj, B. (2018). Application specific microcontroller design for IoT based WSN. *Sensor Letter*, *16*, 374–385.
79. Singh, G., Sarin, R. K., and Raj, B. (2018). Fault-tolerant design and analysis of quantum-dot cellular automata based circuits. *IEEE/IET Circuits, Devices & Systems*, *12*, 638–64.
80. Singh, J. and Raj, B. (2018). Modeling of mean barrier height levying various image forces of metal insulator metal structure to enhance the performance of conductive filament based memristor model. *IEEE Nanotechnology*, *17*(2), 268–267.
81. Jain, A., Sharma, S., and Raj, B. (2018). Analysis of triple metal surrounding gate (TM-SG) III-V nanowire MOSFET for photosensing application. *Opto-electronics Journal*, *26*(2), 141–148.
82. Jain, A., Sharma, S., Raj, B. (2016). Design and analysis of high sensitivity photosensor using cylindrical surrounding gate MOSFET for low power sensor applications. *Engineering Science and Technology, an International Journal*, *19*(4), 1864–1870.
83. Singh, A., Khosla, M., and Raj, B. (2017). Analysis of electrostatic doped Schottky barrier carbon nanotube FET for low power applications. *Journal of Materials Science: Materials in Electronics*, *28*, 1762–1768.
84. Kumar, G. S., Singh, A., and Raj, B. (2018). Design and analysis of gate all around CNTFET based SRAM cell design. *Journal of Computational Electronics*, *17*(1), 138–145.
85. Singh, G. P., Sohi, B. S., Raj, B. (2017). Material properties analysis of graphene base transistor (GBT) for VLSI analog circuits. *Indian Journal of Pure & Applied Physics (IJPAP)*, 55, 896–902.

86. Singh, A., Khosla, M., and Raj, B. (2016). Comparative analysis of carbon nanotube field effect transistor and nanowire transistor for low power circuit design. *Journal of Nanoelectronics and Optoelectronics*, *11*, 388–393.
87. Kumar, S. and Raj, B. (2017). Estimation of stability and performance metric for inward access transistor based 6T SRAM cell design using n-type/p-type DMDG-GDOV TFET. *IEEE VLSI Circuits and Systems Letter*, *3*(2), 25–39.
88. Sharma, S., Kumar, A., Pattanaik M., and Raj, B. (2013). Forward body biased multimode multi-threshold CMOS technique for ground bounce noise reduction in static CMOS adders. *International Journal of Information and Electronics Engineering*, *3*(3), 567–572.
89. Singh, H., Kumar, P., and Raj, B. (2013). Performance analysis of majority gate SET based 1-bit full adder. *International Journal of Computer and Communication Engineering* (IJCCE), *2*(4), ISSN: 2010-3743.
90. Bhardwaj, A. K., Gupta, S., and Raj, B. (2020). Investigation of parameters for Schottky barrier (SB) height for Schottky barrier based carbon nanotube field effect transistor device. *Journal of Nanoelectronics and Optoelectronics*, *15*, 783–791.
91. Bansal, P. and Raj, B. (2017). Memristor: A versatile nonlinear model for dopant drift and boundary issues. *JCTN*, *14(*5), 2319–2325.
92. Jain, N. and Raj, B. (2016). An analog and digital design perspective comprehensive approach on fin-FET (fin-field effect transistor) technology — A review. *Reviews in Advanced Sciences and Engineering (RASE)*, *5*, 1–14.
93. Sharma, S., Raj, B., and Khosla, M. (2017). Subthreshold performance of $In_{1-x}Ga_xAs$ based dual metal with gate stack cylindrical/surrounding gate nanowire MOSFET for low power analog applications. *Journal of Nanoelectronics and Optoelectronics*, *12*, 171–176.
94. Raj, B., Saxena, A. K., and Dasgupta, S. (2009). Analytical modeling for the estimation of leakage current and subthreshold swing factor of nanoscale double gate FinFET device. *Microelectronics International*, *26*, 53–63.
95. Soniya, S. S., Wadhwa, G., and Raj, B. (2019). An analytical modeling for dual source vertical tunnel field effect transistor. *International Journal of Recent Technology and Engineering (IJRTE)*, *8*(2), 603–608.
96. Singh, S. and Raj, B. (2019). Design and analysis of hetrojunction vertical T-shaped tunnel field effect transistor. *Journal of Electronics Material*, *48*(10), 6253–6260.
97. Goyal, C., Ubhi, J. S., and Raj, B. (2019). A low leakage CNTFET based inexact full adder for low power image processing applications. *International Journal of Circuit Theory and Applications*, *47*(9), 1446–1458.
98. Raj, B., Saxena, A. K., and Dasgupta, S. (2008). A compact drain current and threshold voltage quantum mechanical analytical modeling for FinFETs. *Journal of Nanoelectronics and Optoelectronics (JNO)*, *3*(2), 163–170.

99. Wadhwa, G. and Raj, B. (2020). An analytical modeling of charge plasma based tunnel field effect transistor with impacts of gate underlap region. *Superlattices and Microstructures*, *142*, 106512.
100. Singh, Shailendra and Raj, B (2020). Modeling and simulation analysis of SiGe hetrojunction double gatevertical t-shaped tunnel FET. *Superlattices and Microstructures*, *142*, 106496.
101. Singh, A., Saini, D. K., Agarwal, D., Aggarwal, S., Khosla, M., and Raj, B. (2016). Modeling and simulation of carbon nanotube field effect transistor and its circuit application. *Journal of Semiconductors (JoS)*, *37*, 074001-6.
102. Jain, N. and Raj, B. (2016). Device and circuit co-design perspective comprehensive approach on FinFET technology — A review. *Journal of Electron Devices*, *23*(1), 1890–1901.
103. Kumar, S. and Raj, B. (2016). Analysis of ION and Ambipolar current for dual-material gate-drain overlapped DG-TFET. *Journal of Nanoelectronics and Optoelectronics*, *11*, 323–333.
104. Anjum, N., Bali, T., and Raj, B. (2013). Design and simulation of handwritten multiscript character recognition. *International Journal of Advanced Research in Computer and Communication Engineering*, *2*(7), 2544–2549.
105. Gupta, B. B. and Badve, O. P. (2017). Taxonomy of DoS and DDoS attacks and desirable defense mechanism in a cloud computing environment. *Neural Computing and Applications*, *28*, 3655–3682.
106. Srivastava, A., Gupta, B. B., Tyagi, A., Sharma, A., and Mishra, A. (2011). A recent survey on DDoS attacks and defense mechanisms. In *2011 Proceedings of the Advances in Parallel Distributed Computing: First International Conference on Parallel, Distributed Computing Technologies and Applications, PDCTA 2011, Tirunelveli, India, September 23–25*, (pp. 570–580). Springer: Berlin, Heidelberg.
107. Chhabra, M., Gupta, B., and Almomani, A. (2013). A novel solution to handle DDOS attack in MANET. *Journal of Information Security*, *4*(3), 165–179.
108. Gaurav, A., Arya, V., and Santaniello, D. (2022). Analysis of machine learning based DDoS attack detection techniques in software defined network. *Cyber Security Insights Magazine (CSIM)*, *1*(1), 1–6.
109. Dahiya, A. and Gupta, B. B. (2021). How IoT is making DDoS attacks more dangerous? Insights2Techinfo, p. 1. https://insights2techinfo.com/how-iot-is-making-ddos-attacks-more-dangerous/.

https://doi.org/10.1142/9789811279539_0002

Chapter 2

Designing Scalable Event-Driven Systems with Message-Oriented Architecture

Daniel Šandor[‡] and Marina Bagić Babac[*,†]

Faculty of Electrical Engineering and Computing, University of Zagreb, Zagreb, Croatia

[†]marina.bagic@fer.hr

[‡]daniel.sandor@fer.hr

Abstract

This chapter provides an overview of the message-oriented architectural pattern. The core components of this type of system are explained, and an overview of common messaging protocols is provided. Two binary messaging protocols (AMQP and MQTT) and a text-based (STOMP) one are analyzed. The exchange of information in these protocols, the message format they expect, and the message exchange order are described. The core components of the message-oriented architectural pattern are the message and the message broker. Message brokers provide implementations of messaging protocols. Five commonly used message brokers (Google Cloud Pub/Sub, Azure Service Bus, RabbitMQ, NATS, and Apache Kafka) and their components are introduced and explained, and common use cases for each broker are listed. A message broker contains a topic and/or queue as a means of exchanging messages between message producers (publishers) and message consumers. The

[*]Corresponding author.

way messages are exchanged between these components is described. An example of a message-oriented architecture is implemented. The central component of the architecture is a Spring Boot application which communicates with the Apache Kafka broker and additional external components. Finally, two use cases of such a component are explained: a chat application in the React framework, and a high-throughput message producer in the Golang programming language.

Keywords: Message-oriented architecture, message protocol, message broker, Spring Boot

2.1. Introduction

Message-oriented architecture, sometimes referred to as message-oriented middleware (MoM), refers to software architecture that uses a message broker as a "middleman" between layers.

The core component of a message-oriented system is the messages. Messages are pieces of data in a format dictated by the message broker which processes the data. The messages are relayed between publishers and consumers, using the broker as a "middleman."

A message broker is a piece of software that allows systems to communicate with each other and exchange information in the form of a message. The message broker does this by translating messages between formal messaging protocols. There are multiple variants of message brokers, but generally, they are split into two groups: queue-based and topic-based message brokers. There are also hybrids of the two types, though the underlying concept remains the same, and multiple implementations of each variant by different providers [1–5].

A message protocol, sometimes referred to as the wire protocol, defines the format of the message that is transferred in the communication process and the combination of messages, or packets, that are exchanged for a particular operation. Specifically, message protocols are application layer protocols that define data formats, communication endpoints, and features.

When dealing with multiple distributed systems, communication between systems is a challenge. Traditional approaches are the following: an exchange of data through exposed application programming interface (API) services or an exchange of files containing raw data, e.g., XML files. Message-oriented architectures provide a relatively simple and thoroughly effective means of communication between services. One of the biggest advantages of a message-oriented architecture is its flexibility. This kind of architecture involves the asynchronous passing of, preferably, small packets of data. This asynchronous nature of the architecture design allows applications or services to be loosely coupled, meaning that a single service can be split into multiple lightweight services which communicate through a messaging system, i.e., the message broker.

This chapter describes the inner workings of some of the most common message brokers: Google Cloud Pub/Sub, Azure Service Bus, RabbitMQ, NATS, and Apache Kafka. The architectural patterns they use and the messaging protocols they implement are reviewed. The features they provide are examined.

Finally, an example of a message-oriented system is presented. A Java Spring Boot server that communicates with the Apache Kafka messaging broker [6], and Elasticsearch and PostgreSQL storage systems are implemented and exposed through a REST API. The usage of the REST API is shown through a simple React chat application. To showcase the scalability and real-time consumption capabilities of the API, a custom publisher in the Golang programming language is implemented [7–11].

2.2. Related Work

Distributed systems have become a standard in modern software development. High performance, availability, and scalability are necessary requirements for modern applications. These requirements can be met by implementing message-oriented architectural patterns.

Some of the most common problems of distributed software are network unreliability, strongly coupled services, and the heterogeneity of applications. In a paper by Magnoni [12], a solution to this problem is presented, which involves decoupled services communicating through a message-based approach. The problems with connection-oriented communication are defined as a temporal dependency, location, and data structure and representation. That is, in connection-based architectures, all components must be available simultaneously, each component must know the address of the component it is dependent on, and a single data format and binary representation must be agreed upon between components. In message-based architectures, these issues are resolved by a single addressable channel, the message broker, which queues up messages that can be consumed at any time. Also, no assumptions are made concerning the data format [13–17].

Yongguo *et al.* [18] gave a review of the basic structural components of MoM. The core part of the system is the message, which contains the message header and the message body. The message header contains the metadata (e.g., message size in bytes). The message body contains the message payload, i.e., the content to be published. The management of messages is done by the message broker. It controls the message life cycle, meaning it disposes of invalid and expired messages. It also implements a mechanism to give order and priority to each message. In some systems, the broker can compress messages for easier network transfer. The queue is a buffer implementation that stores messages. Two important optional features of queues are message ordering and message persistence. In this chapter, these components and features are reviewed for each analyzed message broker implementation.

Apukhtin *et al.* [3] reviewed different message broker patterns: the publisher–subscriber pattern, the exclusive point-to-point pattern, and the remote procedure call (RPC) pattern. The most common pattern is the publisher–subscriber pattern. This pattern contains multiple communication nodes which communicate in a many-to-many-type relationship over a message queue or topic. Producer nodes create messages and publish them to a queue or topic that retains the messages until they are consumed by the subscriber nodes. A single message can be delivered to multiple subscribers. The message broker has the task of resolving which subscriber can receive which message. An exclusive point-to-point pattern refers to a direct type of communication between nodes. One publisher node communicates with one subscriber (consumer) node over a communication channel. The communication channel is a queue that retains the order of published messages. The remote procedure call pattern extends the exclusive point-to-point pattern with an additional queue that enables two-way communication between nodes. This chapter describes the inner workings of some of the most common publisher–subscriber and point-to-point pattern message brokers.

In a study by Wise and McDermott [17], the advantages of the asynchronous nature of the publisher–subscriber pattern were explained. One of the biggest advantages is performance. By overlapping message processing on the subscriber side with continuous message production on the publisher side, the concurrency of the system is increased. In synchronous systems, to achieve a similar concurrency increase, additional buffering processes are needed. Instead of additional buffer processes, an added complexity in the form of message buffering on the subscriber side could be implemented. This means asynchronous messaging systems are cleaner, more compact, and easier to implement and understand. This type of architecture proves most useful in systems where no response is required from the message consumer to the message producer.

Systems that require real-time data processing have greatly benefited from applications based on message-oriented architectures. A study by Intorruk and Numnonda [9] compared the performance, architecture, and resource utilization between Apache Kafka and Apache Pulsar message brokers. These parameters were compared for the task of data analytics using Big Data as a source. Performance-wise, Apache Pulsar outperformed in throughput, latency, and average resource utilization when the size of messages was less than 1 MB. This implies that Apache Pulsar is a better choice for the task of real-time data analytics. In this chapter, Apache Kafka [2] is explored in detail and used in a practical demonstration of a message-based architecture.

Another example of a message-based architecture is given in an article by Guner and colleagues [8]. A message broker was used as an arbitrator of communication between devices in an Internet of Things (IoT) setting. The broker allows for seamless many-to-many communication between data providers and data requestors. An MQ Telemetry Transport (MQTT, as explained in the following section) broker implementing the publish–subscribe pattern was used. It was responsible for the tasks of receiving messages, filtering them based on subject/topic, content, and message type, and sending the messages to the correct subscribed clients.

2.3. Messaging Protocols

There are two variants of message protocols: binary and text-based. Common binary protocols are MQTT and Advanced Message Queuing Protocol (AMQP). The most common text-based messaging protocol is the Streaming Text Oriented Messaging Protocol (STOMP).

2.3.1. *MQ Telemetry Transport Protocol*

MQTT is a binary, lightweight open network protocol designed for messaging in low-bandwidth environments. It is a low-overhead protocol, meaning that it works well in systems limited in bandwidth and CPU power, e.g., embedded systems [4].

The name stems from an IBM product called IBM MQ, which provides messaging and queuing middleware in several modes of operation: publish–subscribe, file transfer, and point-to-point. It was originally used to transport telemetry data in the oil industry [19–24].

The MQTT protocol implements the publish–subscribe pattern. That is, it considers two entities that communicate over a single server or broker. The two entities are the publisher of the data and the subscriber to the sent data. The MQTT protocol has the standard advantages of the publisher–subscriber pattern. It enables application service decoupling in a spatial and temporal way. The spatial entity which is recognized between publishers and subscribers is the broker, with a known address and port. The asynchronous flow of programs between publishers and subscribers is achieved using the broker component. Brokers are scalable, meaning they can handle a larger number of concurrent connections from clients [4].

In the MQTT publisher–subscriber pattern, there is a single publisher who publishes data to the message broker, and the message broker then forwards the messages to the corresponding subscribers. MQTT protocol relies on the TCP protocol for message transmission [4].

The MQTT broker is an intermediary between publishers and subscribers. It filters messages based on several principles. The most common is subject-based filtering. Subject-based filtering refers to the filtering of messages according to the topic they are published to. Each MQTT message contains the topic name. Topic names are strings that represent a hierarchical structure that can be filtered using regular expressions. A topic can also be interpreted as an abstract buffer of messages with a certain key. A subscriber subscribed to a topic receives all the messages published to that topic [4].

The MQTT protocol operates by exchanging control packets of a certain type. The official documentation [4] describes these packets and how they are exchanged. Control packets are differentiated by the command type, header, and payload. Every packet contains a two-byte fixed header. The first half of the first byte denotes the MQTT control packet type, and the second half contains flags specific to each control packet type. The second byte denotes the remaining length of the packet, which consists of the length of the variable headers and the length of the payload. Variable headers are optional additions whose content varies depending on the packet type. The optional payload carries the data itself. For example, in a packet of type *PUBLISH*, the payload stores the content of the message sent to the broker. The packet types are: *CONNECT, PUBLISH, PUBACK, PUBREC, PUBREL, PUBCOMP, SUBSCRIBE, SUBACK, UNSUBSCRIBE, UNSUBACK, PINGREQ, PINGRES, DISCONNECT,* and *AUTH* [25–33].

An MQTT connection is established upon an existing TCP connection. At the end of the packet transaction, the TCP connection is closed. The communication flows in a request–response manner. For each packet sent to the broker, the client receives a response. There are two general communication flows, one for the publisher and the other for the subscriber.

The publisher communication flow begins with a connection establishment process. To create a connection, a connect packet is sent. The connect packet must be sent exactly once. The packet payload contains encoded client credentials and the connection target. The variable header contains the protocol name ("MQTT"), protocol version (e.g., "5.0"), connect flags (MQTT connection control parameters), keep-alive (maximum time interval which can pass between two published messages), and other properties. After the connect packet is sent, the client waits for the acknowledgment of the connection in the form of a *CONNICK* packet. The connection can be accepted or refused. The status of the connection is denoted by the *Reason Code* (0x00 for success) [34–42]. After the connection is successfully established, a *PUBLISH* packet is sent containing target information in the variable header (topic name), the data in the payload, and additional configuration in the headers. Depending on the quality-of-service (QoS) setting in the fixed headers, the broker will not return a response (set to 0), will return the *PUBREC* packet (set to 1), or will return the *PUBREC* (publish received), *PUBREL* (publish released), and *PUBCOMP* (publish completed) packets in order (set to 2). When the communication is done, a *DISCONNECT* packet is sent to close the connection [4].

Similarly, when subscribing to a topic, a request–response communication flow is required. First, the connection is established using the *CONNECT* packet. After the client receives a successful acknowledgment of the connection, the *SUBSCRIBE* packet is sent. The payload field of the *SUBSCRIBE* packet contains topic filters as a UTF-8-encoded string. One or more topics to subscribe to can be chosen. When the communication is done, the connection is closed using the *DISCONNECT* packet [4].

2.3.2. *Advanced Message Queuing Protocol*

AMQP is a binary, open standard network protocol designed for passing business messages between applications and organizations. It is widely used in financial systems, and its benefits include security, reliability, interoperability, and standardization. AMQP describes both the format of messages exchanged between components and the protocol model. The protocol model contains a specification that is required for different AMQP implementations to be able to communicate [43–48].

In an article by Vinoski [16], the modularity of AMQP is explained. Message brokering is divided between exchanges and message queues. An exchange is a non-persistent, stateless central component that routes messages to queues. The routing is also called binding. A message queue is a persistent message buffer. The messages can persist in memory or on disk. Message queues are open for implementation, and no requirements are made on persistence by the protocol itself. Consumers (subscribers) consume the messages directly from the message queues.

An AMQP network consists of containers, which are abstractions that hold nodes connected by links. For example, a broker is a container with the exchange and queues as nodes. Nodes are entities within an AMQP network that store and/or deliver messages. Links are unidirectional connections between nodes. A link connects the *source* of one node with the *target* of another. The source can apply a filter to the messages, meaning that it transmits a message to the target only if it has certain properties. AMQP, similarly to MQTT, relies on a TCP connection to exchange data packets over a link. A link is comprised of multiple channels over which sequential data frames are transmitted. A channel is a multiplexed virtual connection shared between AMQP peers. Multiple channels share one TCP connection. To allow for efficient reception of data, each frame contains a channel number and frame size [16].

AMQP defines a standard message format that is expected by the broker. The format is described in the official documentation by Godfrey *et al.* [7]. Each message is split into the bare message and the annotated message. The bare message consists of three parts: standard properties, application properties, and application data (the message body). The bare message is the data supplied by the client and is immutable after creation. The annotated message contains annotations added by the messaging infrastructure [49–56].

The header contains standard delivery details about the message transfer. Properties such as durability (persistence), message priority, or time to live (TTL, i.e., duration for which the message is considered valid) are set in the headers. If excluded, their values are set to system defaults. Delivery and message annotations contain information about the message and configuration in an infrastructure-specific format. The standard properties (properties listed above) contain a set of standard properties defined by the protocol. They contain information on the type of content, the encoding of the payload, and information on the sender and receiver of the message. Application properties contain user-defined key–value pairs which are application-specific. The application data contain the payload of the message and binary data. The footer contains details about the message evaluated once it is constructed (e.g., the hash of the message) [7].

The AMQP protocol implements the publish–subscribe pattern. The message producers are publishers. A general communication flow for a publisher would be the following: the publisher connects to exchange over a link, the publisher sends messages with the defined format to the exchange, and the exchange routes the messages to queues according to a message binding defined in the message properties. A general communication flow for the consumer would be the following: the consumer creates the queue and binds it to a named exchange, and the consumer pulls messages directly from the queue. The consumer doesn't need to do additional processing on the messages, as the exchange makes sure the queue contains the correct messages [7].

2.3.3. *Simple Text-Oriented Messaging Protocol*

STOMP is a simple, lightweight, interoperable, and text-based protocol. It is a text-based wire network protocol created with the simplicity of design found in the HTTP protocol. STOMP clients can communicate with STOMP brokers in an interoperable way using different programming languages, platforms, and brokers [14].

STOMP messaging is based on frames modeled on HTTP frames. As in the MQTT and AMQP protocols, the communication of components relies on a TCP connection underneath a STOMP connection. STOMP implements a publish–subscribe pattern. A STOMP client can act as a producer or a consumer. A producer (publisher) publishes messages to a destination on the broker. A consumer (subscriber) subscribes to a destination on the broker. Communication is initiated by *SEND* or *SUBSCRIBE* data frames [14].

One STOMP data frame consists of an obligatory command section and optional headers and body. The headers are UTF-8-encoded key–value pairs. Content length and content type are two common header values. The content length represents the number of octets in the message body. The content type represents the type of data sent in the body. It includes the encoding if the encoding is different from the default type of UTF-8. These header values allow for easier parsing of frames with a body [47–54].

To start a STOMP connection, the client sends the *CONNECT* frame. The connect frame signals the server (broker) to initiate a TCP connection. The message body must be empty for connecting frames. The server replies to the connect frame with a *CONNECTED* frame if the connection was established successfully or an *ERROR* frame if the connection was lost for any reason. In the case where the broker requires client authentication, the username (*login*) and password (*passcode*) should be set in the headers. The protocol version needs to be agreed upon during the connection initialization process. The client sets supported protocol versions in headers, and the broker responds with the accepted version. The highest value for both parties is chosen as the protocol version [14].

To send a message, a *SEND* frame is required. The SEND frame has one obligatory header, the destination on the broker. The protocol itself has no restrictions or specifications made on the destination value. The destination is an opaque string without further semantics in place. It's up to the broker implementation to specify semantics considering destination configuration. The body of the frame is the message to be sent to the broker [14].

To listen to a destination, a client needs to send a *SUBSCRIBE* frame. The *SUBSCRIBE* frame also contains a destination to receive messages from. An ID is required to make a distinction between subscriptions on the same connection to the broker [55–62].

To stop listening to a destination, an *UNSUBSCRIBE* frame is sent. The only required value for the *UNSUBSCRIBE* frame is the ID of the subscription in the headers [14].

Additional support for ensuring message delivery is enabled in the form of message acknowledgments. *ACK* and *NACK* frames are used to acknowledge the receipt of a batch of messages. *ACK* means the batch was processed, and *NACK* means that it was not [63–69].

To process messages atomically, a transaction can be started using the *BEGIN* frame. The transaction is committed when a *COMMIT* frame is sent.

Finally, to ensure all sent messages were successfully received by the broker, a graceful shutdown of the connection is required. The connection should be closed with the *DISCONNECT* frame [14].

2.4. Message Broker Implementations

Message brokers are components of message-oriented systems which act as an intermediary between two or more services that need to communicate. Message brokers provide an implementation of a messaging network protocol and expose the service with a publicly available interface (API). Each message broker implementation expands on the basic network protocol and provides additional features [7].

2.4.1. *Google Cloud Platform Pub/Sub*

Pub/Sub[1] is a fully managed messaging service available on the Google Cloud Platform (GCP). It allows services to communicate asynchronously, with latencies of about 100 ms. It is mostly used for streaming analytics and data ingestion pipelines. Pub/Sub can be used for service integration in distributed systems and as a queue to parallelize tasks [70–76].

Pub/Sub is a managed service. This means that the infrastructure is controlled by the cloud provider itself. After the user creates the required resources in the GCP console (or programmatically), no additional configuration or maintenance is required by the user. This makes Pub/Sub easy and fast to set up and use. Pub/Sub is most useful when used in tandem with other GCP-managed services because of their native integration capabilities. Common Pub/Sub use cases are event gathering by end-user applications and servers, database replication using change data capture events, data streaming from IoT devices, etc.

---

[1] https://cloud.google.com/pubsub/docs.

Pub/Sub directly implements the publish–subscribe architectural pattern. The key components of the Pub/Sub service are the message, the broker containing topics and subscriptions, the publisher, and the subscriber [77–84].

The *message* is the data that is passed through the service. It contains the message payload in byte format, an optional ordering key, and optional key–value attributes.

The *topic* is a named entity that represents a buffer of messages. The *subscription* is a named entity that represents a request to receive messages from a certain topic.

The *publisher*, or the producer, is a component that uses publicly available APIs to publish messages of a defined type to the messaging service on a specific topic.

The *subscriber*, or the consumer, consumes messages from a topic using a named subscription. The ordering of messages is not guaranteed by default. It can be enabled on a per-topic basis. Once enabled, messages published to the same region with the same ordering key can be consumed in order. Pub/Sub offers at-least-once delivery for all subscription types, meaning that messages may be delivered more than once in the event of an error [85–92].

Multiple publishers can publish messages on one topic. One topic can have multiple subscriptions. Multiple consumers can use a single subscription. In this case, messages will be load-balanced across the consumers, meaning that each consumer receives and processes a subset of messages which are available on the subscription.

The infrastructure running the Pub/Sub broker consists of physical servers running in GCP-supported regions around the world. This allows for low-latency data publishing. Publishers in multiple regions can publish on the same topic, but the data for the topic are stored on servers closest to the publisher itself. The Pub/Sub service is controlled by two abstractions, the data plane and the control plane.

The *data plane* comprises servers, which are called *forwarders*. It handles message transmission between publishers and subscribers. Because of this infrastructure component, Pub/Sub ensures message persistence and, with that, message delivery. Each publisher is connected to one or more forwarder servers, depending on the throughput requirements. Each publishing forwarder keeps a list of subscriptions attached to a topic and acknowledgments of messages from each subscription on the topic. When a publisher sends a message to a topic, the message first arrives at the forwarder. The forwarder stores (persists) the message on multiple independent disks in multiple clusters (collection of machines, or servers, in a shared network). This replication ensures message durability. The forwarder sends the messages to the correct subscriptions. Subscribers connect to subscribing forwarders to receive messages. The subscribing forwarders make requests for messages to the publishing forwarders. The publishing forwarder sends unacknowledged messages to the subscribing forwarders, which relay the messages to subscribers. Once the subscriber processes the message, it sends an acknowledgment to the subscribing forwarder, which relays it to the publishing forwarder. Once all subscriptions listening to a topic have acknowledged the message, the message is deleted from storage [93–98].

The *control plane* is a router that directs publishers and consumers to forwarder servers. The distribution is done in a manner that provides scalability, availability, and low latency. The router has two goals, assign a server to a client considering the shortest network distance and load-balance clients on servers. Each client should get the best possible service in terms of latency, but one server cannot be overloaded with traffic. The router sends a list of servers that are optimal for a client to the proxy service, which has the task of connecting a client and a server. The client connects to one or more of these servers, depending on the performance requirements.

Replaying messages is a seek feature enabled when a topic's message retention period is configured. By default, acknowledged messages are immediately discarded from the topic, whereas unacknowledged ones are discarded after a timeout period. When a retention period is set, both acknowledged and unacknowledged messages will persist. Persisted messages can be republished at a certain timestamp, with applied filters. This feature enables recovery from unexpected subscriber problems and testing new and updated subscribers on known data [99–102].

Schema validation is the process of ensuring data are in a previously agreed-upon format. Consumers are often specialized in processing messages in a certain format. The messages need to contain data with specific values and data types. Pub/Sub enables schema validation on a topic. When subscribing to a topic with a schema, the messages sent to a subscriber are guaranteed to be valid messages of a certain type and encoding.

2.4.2. *Azure Service Bus*

Azure Service Bus[2] is a fully managed message broker from Azure that provides an implementation of the AMQP message protocol. It supports message queues as well as topics. It provides load-balancing across competing workers, coordinates transactions that require high reliability, decouples connected services, etc. It is commonly used to transfer business data, invoice data, inventory changes, etc.

The key components of the Service Bus architecture are the message broker, which contains topics and queues, the publishers, and the consumers.

The *broker* contains queues and topics, which are located in a *namespace*. The namespace is an application container that provides the infrastructure for the messaging service offered by Service Bus. It allocates resources for topics and queues on virtual machines running in chosen Azure availability zones (data center locations). Messages are persisted in these resources for a specified duration (TTL) that is defined in the message properties.

Service Bus offers two types of architectural patterns: point-to-point communication and the standard publish–subscribe pattern.

Point-to-point communication is enabled using queues. A queue is a buffer holding messages which are processed in a first-in, first-out (FIFO) manner. This means the order of messages is preserved in the queue. The sender sends messages to a named queue, which gets a timestamp on arrival. One receiver pulls messages from the queue for processing.

Queue-based communication offers durability and redundancy of messages. Messages are stored on three different servers, spread across multiple zones or regions. The messages are kept in memory and in storage until they are acknowledged by the receiver.

The publish–subscribe pattern is enabled using topics. Topics provide a one-to-many-type relationship using subscriptions, which are virtual buffers of messages. Messages are sent to a topic that has one or more subscriptions attached to it. Unlike queue-based communication, messages are not consumed directly from the topic. Consumers use subscriptions to receive messages. Each subscription receives messages which satisfy the filter applied to it. In the case where no additional filter is added to a subscription, it will receive all messages sent to the topic.

---

[2]https://azure.microsoft.com/en-us/products/service-bus/.

The *publishers* publish messages containing the binary payload (bytes) and key–value properties describing the payload.

Consumers can receive messages in two modes: *receive and delete* and *peek lock*. In the receive-and-delete mode, when a request from the consumer is made, the message in the queue or topic is marked as being consumed and returned to the consumer. This mode gives no guarantees in the event of a processing error. If an error occurred during message processing, these messages will not be resent and are lost from the queue or topic. The peek lock mode guarantees that all messages will be processed in the event of an error. Upon request from the consumer, the message is locked to prevent other consumers from receiving it. When the consumer processes the message, it sends a request to Service Bus, which then marks the message as consumed. The message lock can be released in the event of an error. When an *abandon* request is sent, the message is unlocked and can be consumed again. Each message lock has a timeout; when it expires, the message is unlocked and available to be received again. In the event of an error during message processing, but before the message is marked as being consumed, Service Bus resends the message to the consumer to ensure at-least-once processing. Exactly-once processing can be achieved by using an additional duplicate detection feature in Service Bus. Service Bus doesn't provide message ordering guarantees for standard message consumption. Order is guaranteed within a message session.

Message sessions are identifiers for unbounded sequences of messages which are related in some way. A session is created when a message with a unique session ID (group-id in the standard properties of an AMPQ message) is sent to a topic or queue. Sessions don't have a defined timeout. When a session is started, producers and consumers can no longer work with standard messages. A session receiver is created by a client which accepts a session, i.e., it consumes messages with a specified session ID. Only one session receiver is allowed per session. A session receiver holds exclusive locks on all current and future messages within the session in the queue or subscription (when consuming from a topic). The lock is closed when the consumer is closed, or the lock expires. Sessions guarantee message ordering (FIFO) and at least one processing.

Topic filters and *actions* are tools for specifying and customizing messages which are forwarded to subscriptions. Subscribers define subscription rules, which consist of filter conditions and actions that annotate messages. Rules without actions are joined using an OR condition.

Messages which pass a filter with an action are duplicated. Actions add, update, or delete properties from a message. Filters can be SQL-like conditional expressions that are evaluated against the system or user-defined message properties (SQL filters), filters to allow all messages or block all messages (Boolean filters), and a set of values for message properties that need to match exactly (correlation filters).

Dead-letter queues are Azure-managed secondary sub-queues for invalid messages. Invalid messages are messages that couldn't be delivered to any consumer or that couldn't be processed. Dead-letter queues ensure messages aren't lost during application failures. Messages can be received and inspected from the queue. Producers can't publish messages directly to the dead-letter queue.

Message deferral is a mechanism for delayed message processing. When a queue or subscription client receives a message that should be processed but can't be done at that moment, the message is deferred. Deferred messages stay in the queue or subscription and can no longer be received by a consumer using regular operations. Deferred messages need to be received using the exact sequence number of the message. The sequence number is a unique identifier for a message assigned by the broker itself.

2.4.3. *RabbitMQ*

RabbitMQ is a message broker which offers reliability in the form of persistence, delivery acknowledgments, high availability using replication and streams, and flexibility in routing. The core RabbitMQ implementation is an open-source implementation of the AMQP 0-9-1 protocol. Additional support through plugins is provided for the AMQP 1.0, STOMP, MQTT, HTTP, and WebSocket protocols.

An architectural overview of RabbitMQ is given in a guide by Toshev [15]. RabbitMQ consists of the RabbitMQ broker, which contains the exchange and queues, and producers and consumers outside the broker.

The *broker* is a server running on a virtual machine node. All RabbitMQ brokers start out running on a single node. RabbitMQ supports cluster infrastructure with multiple nodes which communicate. Nodes are defined by unique names (*unique-prefix@hostname*). In a multi-node environment, all the data and state of a broker are replicated across all nodes, except standard message queues. All message queues are accessible across all nodes. Clients can connect to any node and have their requests routed to the node containing the queue they are trying to access.

The *exchange* is a RabbitMQ server endpoint to which the producers connect when publishing messages. It is identified by a unique key. The exchange is in charge of distributing message copies to queues using rules called *bindings*, which differ for different exchange types. Bindings are logical links between exchanges and queues which specify the routing semantics of messages. The exchange is declared with a number of attributes. The most important attributes are the name of the exchange, the type of the exchange, durability (if the exchange should survive the broker restart), auto-delete (if the exchange should be deleted if no queues are bound to it), and optional arguments. RabbitMQ contains a default exchange that is pre-declared by the broker using an empty string for a name. Messages sent to the default exchange are routed to queues using the queue name as a routing key. In the case where a routing key isn't set, the messages are sent round-robin across queues. Messages and message queues can be removed after a certain period (TTL/expiration). TTL can be set for messages in a single queue, multiple queues, or on a message-per-message basis. Queues can be deleted after a defined expiration time if they are not used [15].

A direct exchange is an exchange type where messages are delivered to queues based on the message routing key provided in the message header. A queue is bound to the exchange by a routing key. Messages published with the corresponding routing key are sent to the bound queue.

A fanout exchange is an exchange type where the routing key is ignored, and the messages are routed to all queues bound to the exchange. It is used to establish a broadcast pattern. One message is sent to the exchange, and the same is copied for each queue bound to the exchange.

A topic exchange is an exchange type where messages are delivered to a queue based on a routing filter between the topic exchange and the queues. The filter is evaluated based on the message routing key. Filters are regular expressions that need to be satisfied by the message routing key. For example, to filter the messages containing the word "invoice" in the routing key, the filter would be: "*invoice*". Topic exchanges are used to implement the publish–subscribe pattern.

A header exchange is an exchange type where messages are routed based on some values in the message headers, excluding the routing key. A queue is bound to the exchange using one or more header-matching filters. When binding the exchange, the "x-match" argument can be set to *all* or *any*. *All* will enforce an AND condition, and *any* will enforce an OR condition between filters.

The *queue* is a named message buffer connected to an exchange. Queues have additional properties which need to be defined during initialization. The durability property defines if a queue should survive a broker restart. Durable queues persist the messages in storage and place unprocessed messages back in the queue upon broker restart. The exclusiveness property defines a connection lock on the queue, meaning that the queue is used by one connection exclusively and deleted when that connection closes. The auto-delete property defines whether the queue should be deleted when the last consumer stops consuming messages.

The producer publishes messages to the exchange at a rate suitable for consumers. The TCP connection to exchange is wrapped by multiplexer channels. Channels are annotated with channel IDs. Publishing to a non-existent channel results in a channel error. The messages contain the message properties (headers) and the payload as a byte array. The only required message property is the delivery mode. The delivery mode can be transient and persistent. Transient messages are stored on disk only when they need to be evicted from memory during memory pressure. Persistent messages are stored on the disk as soon as they arrive in the queue. The client should also set the routing key to route the message to the correct queue (if the direct exchange type is used). The exchange can be explicitly set; if not, the messages are sent to the default exchange. The delivery tag is a unique identifier for a message set by the broker [15].

The *consumer* is an entity that subscribes to message queues. There are two ways for a consumer to receive messages. In *push* mode, the consumer subscribes to a queue, and the broker delivers the messages to the consumer from the queue. There can be multiple consumers per queue or an exclusive consumer (one consumer has an exclusive lock on the queue). Consumers are differentiated by a unique string, the *consumer tag*. In *pull* mode, the messages are polled from the queue. When a request to pull messages is made, all messages from the queue are returned to the consumer. The latter mode is less efficient and not recommended. RabbitMQ ensures at-least-once message delivery when using message acknowledgments. Without acknowledgments, message loss is possible, and only at-most-once message processing is guaranteed. RabbitMQ ensures messages are received in the order they were published when there is one consumer consuming data from a queue. In the case of multiple consumers in a queue, the guarantee is no longer valid.

Message acknowledgments are confirmations from the consumer to the broker that a message has been processed. Message acknowledgments are used to ensure reliability when an internal processing error or a network error occurs. Messages shouldn't be lost if an error occurs. There are two acknowledgment modes. Automatic acknowledgment mode considers a message successfully delivered immediately after it is sent to the consumer (written to a TCP socket). This is an unsafe mode used when performance is a priority over reliability. The manual acknowledgment mode requires the consumer to, positively or negatively, acknowledge a message or a batch of messages. Positive acknowledgment means the consumer successfully processed the message/messages, and it/they can be removed from the queue. Negative acknowledgment means the message/messages couldn't be processed and should be discarded, dead-lettered, or requeued by the broker. If the acknowledgment is not sent before a consumer is closed, the broker will redeliver the message to another available consumer [15].

Automatic requeuing is a safety mechanism used when manual message acknowledgment is used. Any message that was not acknowledged when a connection between the consumer and the broker is closed is automatically redelivered. The connection can be closed when a TCP connection from the client is lost, the consumer application fails, or during channel-level protocol exceptions. Messages which are redelivered have the *redeliver* property set to *true*.

The *prefetching message* is a feature in which a consumer specifies the number of messages it can receive before sending an acknowledgment for the messages. This feature is a load-balancing technique used to improve the consumption throughput of messages published in batches.

A *quorum queue* is a durable, replicated implementation of an ordered queue. Queues are replicated across nodes in the cluster. A consensus algorithm is used to determine the current state of the queue. This feature is used in systems where fault tolerance and data safety are the highest priority.

2.4.4. *NATS*

NATS is a secure, lightweight connective technology that is responsible for addressing, discovering, and exchanging messages [13]. The official documentation states that NATS is a message broker which offers M:N connectivity. It refers to a different number of two types of microservices that need to communicate. It is a self-healing system that scales up or down and handles topology systems without service downtime. It is easy to deploy on bare metal, virtual machines, container orchestration tools, and embedded systems.

NATS supports plain NATS connections, TLS-encrypted NATS connections, WebSocket NATS connections, and MQTT client connections. NATS is used in cloud messaging services, event streaming (data analytics), IoT, updating legacy messaging systems, etc. NATS implements multiple messaging patterns: the publish-subscribe pattern, the request-reply pattern, and queue groups [13].

NATS client applications connect to a NATS server. NATS server processes are fully interconnected, meaning the infrastructure can be horizontally scaled. Servers can be arranged into clusters of servers, and clusters can be arranged into superclusters. The NATS infrastructure consists of the NATS server, subjects, message producers, and message subscribers.

The *NATS server* is a machine that offers messaging services. The server is accessed by a URL (string), which specifies the IP address and port where the server is exposed and the type of connection (TCP, TLS, or WebSocket). Applications can be requested to authenticate themselves with a username and password, token, TLS certificate, or similar authentication schemes. The simplest version of the NATS server is a locally run *nats-server* process. To increase scalability and ensure fault tolerance, nats-server processes can be organized in a cluster. The client establishes a connection with only one server by using its URL. Multiple clusters can be organized into superclusters with clusters in multiple regions, virtual machines, cloud providers, etc. The client establishes a connection with any server in any cluster. Clusters and superclusters can be extended by leaf nodes, which are servers that act as a proxy between their client applications and the NATS service infrastructure. Messages flow into the leaf node and to the cluster based on leaf node connection permissions [13].

The *subject* is a string of characters that is used for message routing. A subject is used to simulate streams or topics. A producer publishes messages with a subject, and consumers subscribe to messages with certain subjects. The subject consists of tokens divided by the dot character. Wildcards can be used on the consumer side to subscribe to multiple subjects with a single subscription (e.g., *time.us.**).

The *producer* publishes messages to a subject. The message contains the subject, additional header fields, an optional "reply" address field, and the message payload in the form of a byte array. The subject in published messages must not contain wildcard characters. The publisher–subscriber pattern is implemented by default. To implement the request–reply pattern, the producer sends the messages to a certain subject. Listeners on that subject receive the messages and send replies to the "reply" subject. The replies are dynamically sent to the producer. When multiple instances of a subscriber listen on the same subject, one randomly chosen subscriber receives each request. That is, each request receives exactly one response [13].

The *consumer* subscribes to a subject to receive messages. The consumer can define subjects using wildcards, meaning that one consumer can receive messages from multiple subjects. Consumers can be arranged in a queue group. When multiple subscribers subscribe to a queue (same as a subject), each message is forwarded to exactly one subscriber. Queue groups provide a built-in load-balancing feature that enables workload scaling. Core NATS provides at-most-once message delivery because of possible data loss. NATS offers source-ordered delivery per publisher. This means messages from a single publisher will be delivered to all subscribers in the order in which they were published [13].

JetStream is a separate component built into NATS servers that enables distributed persistence. It enables message streaming and consumption at any time. Published messages are captured and stored in persistent storage. A NATS broker without the JetStream component is a fire-and-forget system, meaning that messages sent to a subject without a subscriber will be lost. JetStream allows for message replay, or delayed consumption. This ensures at-least-once and exactly-once message delivery. The messages can be consumed in multiple modes. The methods of consumption are the following: from a specific start time or sequence number, just the last message (from the stream of messages or a specific subject), and all messages.

When a defined limit on a stream of messages is reached (maximum message age, maximum total stream size, etc.), the arrival of new messages should be handled by discarding existing messages. The client can choose to discard new arriving messages and return an error or discard the oldest messages in the stream.

To consume messages from a stream, a special JetStream consumer needs to be created. The consumer defines how, when, and which messages to pull from a stream. The client then subscribes to the consumer to receive messages.

The *Key/Value Store* is a persistent storage for key–value pairs. The client can perform standard storage operations (CRUD operations) on the key–value store (bucket). Limits can be set on the maximum size of the bucket, the maximum size of any value, and the expiration time for values. An additional feature not found in most key–value stores is a subscription on changes in the key–value store. The client receives updates when a specific key or all keys in the bucket are updated [13].

2.4.5. *Apache Kafka*

Apache Kafka is an open-source distributed event-streaming platform [2]. The main features of Kafka include high throughput using a cluster of machines with minimal latencies; horizontal scalability of clusters; distributed, durable, and fault-tolerant message storage; and high availability over multiple geographic regions. It is a message broker which implements its own message protocol. The Kafka message protocol is a binary protocol that relies on the TCP protocol.

Kafka is used for event streaming. Event streaming is the practice of real-time data processing, storing, and routing to different destination technologies. Events are messages which contain a key, value, timestamp, and optional metadata. Example use cases of Kafka would be financial transaction processing, IoT data processing, medical data monitoring, automotive industry data tracking and monitoring, etc.

Apache Kafka implements the publish–subscribe pattern. In the high-level architecture of the Apache Kafka messaging system, the core components are the broker, producers, and consumers. There can be multiple brokers organized in a cluster. Consumers can be organized into consumer groups [6].

The *Kafka broker* is a server that acts as a bridge between producers and consumers. A Kafka cluster is a container for multiple Kafka brokers which act as peers (no master–slave node relationship). The broker contains one or more topics. Topics are partitioned event (message) buffers, and partitions are sub-buffers within the topic. Partitions of a topic can be located on different brokers to increase throughput and scalability. Partitions are replicated across brokers. There is one leader broker for the partition and zero or more follower brokers. All write operations for a partition are executed through the leader broker. The follower partitions consume data from the leader to stay up to date. The order of messages is guaranteed only within a partition (FIFO). Data from the topics is stored in a persistent storage. All topics have a default data retention policy. The data are kept in the storage for seven days, then the oldest messages are deleted. Other limits can be defined (e.g., stored message size), on a per-topic basis [6].

Kafka brokers are maintained by Zookeeper nodes. Zookeeper is a centralized service for maintaining configuration information, naming, and providing distributed synchronization in distributed systems. As stated in the book by Junqueira and Reed [10], Zookeeper detects crashes, implements topic discovery, and maintains the production and consumption states for Apache Kafka [2].

The *producer* publishes events to a topic on the broker. The client controls which partition the data is published through a partitioning function. The default partitioning function is the hash function. The client specifies the key to the event, and the partition is chosen based on the hash of the key. The partitioning function can be overwritten as well. The producer can work in an asynchronous mode where the messages (events) are batched to decrease the number of requests to the broker.

Messages that are published to the broker consist of a header, a key byte array, and a value byte array. Messages (Records) are always written in batches (record batches). Record batches contain all the records to be published and the metadata. The metadata contains the number of records, message compression type, timestamps, producer identifier, etc. Each record has its own optional metadata (headers) as key–value pairs. The record also contains the key, value, and timestamp. Each message is given a sequence number to prevent duplicate messages. Client applications usually provide wrappers around the record type [2].

The *consumer* receives data from the broker in a pull-based system. It issues a request to pull messages, containing the offset, to the leader of the partition it wants to consume from. The offset is an incremental integer value for messages arriving in the partition. The consumer stores the offset of the last processed message in Zookeeper. The broker responds to the consumers' request with all the messages in the partition, beginning from the offset, in the order they were written.

Consumers are arranged into consumer groups, which consist of multiple consumers consuming from the same topic. One partition can be consumed by exactly one consumer from the group, but one consumer from the group can consume from multiple partitions.

When publishing messages, the broker stores them on disk (also called, "committing to a log"). The message is marked as committed when a defined number of partition replicas have committed the messages. Consumers can read only the committed messages. This process guarantees at-least-once message delivery. Kafka also enables exactly one message delivery [6].

Kafka Connect is a tool built upon the Apache Kafka message broker. It is used for scalable and reliable streaming of data between Apache Kafka and other systems, such as databases, key–value stores, and search indexes [6]. It can run in standalone mode (single process) or distributed mode (multi-node cluster). It contains standardized connectors for integration with other systems. Connectors are added as properties files using the command line interface (standalone) or as a configuration in JavaScript Object Notation (JSON) format using an exposed REST API (distributed).

Kafka Connect requires a source connector and a sink connector. The source connector reads from a data source (e.g., PostgreSQL database, message broker) and publishes the data to a Kafka topic. The sink connector consumes the data from a Kafka topic and forwards it to other systems (e.g., Elasticsearch, Hadoop). The framework automatically commits the consumer offset at a defined rate [2].

The connectors can be configured to perform data transformations. The transformations are lightweight and used for simple data enrichment, modification, and event routing. Transformations are executed on every message which passes through the framework (e.g., insert an additional static field in each message).

Kafka Streams is a client library for fast and scalable processing of data stored in Kafka topics. It is based on streams, an abstraction that represents an unbounded, continuous flow of data records. A stream is ordered, repayable, and fault tolerant. The stream of data is processed by one or more stream processors. Stream processors are nodes that receive records from upstream processors, apply transformations to the data, and, potentially, produce records for downstream processors.

Source processors represent the entry point into the stream processor topology. The source processor consumes data from one or more Kafka topics and forwards it to downstream processors. Sink processors receive data from upstream processors and publish them to one or more Kafka topics. Common data transformations include filtering, joining, and aggregating data at certain time intervals (windows) or by a unique key [2].

2.4.6. *Message Broker Comparison*

Table 2.1 provides a comparison of five features of the reviewed message brokers.

Every broker implements the publish–subscribe pattern, meaning that messages can be sent from one publisher to multiple subscribers (one-to-many pattern). Service Bus and RabbitMQ natively enable point-to-point communication. In the point-to-point pattern, one publisher communicates directly with one subscriber. NATS offers two additional patterns specific to its architecture: the request–reply pattern, where a publisher receives responses from multiple consumers, and the queue groups pattern, where multiple consumers receive data from one publisher for load-balancing purposes [93–97].

Pub/Sub and Kafka use topics as a data structure that stores messages (in memory). RabbitMQ uses FIFO queues. Service Bus has the advantage of containing both topics and queues as message buffers. This provides users with two distinct options for two distinct use cases

Table 2.1. Message Broker Feature Comparison

	Pub/Sub	Service Bus	RabbitMQ	NATS	Kafka
Architectural pattern	Publish–subscribe	Publish–subscribe/point-to-point	Publish–subscribe/point-to-point	Publish–subscribe/request–reply, queue groups	Publish–subscribe
Message buffer	Topic	Topic/Queue	Queue	Subject	Topic
Message retention	Topic message retention	Message TTL	Message/queue TTL	Fire-and-forget	Topic message retention
Message ordering	Ordering key	Message sessions	By default, with one consumer	By default, with one publisher	Ordering key, within a partition
Reliability	At-least-once	At-least-once/exactly-once	At-most-once/exactly-once	At-most-once/at-least-once/exactly-once	At-least-once/exactly-once

(one-to-many or one-to-one communication). NATS uses an in-memory buffer of messages from which the data is immediately forwarded to subscribers based on a string called the "subject."

Message retention defines how long messages should persist in the internal buffer of messages. After the defined period has passed, messages are marked "dead" and are removed from the buffer. Pub/Sub and Kafka allow for a message retention period to be defined on a per-topic basis. For Pub/Sub, this is not defined by default. Kafka topics have a default retention period of seven days. Service Bus removes messages based on a TTL period defined on a per-message basis. RabbitMQ can remove messages based on a defined message TTL, or it can remove the entire queue (with messages) after a defined period. NATS doesn't hold messages in the internal buffer.

Pub/Sub and Kafka enable message order by defining a message ordering key. Pub/Sub requires all messages to be sent to the same server (geographical region) to guarantee order. Kafka guarantees order only within partitions of a topic. Service Bus does not guarantee orders by default. Order is established only within a message session. RabbitMQ and NATS have ordering guarantees by default in certain use cases. RabbitMQ requires a single consumer, and NATS requires a single producer.

Reliability refers to the reliability of message delivery, even when encountering failures of various kinds. RabbitMQ and NATS, in their default working modes, offer the least reliable message delivery (at most once), meaning that messages can be lost in the event of failure. However, they offer additional features to increase reliability. RabbitMQ offers exactly one delivery (highest reliability) when message acknowledgments are used, meaning that all sent messages are received exactly once. NATS offers at-least-once (messages can be duplicated) and exactly-once delivery when using the JetStream feature. Service Bus and Kafka offer at-least-once and exactly-once message deliveries. Pub/Sub is the only broker which does not provide exactly-once reliability.

2.5. Message-Oriented Architecture in Practice: A Worked Example

To showcase the usage of a message-oriented architecture, a Spring Boot application that uses Apache Kafka for message handling was implemented.

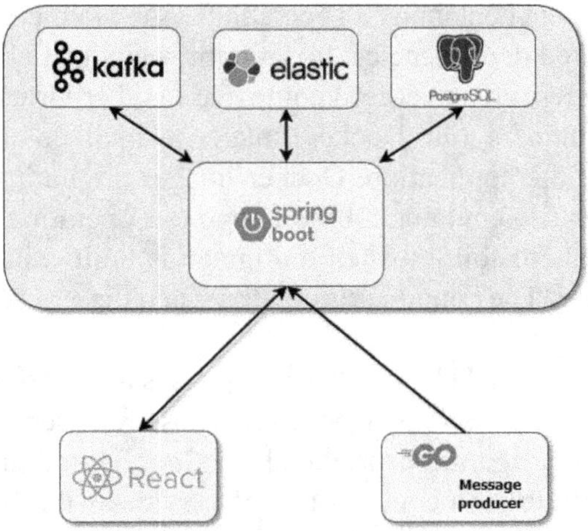

Figure 2.1. Message-oriented Architecture Example

The Spring Boot server communicates with Apache Kafka to asynchronously publish messages to Kafka topics. These messages are consumed in real-time. Messages published to a certain API endpoint are stored in the Elasticsearch search engine for easier querying. The server exposes a REST API, which enables client applications to produce and consume data from the server.

To showcase the usage of the Spring Boot server, two additional applications were implemented: A simple React chat application and a high throughput message producer in the Golang programming language. For the purposes of the React application, a PostgreSQL database was used to store mock users. Figure 2.1 shows a diagram representing the described architecture.

2.5.1. *Infrastructure*

The React application, the message producer, and the Spring Boot server were deployed as Docker containers from Docker files. External components for the Spring Boot server were deployed as Docker containers using Docker Compose.

As explained in a paper by Rad *et al.* [5], Docker is an open-source platform that is used to build, distribute, and run applications. Software is packaged in a standard form called a container. The container provides an isolated environment containing an operating system kernel, the application, and all the required dependencies. It is a lightweight and efficient environment with a low resource overhead compared to other virtualization tools.

Docker containers run Docker images, which contain the virtual environment for the application. Docker images are built from a Docker file. The Docker file contains a base image (an operating system) and a list of instructions to follow when the image is built using the `docker build` command. The container is created when the `docker run` command is executed [].

As stated by Jangla [11], Docker-compose is a tool for running multi-container Docker applications. Docker-compose runs services in containers based on a configuration provided in a configuration file in *YAML* format. The configuration contains named services with defined environments and Docker images. The containers are created by running the `docker-compose up` command.

Docker files are similar for all components, so one example is provided. The following example of a Docker file is used to create the Spring Boot server container (Code 2.1). The base image is defined with the `FROM` command. The base image is the Alpine distribution of Linux, which contains pre-installed Java Development Kit 16. The next three commands specify the working directory inside the image and copy *Maven* (a build tool for Java applications) related files from the local machine to the working directory of the image. The `RUN` command ensures that all the application dependencies are installed locally in the image. The next command copies the source code into the image. The following `EXPOSE` command exposes the port in the container which the Spring Boot application uses. The last command starts the application.

```
# syntax=docker/dockerfile:1
FROM openjdk:16-alpine3.13
WORKDIR src/
COPY .mvn/ .mvn
COPY mvnw pom.xml ./
RUN ./mvnw dependency:go-offline
COPY src ./src
EXPOSE 8080
CMD ["./mvnw", "spring-boot:run"]
```

Code 2.1. Spring Boot Application Docker File

The following example is a simplified Docker compose file used to create the external components for the Spring Boot server (Code 2.2). The Docker compose file contains configuration for Apache Kafka, Elasticsearch, and a PostgreSQL database. For each service, a container name, Docker image, and port mapping are required. Port mapping allows services outside of the Docker container to access the service provided by the container on a specific port.

```
version: '2'
services:
  zookeeper:
    container_name: zookeeper
    image: wurstmeister/zookeeper
    ports:
      - "2181:2181"
  kafka:
    container_name: kafka
    image: wurstmeister/kafka
    ports:
      - "9092:9092"
    depends_on:
      - zookeeper
  db:
    container_name: postgres
    image: postgres
    environment:
      POSTGRES_USER: ****
      POSTGRES_PASSWORD: ****
      POSTGRES_DB: users
    ports:
      - "5433:5432"
  elastic:
    container_name: es791
    image: elasticsearch:7.9.1
    ports:
      - "9200:9200"
```

Code 2.2. Docker Compose File Example

2.5.2. *Spring Boot Application*

Spring Boot is an open-source Java-based framework for creating standalone and production-ready Spring applications. It contains an embedded Tomcat server, which is used to serve HTTP requests from clients. The framework provides libraries that enable easy communication with external services, such as Elasticsearch, PostgreSQL, and Apache Kafka.

The Spring Boot application is the core of the message-oriented architecture. It communicates with Apache Kafka using the Spring Apache Kafka library [6]. Messages which are received from the clients are published to Kafka topics. The application requires user configuration for the location of the Kafka broker. Kafka topics and consumer groups, for handling messages from different client applications, have default values that can be overridden. The configuration is provided in the form of environment variables. The two Kafka topics are automatically created if they do not exist. The topics have three partitions, and each partition is replicated three times. The messages will not have specific keys because ordering is not a requirement. Messages will be load-balanced in a round-robin manner across the three partitions. When the application is started, the Kafka producers and consumers are configured and initialized.

Code 2.3 contains the configuration for both Kafka producers. The producer is configured to wait for acknowledgment from the broker that the message was successfully published to all replicas of the topic. No duplicate messages are tolerated. The producer will attempt to republish messages 10 times in the event of a recoverable error. Messages are sent to the broker in batches of a maximum of 5,000 bytes in size. The broker is defined dynamically from a variable containing the *KAFKA_BROKER* environment variable set by the user. Finally, as messages arrive at the

```
configurations.put(ProducerConfig.ACKS_CONFIG, "all");

configurations.put(ProducerConfig.ENABLE_IDEMPOTENCE_CONFIG, true);

configurations.put(ProducerConfig.RETRIES_CONFIG, 10);

configurations.put(ProducerConfig.BATCH_SIZE_CONFIG, 5000);

configurations.put(ProducerConfig.BOOTSTRAP_SERVERS_CONFIG,
KafkaConfiguration.KAFKA_BROKER);

configurations.put(ProducerConfig.KEY_SERIALIZER_CLASS_CONFIG,
StringSerializer.class);

configurations.put(ProducerConfig.VALUE_SERIALIZER_CLASS_CONFIG,
JsonSerializer.class);
```

Code 2.3. Kafka Producer Configuration

```
configurations.put(ConsumerConfig.BOOTSTRAP_SERVERS_CONFIG,
KafkaConfiguration.KAFKA_BROKER);

configurations.put(ConsumerConfig.GROUP_ID_CONFIG,
KafkaConfiguration.CHAT_GROUP_ID);

configurations.put(ConsumerConfig.KEY_DESERIALIZER_CLASS_CONFIG,
StringDeserializer.class);

configurations.put(ConsumerConfig.VALUE_DESERIALIZER_CLASS_CONFIG,
JsonDeserializer.class);

configurations.put(ConsumerConfig.AUTO_OFFSET_RESET_CONFIG, "latest");

configurations.put(ConsumerConfig.ENABLE_AUTO_COMMIT_CONFIG, true);
```

Code 2.4. Kafka Consumer Configuration

broker in byte format, a conversion method for the message key and the data is required. The data are converted to bytes using Java classes, which implement this functionality based on the type of data. The key is in string format, and a message is a JSON object.

Code 2.4 contains the configuration for both consumers. The broker is defined by the user. The consumer group defaults to “chat” if not defined by the user. The messages consumed by the broker are in byte format. The conversion to regular data types is done using the same Java classes as the producer. The consumer will start consuming data from the “end” of the topic, meaning that it will not consume messages from before it was started. The consumer will periodically commit the offset for each partition of the topic.

The server exposes three groups of REST API endpoints, intended for use by two different client applications.

The first and second groups of endpoints are intended for use with the React chat application. The first group exposes API endpoints to handle users used in the application. Standard *CRUD* operations are supported. The users are stored in a PostgreSQL database. The users are defined by a unique identifier, username, and password. The application communicates with the database using the Spring PostgreSQL library. The second group of endpoints enables clients to send messages in a specified format. The messages need to contain an identifier as a string (`id`), the name of the sender as a string (`sender`), the identifier of the sender as a numerical value (`senderId`), the identifier of the user receiving the message as a numerical value (`receiverId`), the message content as a string (`content`), and an optional timestamp as a string (`timestamp`). If the timestamp is not set on the client side, the server will set it as the time when the message was processed. Messages in the specified format are sent as

POST requests on the `HTTP` endpoint, containing the message content in the request body. The host varies depending on the environment the Spring Boot container is running in. Messages sent to this endpoint are published to a Kafka topic. The messages are concurrently consumed and sent to a WebSocket connection established with the React application [98–103].

WebSocket is a tool that enables a two-way interactive communication session between a browser and a server. When a connection is established on a certain endpoint, the browser can receive event-driven responses without having to send explicit requests.

The second group of endpoints is intended for use with the message producer. It exposes an endpoint to send messages and publish them to a Kafka topic (`http://<host>/api/chat/send`) and multiple endpoints to retrieve messages based on defined parameters (sender, sender–receiver combination, message ID, or all messages). The messages which are published to the Kafka topic are consumed in real-time and stored in Elasticsearch using the Spring Elasticsearch library.

Andhavarapu [1] described Elasticsearch as a highly scalable, distributed search engine. It supports simple search queries and complex filtering and aggregations of data. The data are stored in JSON format. The data are split across multiple servers (nodes), which work together to load-balance requests to the exposed REST API. The API is usually used over the provided client libraries.

The messages are stored in Elasticsearch for their querying possibilities. Apache Kafka does not enable message querying based on defined parameters [6]. The data can only be processed by consumers which read data from a certain offset in a partition. Also, the messages are kept on the disk only for a certain retention period or until a certain memory limit is reached. Storing data in Elasticsearch ensures its durability of data.

2.5.3. *React Chat Application*

React is an open-source front-end JavaScript library. It enables the building of user interfaces based on UI components. UI components are built from one or more encapsulated components that manage their state. The React application built as part of this study is a simple usage example of the implemented Spring Boot application.

The core component of chat applications is the message. Messages are passed between users in a one-to-one or one-to-many mapping. This implicates two architectural patterns: point-to-point and publish–subscribe.

User applications are the publishers and the subscribers, and a central server is the message broker. The message broker in the center of the architecture allows for a temporal and geographical decoupling between publishers and subscribers, i.e., the users who are communicating. This React application demonstrates this functionality on a smaller scale.

The Spring Boot server and the React application communicate over the exposed API. The messages from users are sent to the `http://<host>:8080/api/chat/send` endpoint. This endpoint publishes the messages to a Kafka topic. This topic is concurrently consumed, and consumed messages are sent to the `http://<host>:8080/api/chat/socket/messages` endpoint. In the React application, messages are received using the React *SockJsClient* component. This component opens a WebSocket connection on the aforementioned endpoint and consumes messages arriving through the connection in real-time. The messages contain the identifications of the sender and receiver. Using this information, the messages are displayed to the correct users.

The user first needs to log in to the application with a username and password. The user can be registered using a simple form. The username must be unique. The users are stored in the PostgreSQL database. Figure 2.2 shows the login screen of the application. The username is required for displaying and routing messages.

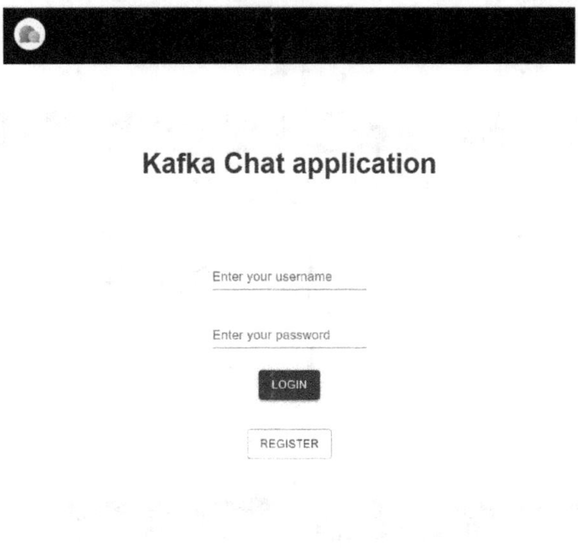

Figure 2.2. User Login

The user can then pick whether to talk to all users or a specific user. Figure 2.3 shows the menu to pick a group (publish–subscribe pattern) or a single user (point-to-point pattern) for two different users. The menu is filled with data from the database. When the user is picked, the ID of the user is stored. This ID becomes the receiverId field in the messages sent to this user. If the user chooses to talk to the group, the `receiver ID` is `-1`.

Figure 2.4 shows two messages exchanged between user Peter and user Daniel. The messages are displayed almost instantly to both users.

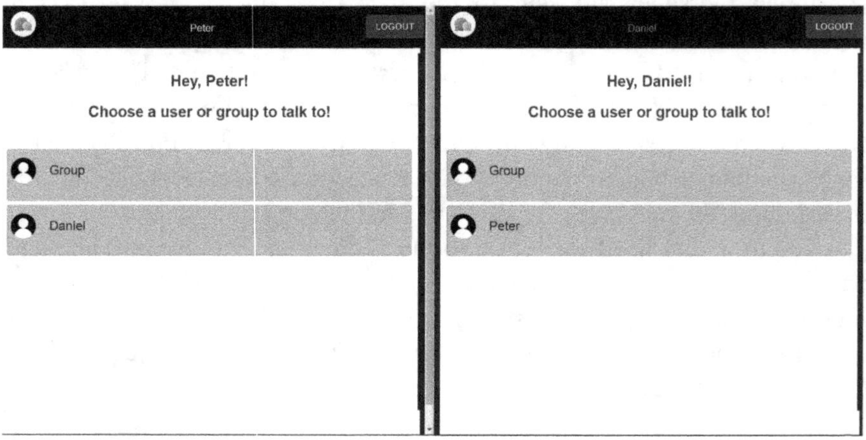

Figure 2.3 User Drop-down Menu

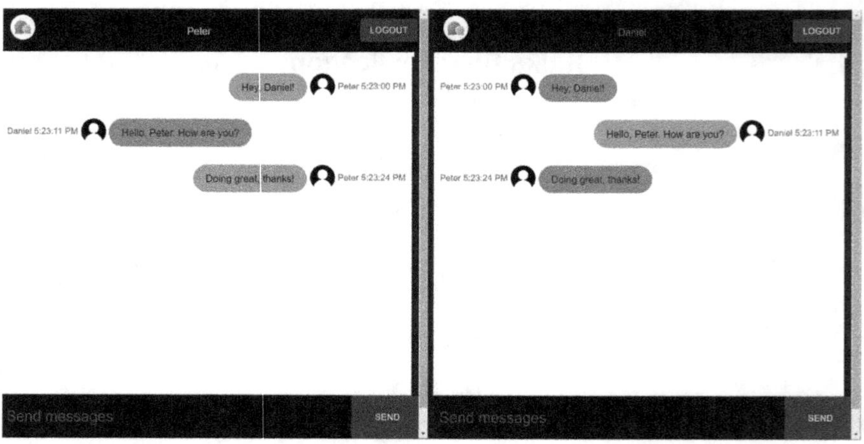

Figure 2.4. Point-to-point Communication

Figure 2.5. User Messages in the Topic

The messages contain the timestamp added by the Spring server. The timestamp represents the time the messages were published to the Kafka topic. Figure 2.5 shows these messages in the *messages* Kafka topic. The messages were consumed in an open-source tool by Apache called *Conduktor*. Conduktor enables the creation of a consumer on the Kafka topic. Since reading messages from a Kafka topic is not destructive, the messages can be consumed by the React application and the consumer in Conduktor simultaneously.

2.5.4. *Message Producer*

Message-oriented systems operate with messages on a large scale. One such example is an extract, transform, and load (ETL) pipeline. ETL is a data integration process that combines data from multiple data sources, applies transformations to the data, and stores the data in target systems. Data can be extracted from databases, files, or generated through message producers. The transformations include data filtering, cleansing, de-deduplicating, validating, etc. The target systems include databases, data warehouses, message brokers, etc. Message brokers are standard components of ETL transformations. In addition to their use as target systems, they can be used to decouple steps of the ETL pipeline, extract data that needs to be processed by another system, etc.

A high-throughput message producer in the Golang programming language was implemented to simulate a source of data. The data are sent to the Spring Boot server, which acts as an intermediary between the source and the message broker. The server stores the data in the Elasticsearch search engine. The server is in charge of the data transformation and data loading steps of the ETL pipeline.

The message generator is an interface that contains a function that generates an array of messages to be sent to the server. The number of

```
type Generator interface {
Generate(n int) ([]*publisher.Message, error)
}
```

Code 2.5. Generator Interface

```
type Publisher interface {
      Publish(messages []*Message) error
}
```

Code 2.6. Publisher Interface

messages to be generated is a function argument. The interface definition is given in Code 2.5.

The interface was implemented with a generator that creates mock messages in the same format as for the React application. The number of messages to be generated is supplied to the application as an environment variable (*NUM_OF_MESSAGES*).

The *id* field is a unique integer in string format. The usernames are chosen from a file containing common American names. Every username (the *senderId* field) is given a unique ID (*senderId*). The content is a string containing randomly generated characters. The *receiverId* field is a randomly selected *senderId* from the pool of created users, without the *senderId* of the user who is sending the message. The message is given the current timestamp as the time of message creation.

The message publisher is an interface that contains a function that publishes messages to any target. The message array is passed as a function argument. The function definition is given in Code 2.6.

The implementation of the interface publishes messages to the Spring Boot server API. The API endpoint is supplied to the application as an environment variable (*API_URL*).

A high throughput of messages is realized by parallelizing message publishing. Each message is transformed into bytes and sent as the body of a *POST* request to the server's API in a separate goroutine. *Goroutines* are lightweight execution threads that execute functions concurrently with the rest of the program. The resource cost of creating a goroutine is very low compared to traditional threads. Because of this, a large number of requests can be made to the server without waiting on a response from the server for each consecutive request. Messages sent to the API are immediately published to a Kafka topic. On average, 5,000 messages can be published to the API in 3.69 s [104–108].

The messages published to the API and, through that, to the Kafka topic are concurrently consumed. After the message is consumed, a transformation in the form of data enrichment is done. The new message contains all the received data and the processing timestamp added by the consumer. The new data are then loaded into the Elasticsearch search engine. The publishing and consumption of messages are temporally decoupled by using the message broker as an intermediary. The asynchronous nature of the publisher and consumer components enhances performance since the consumption of data is a performance bottleneck. Publishing can be faster because it does not wait for the response from the consumption process.

Data can be queried using the exposed API. Figure 2.6 shows a request made from the Postman tool. The response is the message extracted from Elasticsearch with message ID 2728.

Figure 2.7 shows the request to count messages by the user with the username "James." Figure 2.8 shows the request to receive messages by the user with the username "James."

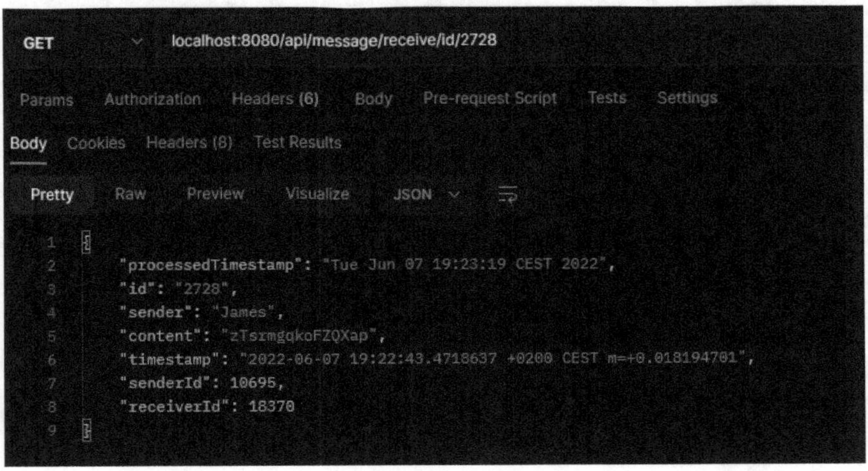

Figure 2.6. Postman Query for a Specific Message

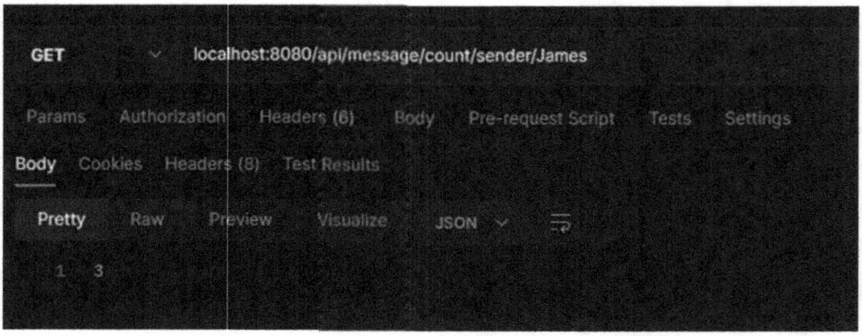

Figure 2.7. Postman Query for a Count of Messages by User

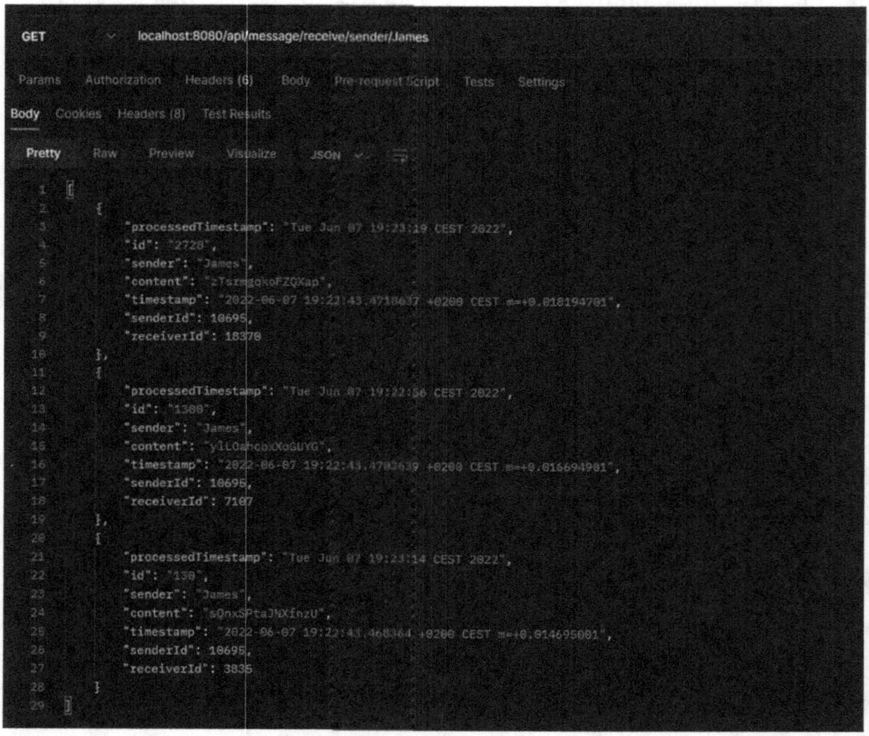

Figure 2.8. Postman Query for Messages by a Specific User

2.6. Conclusion

This chapter provided an overview of the components of message-oriented architectures. Message-oriented architectures provide an effective means of communication between components in distributed systems.

Multiple components (microservices) can communicate with each other in an asynchronous and temporally and spatially decoupled way.

The central components of a message-oriented architecture are messages and message brokers. Message brokers provide a central interface for publishing and consuming messages. The messages are, preferably, small data packets in a certain format. The messages contain a problem-specific payload and the metadata which describes the payload. Message brokers provide messaging protocol implementations.

Messaging protocols, or wire protocols, are application-level protocols that describe the format of the data which is exchanged over a network connection. The data packets which are exchanged between components can be in byte format (binary protocols) or textual format (text-based protocols). Message brokers provide a level of abstraction above the messaging protocols. They provide their own message formats, on top of the protocol messages, which enable additional messaging features.

The basic component of a broker is a message buffer. Depending on the architecture type, the buffer can be a queue (point-to-point architecture) or a topic (publish–subscribe architecture). The basic features of message brokers are message publishing and consumption. Publishers create data and send it to the broker. The broker stores data in message buffers and passes them to the listening consumers. The broker can remember the state of its buffers (persistent storage) or immediately pass data to consumers.

While the underlying concept remains the same, different message broker implementations provide distinct features and varying performances. Depending on the use case and business requirements, a system architect needs to choose the optimal message broker.

References

1. Andhavarapu, A. (2017). *Learning Elasticsearch.* Packt Publishing Ltd.: Birmingham.
2. *Apache Kafka* (2022). https://kafka.apache.org/ (accessed 30 May 2022).
3. Apukhtin, V., Shirokopetleva, M., and Skovorodnikova, V. (2019). The relevance of using message brokers in robust enterprise applications. In *2019 IEEE International Scientific-Practical Conference Problems of Infocommunications, Science and Technology (PIC S&T)*, Kyiv, pp. 305–309.
4. Banks, A., Briggs, E., Borgendale, K., and Gupta, R. (2019). *MQTT Version 5.0.* OASIS Standard. https://docs.oasis-open.org/mqtt/mqtt/v5.0/mqtt-v5.0.html.

5. Bashari Rad, B., Bhatti, H. J., and Ahmadi, M. (2017). An introduction to Docker and analysis of its performance. *IJCSNS International Journal of Computer Science and Network Security*, *17*(3), 228–235.
6. Garg, N. (2013). *Apache Kafka.* Packt Publishing Ltd.: Birmingham.
7. Godfrey, R., Ingham, D., and Schloming, R. (2019). *OASIS Advanced Message Queuing Protocol (AMQP) Version 1.0*. OASIS Standard. http://docs.oasis-open.org/amqp/core/v1.0/amqp-core-messaging-v1.0.html.
8. Guner, A., Kurtel, K., and Celikkan, U. (2017). A message broker based architecture for context-aware IoT application development. In *2017 International Conference on Computer Science and Engineering (UBMK)*, Antalya, pp. 233–238.
9. Intorruk, S. and Numnonda, T. (2019). A comparative study on performance and resource utilization of real-time distributed messaging systems for big dana. In *20th IEEE/ACIS International Conference on Software Engineering, Artificial Intelligence, Networking and Parallel/Distributed Computing (SNPD)*, Toyama, pp. 102–107.
10. Junqueira, F. and Reed, B. (2013). *Zookeeper*. O'Reilly Media, Inc.: Sebastopol.
11. Jangla, K. (2018). *Accelerating Development Velocity Using Docker: Docker Across Microservices*. Apress: Berkeley.
12. Magnoni, L. (2015). Modern messaging for distributed systems. *Journal of Physics: Conference Series, Vol. 608, 16th International workshop on Advanced Computing and Analysis Techniques in physics research (ACAT2014)*. 1–5 September 2014, Prague.
13. *Nats.io*. (2022). https://nats.io/ (accessed 29 May 2022).
14. *STOMP*. (2022). https://stomp.github.io/ (accessed 21 May 2022).
15. Toshev, M. (2015). *Learning RabbitMQ*. Packt Publishing Ltd.: Birmingham.
16. Vinoski, S. (2006). Advanced message queuing protocol. *IEEE Internet Computing*, *10*(6), 87–89.
17. Wise, M. J. and McDermott, T. (1993). Message-brokers: A novel interprocess communications primitive. In *Proceedings of the Twenty-sixth Hawaii International Conference on System Sciences*, Wailea, 8 January 1993, pp. 184–193.
18. Yongguo, J., Qiang, L., Changshuai, Q., Jian, S., and Qianqian, L. (2019). Message-oriented middleware: A review. In *2019 5th International Conference on Big Data Computing and Communications (BIGCOM)*, QinqDao, 9–11 August, 2019, pp. 88–97.
19. Kandasamy, N., Ahmad, F., Ajitha, D., Raj, B., and Telagam, N. (2020). Quantum dot cellular automata based scan flip flop and boundary scan register. *IETE Journal of Research*, *66*, 535–548.

20. Sharma, S. K., Raj, B., and Khosla, M. (2019). Enhanced photosensivity of highly spectrum selective cylindrical gate $In_{1-x}Ga_xAs$ nanowire MOSFET photodetector. *Modern Physics Letter B, 33*(12), 1950144.
21. Singh, J. and Raj, B. (2019). Design and investigation of 7T2M NVSARM with enhanced stability and temperature impact on store/restore energy. *IEEE Transactions on Very Large Scale Integration Systems, 27*(6), 1322–1328.
22. Bhardwaj, A. K., Gupta, S., Raj, B., and Singh, A. (2019). Impact of double gate geometry on the performance of carbon nanotube field effect transistor structures for low power digital design. *Computational and Theoretical Nanoscience, 16*, 1813–1820.
23. Jain, N. and Raj, B. (2019). Thermal stability analysis and performance exploration of asymmetrical dual-k underlap spacer (ADKUS) SOI FinFET for security and privacy applications. *Indian Journal of Pure & Applied Physics (IJPAP), 57*, 352–360.
24. Singh, A., Khosla, M., and Raj, B. (2019). Design and analysis of dynamically configurable electrostatic doped carbon nanotube tunnel FET. *Microelectronics Journal, 85*, 17–24.
25. Jain, N. and Raj, B. (2019). Dual-k spacer region variation at the drain side of asymmetric SOI FinFET structure: Performance analysis towards the analog/RF design applications. *Journal of Nanoelectronics and Optoelectronics, 14*, 349–359.
26. Singh, J., Sharma, S., Raj, B., and Khosla, M. (2018). Analysis of barrier layer thickness on performance of $In_{1-x}Ga_xAs$ based gate stack cylindrical gate nanowire MOSFET. *JNO, 13*, 1473–1477.
27. Jain, N. and Raj, B. (2018). Analysis and performance exploration of high-k SOI FinFETs over the conventional low-k SOI FinFET toward analog/RF design. *Journal of Semiconductors (JoS), 39*(12), 124002-1-7.
28. Goyal, C., Ubhi, J. S., and Raj, B. (2018). A reliable leakage reduction technique for approximate full adder with reduced ground bounce noise. *Journal of Mathematical Problems in Engineering, 2018*, 16. Article ID 3501041.
29. Anuradha, J. S., Raj, B., and Khosla, M. (2018). Design and performance analysis of nano-scale memristor-based nonvolatile SRAM. *Journal of Sensor Letter, 16*, 798–805.
30. Wadhwa, G. and Raj, B. (2018). Parametric variation analysis of charge-plasma-based dielectric modulated JLTFET for biosensor application. *IEEE Sensor Journal, 18*(15), 6070–6077.
31. Singh, J. and Raj, B. (2018). Comparative analysis of memristor models for memories design. *JoS, 39*(7), 074006.
32. Yadav, D., Chouhan, S. S., Vishvakarma, S. K., and Raj, B. (2018). Application specific microcontroller design for IoT based WSN. *Sensor Letter, 16*, 374–385.

33. Singh, G., Sarin, R. K., and Raj, B. (2018). Fault-tolerant design and analysis of quantum-dot cellular automata based circuits. *IEEE/IET Circuits, Devices & Systems*, *12*, 638–64.
34. Singh, J. and Raj, B. (2018). Modeling of mean barrier height levying various image forces of metal insulator metal structure to enhance the performance of conductive filament based memristor model. *IEEE Nanotechnology*, *17*(2), 268–267.
35. Jain, A., Sharma, S., and Raj, B. (2018). Analysis of triple metal surrounding gate (TM-SG) III-V nanowire MOSFET for photosensing application. *Opto-electronics Journal, 26*(2), 141–148.
36. Jain, A., Sharma, S., and Raj, B. (2016). Design and analysis of high sensitivity photosensor using cylindrical surrounding gate MOSFET for low power sensor applications. *Engineering Science and Technology, an International Journal*, *19*(4), 1864–1870.
37. Singh, A., Khosla, M., and Raj, B. (2017). Analysis of electrostatic doped Schottky barrier carbon nanotube FET for low power applications. *Journal of Materials Science: Materials in Electronics*, *28*, 1762–1768.
38. Kumar, G. S., Singh, A., and Raj, B. (2018). Design and analysis of gate all around CNTFET based SRAM cell design. *Journal of Computational Electronics*, *17*(1), 138–145.
39. Singh, G. P., Sohi, B. S., and Raj, B. (2017). Material properties analysis of graphene base transistor (GBT) for VLSI analog circuits. *Indian Journal of Pure & Applied Physics (IJPAP)*, *55*, 896–902.
40. Singh, A., Khosla, M., and Raj, B. (2016). Comparative analysis of carbon nanotube field effect transistor and nanowire transistor for low power circuit design. *Journal of Nanoelectronics and Optoelectronics*, *11*, 388–393.
41. Kumar, S. and Raj, B. (2017). Estimation of stability and performance metric for inward access transistor based 6T SRAM cell design using n-type/p-type DMDG-GDOV TFET. *IEEE VLSI Circuits and Systems Letter*, *3*(2), 25–39.
42. Sharma, S., Kumar, A., Pattanaik, M., and Raj, B. (2013). Forward body biased multimode multi-threshold CMOS technique for ground bounce noise reduction in static CMOS adders. *International Journal of Information and Electronics Engineering*, *3*(3), 567–572.
43 Singh, H., Kumar, P., and Raj, B. (2013). Performance analysis of majority gate SET based 1-bit full adder. *International Journal of Computer and Communication Engineering* (IJCCE), *2*(4), 1–8.
44. Bhardwaj, A. K., Gupta, S., and Raj, B. (2020). Investigation of parameters for Schottky barrier (SB) height for Schottky barrier based carbon nanotube field effect transistor device. *Journal of Nanoelectronics and Optoelectronics*, *15*, 783–791.

45. Bansal, P. and Raj, B. (2017). Memristor: A versatile nonlinear model for dopant drift and boundary issues. *JCTN*, *14*(5), 2319–2325.
46. Jain, N. and Raj, B. (2016). An analog and digital design perspective comprehensive approach on fin-FET (fin-field effect transistor) technology — A review. *Reviews in Advanced Sciences and Engineering (RASE)*, *5*, 1–14.
47 Sharma, S. Raj, B., and Khosla, M. (2017). Subthreshold performance of $In_{1-x}Ga_xAs$ based dual metal with gate stack cylindrical/surrounding gate nanowire MOSFET for low power analog applications. *Journal of Nanoelectronics and Optoelectronics*, *12*, 171–176.
48. Raj, B., Saxena, A. K., and Dasgupta, S. (2009). Analytical modeling for the estimation of leakage current and subthreshold swing factor of nanoscale double gate FinFET device. *Microelectronics International*, *26*, 53–63.
49. Soniya, S. S., Wadhwa, G., and Raj, B. (2019). An analytical modeling for dual source vertical tunnel field effect transistor. *International Journal of Recent Technology and Engineering (IJRTE)*, *8*(2), 603–608.
50. Singh, S. and Raj, B. (2019). Design and analysis of hetrojunction vertical T-shaped tunnel field effect transistor. *Journal of Electronics Material*, *48*(10), 6253–6260.
51. Goyal, C., Ubhi, J. S., and Raj, B. (2019). A low leakage CNTFET based inexact full adder for low power image processing applications. *International Journal of Circuit Theory and Applications*, *47*(9), 1446–1458.
52. Raj, B., Saxena, A. K., and Dasgupta, S. (2008). A compact drain current and threshold voltage quantum mechanical analytical modeling for FinFETs. *Journal of Nanoelectronics and Optoelectronics (JNO)*, *3*(2), 163–170.
53. Wadhwa, G. and Raj, B. (2020). An analytical modeling of charge plasma based tunnel field effect transistor with impacts of gate underlap region. *Superlattices and Microstructures*, *142*, 106512.
54. Singh, S. and Raj, B. (2020). Modeling and simulation analysis of SiGe hetrojunction double gatevertical t-shaped tunnel FET. *Superlattices and Microstructures*, *142*, 106496.
55 Singh, A., Saini, D. K., Agarwal, D., Aggarwal, S., Khosla, M., and Raj, B. (2016). Modeling and simulation of carbon nanotube field effect transistor and its circuit application. *Journal of Semiconductors (JoS)*, *37*, 074001-6.
56. Jain, N. and Raj, B. (2016). Device and circuit co-design perspective comprehensive approach on FinFET technology — A review. *Journal of Electron Devices*, *23*(1), 1890–1901.
57. Kumar, S. and Raj, B. (2016). Analysis of I_{ON} and Ambipolar current for dual-material gate-drain overlapped DG-TFET. *Journal of Nanoelectronics and Optoelectronics*, *11*, 323–333.

58. Anjum, N., Bali, T., and Raj, B. (2013). Design and simulation of handwritten multiscript character recognition. *International Journal of Advanced Research in Computer and Communication Engineering, 2*(7), 2544–2549.
59. Sharma, S., Raj, B., and Khosla, M. (2016). A Gaussian approach for analytical subthreshold current model of cylindrical nanowire FET with quantum mechanical effects. *Microelectronics Journal, 53*, 65–72.
60. Singh, K. and Raj, B. (2015). Performance and analysis of temperature dependent Multi-walled Carbon Nanotubes as Global Interconnects at different technology nodes. *Journal of Computational Electronics, 14*(2), 469–476.
61. Kumar, S. and Raj, B. (2015). Compact channel potential analytical modeling of DG-TFET based on evanescent–mode approach. *Journal of Computational Electronics, 14*(2), 820–827.
62. Singh, K. and Raj, B. (2015). Temperature dependent modeling and performance evaluation of multi-walled CNT and single-walled CNT as global interconnects. *Journal of Electronic Materials, 44*(12), 4825–4835.
63. Sharma, V. K., Pattanaik, M., and Raj, B. (2014). INDEP approach for leakage reduction in nanoscale CMOS circuits. *International Journal of Electronics, 102*(2), 200–215.
64. Singh, K. and Raj, B. (2015). Influence of temperature on MWCNT bundle, SWCNT bundle and copper interconnects for nanoscaled technology nodes. *Journal of Materials Science: Materials in Electronics, 26*(8), 6134–6142.
65. Anjum, N., Bali, T., and Raj, B. (2013). Design and simulation of handwritten Gurumukhi and Devanagri numerical recognition. *International Journal of Computer Applications, 73*(12), 16–21.
66. Khandelwal, S., Gupta, V., Raj, B., and Gupta, R. D. (2015). Process variability aware low leakage reliable nano scale DG-FinFET SRAM cell design technique. *Journal of Nanoelectronics and Optoelectronics, 10*(6), 810–817.
67. Sharma, V. K., Pattanaik, M., and Raj, B. (2014). ONOFIC approach: Low power high speed nanoscale VLSI circuits design. *International Journal of Electronics, 101*(1), 61–73.
68. Khandelwal, S., Raj, B., and Gupta, R. D. (2015). FinFET based 6T SRAM cell design: Analysis of performance metric, process variation and temperature effect. *Journal of Computational and Theoretical Nanoscience, 12*, 2500–2506.
69. Singh, S., Yadav, S., Rahul, J., Srivastava, A., and Raj, B. (2015). Impact of HfO_2 in graded channel dual insulator double gate MOSFET. *Journal of Computational and Theoretical Nanoscience, 12*(6), 950–953.
70. Sharma, V. K., Pattanaik, M., and Raj, B. (2014). PVT variations aware low leakage INDEP approach for nanoscale CMOS circuits. *Microelectronics Reliability, 54*, 90–99.

71. Raj, B., Saxena, A. K., and Dasgupta, S. (2013). Quantum mechanical analytical modeling of nanoscale DG FinFET: Evaluation of potential, threshold voltage and source/drain resistance. *Journal of Material Science in Semiconductor Processing, 16*(4), 1131–1137.
72. Gopal, M., Prasad, S. S. D., and Raj, B. (2013). 8T SRAM cell design for dynamic and leakage power reduction. *International Journal of Computer Applications, 71*(9), 43–48.
73. Pattanaik, M., Raj, B., Sharma, S., and Kumar, A. (2012). Diode based trimode multi-threshold CMOS technique for ground bounce noise reduction in static CMOS adders. *Advanced Materials Research, 548*, 885–889.
74. Raj, B., Saxena, A. K., and Dasgupta, S. (2011). Nanoscale FinFET based SRAM cell design: Analysis of performance metric, process variation, underlapped FinFET and temperature effect. *IEEE Circuits and System Magazine, 11*(2), 38–50.
75. Sharma, V. K., Pattanaik, M., and Raj, B. (2013). Leakage current ONOFIC approach for deep submicron VLSI circuit design. *International Journal of Electrical, Computer, Electronics and Communication Engineering, 7*(4), 239–244.
76. Chawla, T., Khosla, M., and Raj, B. (2021). Design and simulation of triple metal double-gate germanium on insulator vertical tunnel field effect transistor. *Microelectronics Journal, 114*, 105125.
77. Kaur, P., Gill, S. S., and Raj, B. (2022). Comparative analysis of OFETs materials and devices for sensor applications. *Journal of Silicon, 14*, 4463–4471.
78. Sharma, S. K., Kumar, P., Raj, B., and Raj, B. (2022). $In_{1-x}Ga_xAs$ double metal gate-stacking cylindrical nanowire MOSFET for highly sensitive photo detector. *Journal of Silicon, 14*, 3535–3541.
79. Raj, B., Saxena, A. K., and Dasgupta, S. (2009). Analytical modeling of quasi planar nanoscale double gate FinFET with source/drain resistance and field dependent carrier mobility: A quantum mechanical study. *Journal of Computer (JCP), 4*(9), 1–8.
80. Bhushan, S., Khandelwal, S., and Raj, B. (2012). Analyzing different mode FinFET based memory cell at different power supply for leakage reduction. In *Seventh International Conference on Bio-Inspired Computing: Theories and Application, (BIC-TA 2012) Advances in Intelligent Systems and Computing*, 14–16 December 2012, IIITM Gwalior, India, 202, pp. 89–100.
81. Singh, J. and Raj, B. (2018). Temperature dependent analytical modeling and simulations of nanoscale memristor. *Engineering Science and Technology, an International Journal, 21*, 862–868.
82. Singh, S., Bala, S., Raj, B., and Raj, B. (2020). Improved sensitivity of dielectric modulated junctionless transistor for nanoscale biosensor design. *Sensor Letter, 18*, 328–333.

83. Kumar, V., Vishvakarma, S. K., and Raj, B. (2020). Design and performance analysis of ASIC for IoT applications. *Sensor Letter ASP*, *18*, 31–38.
84. Jaiswal, A., Sarin, R. K., Raj, B., and Sukhija, S. (2018). A novel circular slotted microstrip-fed patch antenna with three triangle shape defected ground structure for multiband applications. *Advanced Electromagnetic (AEM)*, *7*(3), 56–63.
85. Wadhwa, G. and Raj, B. (2018). Label free detection of biomolecules using charge-plasma-based gate underlap dielectric modulated junctionless TFET. *Journal of Electronic Materials (JEMS)*, *47*(8), 4683–4693.
86. Singh, G., Sarin, R. K., and Raj, B. (2017). Design and performance analysis of a new efficient coplanar quantum-dot cellular automata adder. *Indian Journal of Pure & Applied Physics (IJPAP)*, *55*, 97–103.
87. Singh, A., Khosla, M., and Raj, B. (2017). Design and analysis of electrostatic doped Schottky barrier CNTFET based low power SRAM. *International Journal of Electronics and Communications*, *80*, 67–72.
88. Kaur, P., Pandey, V., and Raj, B. (2020). Comparative study of efficient design, control and monitoring of solar power using IoT. *Sensor Letter*, *18*, 419–426.
89. Bhardwaj, A. K., Gupta, S., and Raj, B. (2020). Development & analysis of compact model for double gate Schottky barrier CNTFET. *Journal of Nanoelectronics and Optoelectronics*, *15*, 1199–1208.
90. Wadhwa, G., Kamboj, P., Singh, J., and Raj, B. (2020). Design and investigation of junctionless DGTFET for biological molecule recognition. *Transactions on Electrical and Electronic Materials*, *22*, 8.
91. Chawla, T., Khosla, M., and Raj, B. (2020). Optimization of double-gate dual material GeOI-vertical TFET for VLSI circuit design. *IEEE VLSI Circuits and Systems Letter*, *6*(2), 13–25.
92. Verma, S. K., Singh, S., Wadhwa, G., and Raj, B. (2020). Detection of biomolecules using charge-plasma based gate underlap dielectric modulated dopingless TFET. *Transactions on Electrical and Electronic Materials (TEEM)*, *21*, 528–535.
93. Jain, N. and Raj, B. (2017). Impact of underlap spacer region variation on Electrostatic and analog/RF performance of symmetrical high-k SOI FinFET at 20 nm channel length. *Journal of Semiconductors (JoS)*, *38*(12), 122002.
94. Singh, S. and Raj, B. (2020). Analytical modeling and simulation analysis of T-shaped III-V heterojunction Vertical T-FET. *Superlattices and Microstructures*, *147*, 106717.
95. Singh, G., Sarin, R. K., and Raj, B. (2017). Design and analysis of area efficient QCA based reversible logic gates. *Journal of Microprocessors and Microsystems*, *52*, 59–68.

96. Singh, A., Khosla, M., and Raj, B. (2016). Compact model for ballistic single wall CNTFET under quantum capacitance limit. *Journal of Semiconductors (JoS)*, *37*, 104001–104008.
97. Singh, S., Khosla, M., Wadhwa, G., and Raj, B. (2021). Design and analysis of double-gate junctionless vertical TFET for gas sensing applications. *Applied Physics A*, *127*(16).
98. Singh, I., Raj, B., Khosla, M., and Kaushik, B. K. (2020). Potential MRAM technologies for low power SoCs. *SPIN, 10*(4), 2050027.
99. Singh, S. and Raj, B. (2020). Parametric variation analysis on hetero-junction vertical t-shape TFET for Supressing ambipolar conduction. *Indian Journal of Pure and Applied Physics*, *58*, 478–485.
100. Soniya, S. S., Wadhwa, G., and Raj, B. (2019). Design and analysis of dual source vertical tunnel field effect transistor for high performance. *Transactions on Electrical and Electronics Materials*, *21*, 74–82.
101. Kaur, M., Gupta N., Kumar, S., Raj, B. and Singh, A. K. (2020). RF performance analysis of intercalated graphene nanoribbon based global level interconnects. *Journal of Computational Electronics*, *19*, 1002–1013.
102. Wadhwa, G. and Raj, B. (2019). Design and performance analysis of junctionless TFET biosensor for high sensitivity. *IEEE Nanotechnology*, *18*, 567–574.
103. Singh, J. and Raj, B. (2019). Enhanced nonlinear memristor model encapsulating stochastic dopant drift. *JNO*, *14*, 958–963.
104. Quamara, M., Gupta, B. B., and Yamaguchi, S. (2019). MQTT-driven remote temperature monitoring system for IoT-based smart homes. In *2019 IEEE 8th Global Conference on Consumer Electronics (GCCE)*, (pp. 968–970). IEEE, Osaka, Japan.
105. Stergiou, C., Psannis, K. E., Kim, B. G., and Gupta, B. (2018). Secure integration of IoT and cloud computing. *Future Generation Computer Systems*, *78*, 964–975.
106. Memos, V. A., Psannis, K. E., Ishibashi, Y., Kim, B. G., and Gupta, B. B. (2018). An efficient algorithm for media-based surveillance system (EAMSuS) in IoT smart city framework. *Future Generation Computer Systems*, *83*, 619–628.
107. Sharma, S. and Singh, S. K. (2022). IoT and its uses in Security surveillance, Insights2Techinfo, p. 1. https://insights2techinfo.com/iot-and-its-uses-in-security-surveillance/.
108. Chattopadhyay S., Banerjee S., Pal A., Adi N. S., and Rahaman M. (2022). Secure smart socket system for energy monitoring embedded with IoT. *Cyber Security Insights Magazine*, *3*, 1–5.

https://doi.org/10.1142/9789811279539_0003

Chapter 3

Bone Fracture Detection Using Machine Learning

Chirag Sethi[†,§], Shailendra Singh[*,†,¶], Puspraj Singh Chauhan[†,||], and Balwinder Raj[‡,]**

[†]*Department of Electronics and Communication Engineering, Pranveer Singh Institute of Technology, Kanpur, India*

[‡]*Department of Electronics and Communication Engineering, NITTTR Chandigarh, India*

[§]*sethichirag646@gmail.com*

[¶]*shailendras.nitj@gmail.com*

[||]*puspraj.chauhan@gmail.com*

[**]*balwinderraj@gmail.com*

Abstract

Computer-aided diagnostic (CAD) systems are gaining popularity because they enhance the analysis and comprehension of clinical images, as opposed to the preliminary identification of different illnesses by physicians and medical experts. Likewise, bone fractures are a frequent occurrence owing to strain, accidents, and osteoporosis. In addition, bone is a solid structure that supports the whole body. Therefore, bone fractures are regarded as one of the most pressing issues in the later stages of life. The detection of bone

[*]Corresponding author.

fractures using computer vision is gaining importance in CAD systems since it has the potential to minimize the doctor's burden by eliminating simple cases. In this chapter, the identification of tibia fracture types is analyzed utilizing several techniques of image processing. The main purposes of this investigation are to determine whether or not there is a fracture in the lower leg bone from the X-ray scans and to characterize the kind of fracture that is present (tibia). There are three primary stages involved in the process of developing a system to detect tibia fractures. In order to classify the different kinds of fractures and determine the locations of fractures, preprocessing, feature extraction, and classification are used. During preprocessing, the sharpening technique known as unsharp masking is used on the image in order to enhance it and draw attention to the image's edges. After the image has been sharpened, the Harris corner detection approach is used on it to generate corner feature points for further feature extraction. Then, in order to detect fractures or non-fractures and, if detected, classify the fracture, two different classification techniques are used. A simple decision tree is employed for the classification of fractures, while the *K*-nearest neighbor algorithm is used for the classification of the fracture type. In this work, many forms of fractures, including normal, transverse, oblique, and comminute, are described. In addition, fracture locations may be determined by examining the Harris corner points that have been formed. Last but not least, the outputs of the system are evaluated utilizing two different performance assessment approaches. The first component is a performance evaluation that may be conducted in fracture or non-fracture (normal) conditions, and it can provide one of four possible results: TP, TN, FP, or FN. The second goal is to use the Kappa approach in order to conduct an analysis of the accuracy of each fracture type under various error settings. The MATLAB programming environment was used throughout the building of the system, which offers a vast array of image processing capabilities. The technique has an 82% classification accuracy for fracture types. This image processing approach is extraordinarily useful for several applications, including biomedical, security, satellite imaging, personal imaging, and medicine. Fracture detection employs image processing techniques such as picture augmentation, feature segmentation, and feature excitation. This study employs an astute edge detection technique for segmentation. Using a shrewd technique, accurate information can be extracted from the bone image. This research's primary objective is to identify bone fractures using image processing in MATLAB. Preprocessing, segmentation, and fracture identification are the phases that comprise the proposed system. In the feature excitation stage, the Hough transform method is used for the line detection of inline images. The system's primary function is the extraction of features. Multiple experiments demonstrate that the suggested approach is very precise and effective.

Keywords: Bone fracture, machine learning, MATLAB

3.1. Introduction

As a consequence of developments in both technology and software, the subject of medical image enhancement is becoming more popular and gaining significant traction in the medical field. This popularity is largely attributable to the fact that it offers several career opportunities. It is of critical importance in both the diagnosis of illnesses and the improvement of patient care, and it provides assistance to medical practitioners in making accurate decisions about therapy. In today's society, it is becoming more important to diagnose and treat bone fractures, which affect a large number of people of all ages. Bone fractures may occur at any age. Even in the vast majority of countries that have undergone industrialization, bone fractures remain a common problem, and the incidence of fractures is considerably on the rise. Fractures of the bone may be brought on by anything from a minor mishap to a number of different diseases. As a consequence, it is possible for a rapid and accurate diagnosis to be vital to the effectiveness of any treatment that is advised [1–5]. When determining whether or not a fracture has occurred and the precise nature of the fracture, radiologists and other medical professionals mainly depend on X-ray images. The identification of fractures by physical inspection or through the use of standard X-ray technologies is a difficult and time-consuming task. Radiologists who are overworked may fail to see a fracture in an image among other images showing no abnormalities. It is possible for an X-ray image to be flagged by a computer vision system, which would then inform the appropriate medical professionals. When it comes to something as important as this, relying only on experts has at times resulted in intolerable blunders; for this reason, the idea of an automated diagnostic procedure has been alluring for a very long time. Myint *et al.* presented an algorithm for the detection of leg bone fractures in X-ray images by performing preprocessing, segmentation, fracture detection, and classification of the images [6–10]. It provides information about the process by which the canny edge detector creates faultless segmentation information based on the bone image. The Hough transform approach is used for line identification in the feature extraction process. Mallikarjunaswamy and Raman focused on developing a system based on image processing for the quick and accurate categorization of broken bones based on X-ray/CT image data. They did this by analyzing the fractures seen in patients' X-ray and CT scans. Pre-processing, separation, edge detection, and extraction of features are all examples of approaches to image processing that were used by the authors of the study. This system was able to determine whether or not a bone was broken by using the tools available in the MATLAB 7.8.0 programming language. Additionally, it was able to evaluate the degree to which different methods were accurate. The authors determined that the system was accurate 85% of the time but that its performance had some limitations [2]. X-ray images were analyzed by Vegi *et al.* to identify bone fractures. This work discussed the method of importing images into the system, as well as preprocessing, segmentation, and the identification of bone fractures. In order to improve the efficiency and accessibility of this platform, the authors opted to use a graphical user interface design [3]. The author recognizes the value of this approach and its easy accessibility. The authors investigated the issue of differentiating between different types of fractures by utilizing a variety of classification algorithms to determine whether or not a fracture was present and identify its type. The following procedure was used by the authors in the implementation of the system. Following the resolution of the problem, distinguishing traits are recovered at the end of the preprocessing step. In order to determine whether or not fractures are present, many categorization methods are used [4]. The research by Mahendran and Baboo aimed at improving the diagnosis of tibia fractures using X-ray images through image processing tools and fusion classification algorithms. Preprocessing, segmentation, feature extraction, and bone recognition were the four operations that were proposed for inclusion in the system. In the fusion classification phase of bone detection, three different classifiers, namely naive Bayes (NB), support vector machine (SVM) classifier, and feed-forward backpropagation neural network (BPNN), were applied. They reported a significant increase in the rate of detection as well as a rapid rate of classification [5].

Martin and Greg came up with a process that can autonomously determine whether or not long bones have been fractured. There are two different processes to consider in this study. In the beginning, a nonlinear anisotropic diffusion approach was used in order to extract the bone edge from the X-ray image. The authors then constructed a modified Hough transform that includes automated peak recognition in addition to the intensity and orientation of the gradient with the help of the compute line parameter. Using this methodology, midshaft fractures in long bones may be accurately identified [6]. Baboo and Mahendran retrieved and analyzed fracture X-ray photographs to identify properties like connection, homogeneity, power, entropy, mean, variation, Markov Random Field, and the direction of an intensity gradient direction are some of the concepts that will be covered in this section. Using these features, classifiers were trained to detect fractures in X-ray images, and then, their performance was assessed. A total of three classifiers were used: BPNN, SVM, and NB. These features and classifiers were used as inputs in the development of three single and four multiple classifiers, respectively. As indices of performance, the following attributes were utilized: sensitivity, specificity, positive predictive value, negative predictive value, accuracy, and execution time. In addition, they suggested that the use of fusion classifiers would allow for increasing the capacity of detection, with the combination of SVM and BPNN offering the best results. Wei and Huisheng presented the outcomes of their research on the classification and feature extraction of X-ray fracture images [7]. In order to segment X-ray fracture images, they started by addressing the marker-controlled watershed transform, which is based on gradient and homotopy modification. This allowed them to begin the segmentation process. After that, marker processing and the "regionprops" function were used to extract information from the fracture images [11–16]. This information included the region number, the region area, the region centroid, and the protuberant polygon. After that, the Hough transform was used to locate and extract the information within the protuberant polygon of the X-ray fracture image. Subsequently, the angle between the fracture line and the line perpendicular to the centerline was calculated [8]. Hmeidi and Ayyoub are credited with the development of a technique that, when applied to X-ray images of the hand, provides a precise diagnosis of the presence of hand bone fractures. Certain identifying qualities were recoverable after the images were subjected to processing that included noise reduction and enhancement. In addition to that, they characterized the functioning of the system as having an accuracy rate of greater than 86% [9]. A computer-aided design (CAD) system for the identification of fractures was created by Chai *et al.* The system is based on GLCM identification, which enhances the accuracy of human examination of X-ray images. To identify broken and non-fractured bones, the GLCM computation features consist of homogenization, contrasts, energy, and autocorrelation. They evaluated 30 images of femur fractures and found that the technique was accurate 86.67% of the time. However, they also noted that the effectiveness of this system might be enhanced by utilizing various GLCM characteristics and that future research could focus on categorizing bone fractures according to their degree [10]. Skeletal fractures are a widespread occurrence even in industrialized nations, with their incidence growing daily at an alarming rate. Bone fractures may be caused by simple accidents or other disorders. Therefore, a prompt and precise diagnosis is essential for the efficacy of any given therapy. Relying solely on human specialists in a situation of such critical importance has resulted in terrible blunders. Consequently, the concept of an automated identification technique has always been interesting. The primary objective of this research is to use MATLAB software to identify bone fractures from X-ray images [17–23].

The lower leg bone is the human body's second largest bone after the sternum. The tibia and the fibula are the bones that make up this structure. When compared to the tibia, the fibula is a more diminutive and fragile bone. However, since it supports a significant amount of the body's weight, tibial fractures occur most often. Figure 3.1 shows bones that are normal in structure. X-ray diagnostics is the most commonly used modality to detect bone fractures due to its low cost, high speed, and wide availability. Other diagnostic modalities include computed tomography (CT), magnetic resonance imaging (MRI), and ultrasonography.

Although CT and MRI provide images of human organs of higher quality, X-ray images are faster, less expensive, more widely accessible, easier to use, and have fewer constraints. In addition, the quality of the images produced by an X-ray machine is sufficient for identifying bone

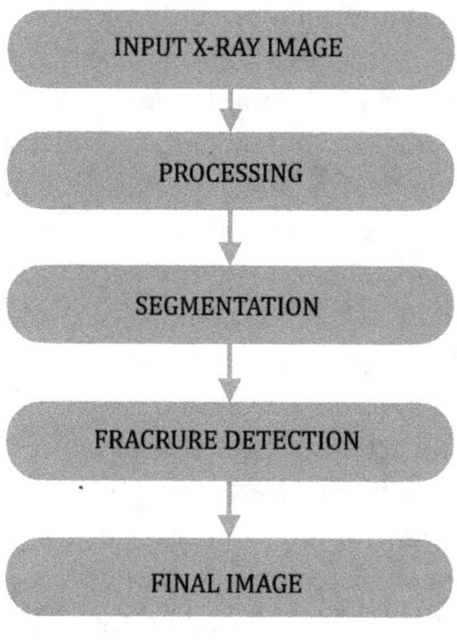

Figure 3.1. Block Diagram of the Proposed System

fractures [11]. By avoiding the treatment of straightforward conditions, this strategy hopes to accomplish two things: cut down on wait times for patients and ease the workload of medical professionals. It intends to reduce human error resulting from hospital physicians manually examining an excessive number of X-ray images for fractures [24–32].

3.2. Building of Framework

A fracture detection method for the tibia is broken down into its component parts and examined in this research. Image capture, feature extraction, preprocessing, and classification are the components that make up this system. The X-ray images are obtained from several hospitals that specialize in orthopedics and radiology, and they include both images of healthy bones and images of bones that have been shattered. When it comes to the effectiveness of computer-assisted diagnostic technologies that are reliant on medical photos and image processing technologies, noise reduction, augmentation, and feature extraction play a crucial role. There are many different kinds of bones [12], and a variety of fractures, including oblique, complicated, comminuted, spiral, greenstick, and transverse, may be detected. Using imaging techniques, fractures can be categorized into several types, such as comminuted, transverse, and oblique. To improve the appearance of the bone areas in the photographs, the initial step in the procedure involves using preprocessing techniques, such as converting the images from RGB to grayscale and sharpening them. The unsharp mask (USM) filter is a technique of image sharpening that is used to improve either the image's sharpness or (local) contrast resulting from value disparities between the original image and the blurred image [13]. This may be accomplished by comparing the blurred image to the original image. It is possible to create a picture with more defined edges by using the mask. Following the completion of the preprocessing step, each image is then extracted using a feature extraction method, such as the Harris corner detection technique, to create a collection of features that represent the image's edge points, break points, and other locations of interest [14]. After that, the categorization procedure is applied to the information that was obtained. During this phase of the process, a decision tree (DT) classifier and a K-nearest neighbor (KNN) classifier were used to categorize the fracture, non-fracture, and fracture types. A block diagram of the proposed system is shown in Figure 3.1, and the following sections provide an in-depth discussion of each step in the process.

3.2.1. *Input X-ray*

The process of capturing an image as input represents the initial stage of the proposed process. In this system, X-ray images are used in the process of identifying long bone fractures [15]. Here, the gathering of relevant information in the form of digital X-ray-scanned images represents the first step in this process. In most contexts, the term "picture acquisition" refers to the process of retrieving an image from a hardware source. For the input of X-ray images, this research makes use of the JPG format since it is very simple for image processing algorithms to work with. In addition, the current imaging equipment used for X-rays supports both the JPG and DICOM file formats simultaneously [33–38]. Because of this, there is no need to convert DICOM files into the JPG format. After this stage, image preprocessing needs to be performed in order to improve the raw X-ray image by removing any noise or artefacts that may be present. In preparation for additional processing, the image edges are being sharpened and soft focus (blurring) is achieved. Figure 3.2 displays an X-ray photograph of the bone in the lower leg.

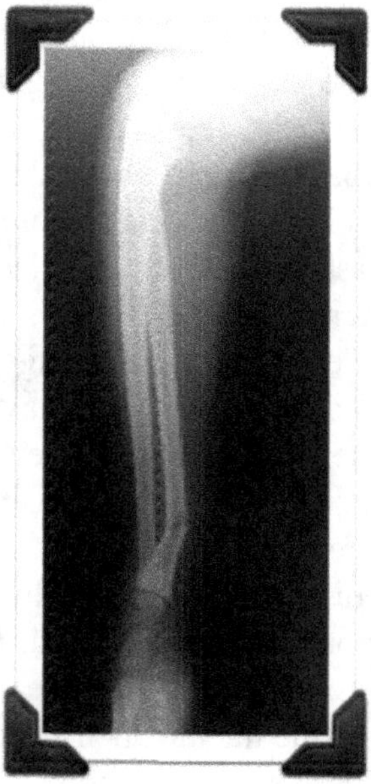

Figure 3.2. Lower Leg Bone X-ray

3.2.2. *Image Filtration Process*

Filtering images is a critical phase since it decides if the results are suitable for the next steps. A technique for improving images may be used as a preprocessing or postprocessing step. Any image enhancing technique that emphasizes edges and minute details is referred to as sharpening [16–18]. The process of USM involves the blurring of the original image, followed by the subtraction of the blurred image from the actual picture. Introducing a distinction into the image that was created in the previous phase as the last step. Equation (3.1) expresses the method used to produce a sharper picture:

$$\text{Sharpened} = \text{Original} + (\text{Initial} - \text{Blurred}) \times \text{Number of units.} \quad (3.1)$$

In this stage, the following approach is used to perform image preprocessing. The X-ray picture input is an RGB image. This image is first transformed to a grayscale, single-layer image in order to reduce processing time and calculations [39–46]. Then, the USM algorithm is used on the gray picture to accentuate it. For display and analysis, enhanced or smoothed picture elements are used to produce the image with edge enhancements. In accordance with the three-setting control of USM, negative effects may be mitigated by making use of a mask to restrict the application of sharpening to the areas in which it is really required. This method is often referred to as "smart sharpening," and for good reason. This method quantifies how much the edge borders should darken or lighten, how big the radius of the edges should be, how wide the edge rims should get, and the threshold for the least amount of brightness change that will be sharpened [19]. The quantity of how much the edge borders darken or lighten can be found in the edge sharpening settings. The step of feature extraction is carried out using the enhanced image. The preprocessing flowchart is depicted in Figure 3.3.

3.2.3. *Data Acquisition*

The properties of an image that capture a specific aesthetic attribute constitute the data. As sharpening a picture increases the contrast between light and dark regions, the process of feature extraction may be skipped [47–54]. Edge detection is a method used for extracting various types of characteristics. Researchers have developed many corner detectors from the Kanade–Lucas–Tomasi operator and the Harris operator that are

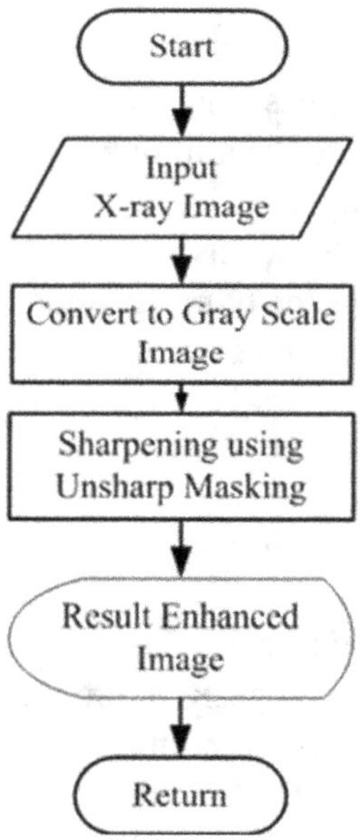

Figure 3.3. Image Preprocessing Flowchart

straightforward to understand, widely implemented, and trustworthy for extracting picture corners. The Harris algorithm is used for the process of discovering edge points as features and for feature extraction [20–22]. This technique detects corners that are more evenly distributed and more informative. The Harris corner identification technique can examine the *X*- and *Y*-derivatives of a liner edge, flat, and corner. Using a mathematical technique, it determines the energy gradient values to acquire two eigenvalues:

$$E = \Sigma\, w\,(x, y)\,[I(x + u, y + v) - I\,(x, y)]^2\,. \quad (3.2)$$

Here, *E* indicates the degree to which the gradient has been altered between the original and the relocated window. The correlation matrix, which may be described as follows, is the foundation upon which the Harris detector builds its corner conclusions:

$$M = \Sigma_{(x,y)}\, w(x, y)\begin{bmatrix} I_x^2 & I_x I_y \\ I_x I_y & I_y^2 \end{bmatrix}, \quad (3.3)$$

where M is the 2×2 matrix of image derivatives. The eigenvalues (λ_1, λ_2) of the matrix M can indicate whether or not a window is necessary. The response function, denoted by R, determines whether a point is a corner and is defined by Harris as $R = \det M - k(\text{trace } M_2)$. R is computed for each window and only makes use of the matrix's eigenvalues, which are (λ_1, λ_2) for the matrix M. In the feature extraction phase of the process, the improved picture is processed using the Harris corner algorithm in order to extract broken points as corner features [55–63]. The Harris algorithm can generate no more than 10 corner points due to the way it is set up. After that, two threshold values are used in order to get rid of the undesired areas at the knee and the ankle. After the feature extraction process, a precise diaphysis shape of the leg bone, which is important for determining whether or not a fracture has occurred, may be determined [23]. Thus, feature extraction constitutes the initial step in processing images for categorization purposes. The Harris corner detection technique is an effective feature point algorithm. This research uses the Harris corner detection method since the current algorithm meets the requirements directly. Figure 3.4 presents a flowchart of the feature extraction process.

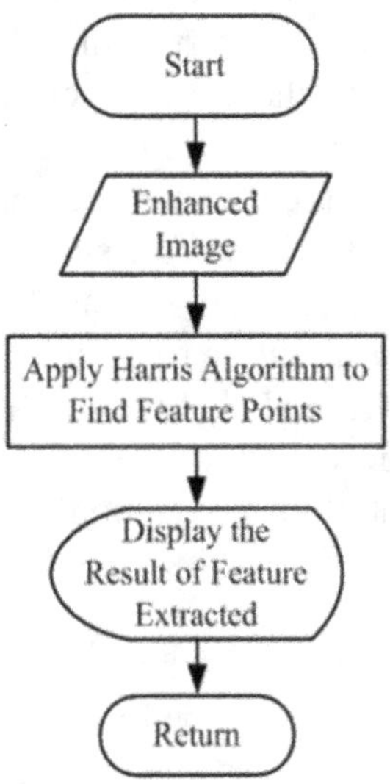

Figure 3.4. Image Retrieval Algorithm Flowchart

3.2.4. *Segmentation*

The strategies that have been presented for detecting fractures may be informally divided into two categories: categorization and transformation. In order to complete the process of recognizing the broken bone appearing on the X-ray image, the classification-based strategy is used in this study [64–70]. This step marks the end of the process in which information analysis, known as classification, is performed. It involves analyzing a collection of data and organizing that data into a number of different categories. In addition to that, it incorporates a wide variety of decision-theoretic techniques to identify the images [24]. There are two criteria that may be considered in relation to classification: binary classification and classification with more than two groups. In a two-category system, which is well understood, there are only two classes involved; however, in multiclass classification, an item is assigned to one of multiple classes. When it comes to image classification, there are many different kinds of classifiers, each of which was examined in greater depth in the previous chapter. The Bayesian classifier has been shown to always have the lowest possible error rate, but in order to use it, one must have complete information on the classes [25]. When using a neural network, classification happens relatively quickly; nevertheless, training may be restrictively slow, which necessitates large and complex artificial neural networks. SVM characterization may be used to overcome the limitations of the above two methods, but it performs poorly when the number of features is much larger than the number of samples [26]. The DT and the KNN classifiers are the two methods used in this study. DT is a very effective model that, in a relatively short amount of time, can provide models that are accurate and simple to comprehend. A tree is referred to as a DT or a classification tree when each internal node, which is a node that is not a leaf, is labeled with an input characteristic. They have applications in a wide variety of fields, including diagnostics, cognitive science, artificial intelligence, game theory, engineering, and data mining. In the context of this task, the system has to make a straightforward choice about whether the state of the bone is fractured or normal. As a result, the DT method is used for both of these scenarios [27]. The Harris algorithm, which was developed to capture the 10 broken points following feature extraction, is responsible for producing the features, also known as break points (n_b). Utilizing edge points, the DT classifier comes to the conclusion that a bone condition has the potential to fracture if there are one or more of these corner points, but if there are none of these corner points, the bone condition is considered to be normal. KNN has the potential to provide good performance for any class condition, but it might be too slow for real-time prediction. KNN needs a lot of training examples, and it does not collaborate well with imbalanced datasets. In contrast, the X-ray input picture undergoes enhancement initially as part of the preprocessing stage for the vast majority of image processing algorithms. Because of this, the adoption of the KNN classifier makes it possible to readily manage noisy data in X-ray images. The only requirement is that, during the "training" phase of the method, the feature vectors and class labels that correspond to the training examples should be saved [71–78]. The phase of classification is represented by the letter *k*, and during this phase, the number of closest neighbors that are used in the classification is given by *k*. If *k* is equal to one, which is the default value, the case is simply assigned to the class that is closest to the query point. If *k* is not equal to one, the case is not allocated to any class. A type of distance metric that is often used for continuous variables is the Euclidean distance [29]. In this study, the different kinds of fractures that occur throughout the training and testing procedures are categorized using the KNN classification system. The four primary stages involved are obtaining an image, preparing it for analysis, determining what it is, and putting it into a category. Once the images are acquired and the input X-ray image goes through four steps of processing, the defected bone area is captured from the final X-ray image. This occurs throughout both the training and testing phases. After that, Otsu's approach is used to apply image segmentation by employing a range filter and a global image threshold. In addition to this, the picture is changed to a black-and-white image, and then, using the command "bwareaopen," tiny components are removed from the black-and-white image to produce a new improved black-and-white image [30]. Finally, the improved black-and-white image is downsized using the immersive function to set its dimensions as 100×100 pixels in order to standardize the image of the fracture outlining to identify the damaged area. After the picture is normalized, it is stored as an image in the training dataset as a ".mat" file. The ".mat" file format is used to store the three different forms of fracture for group categorization considered in this study: transverse, oblique, and comminuted. The KNN classification method involves comparing the pattern features of a test X-ray image with the pattern training feature of each using the Euclidean distance to categorize the image into the three types

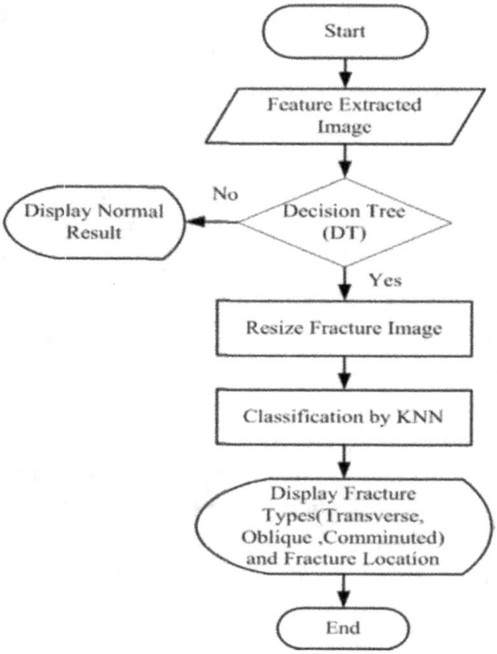

Figure 3.5. Image Retrieval Algorithm Flowchart

of fracture. KNN determines the test X-ray image to be the longitudinal fracture type if the distance between both the query pattern feature and the transverse pattern feature is the nearest distance. Figure 3.5 presents a data flow diagram that shows how fractures are put into groups and how the fracture is located in the image.

3.3. Observation and Computation Outcomes

The output findings from the system are split into two halves, one each for the benefit of orthopedic physicians and radiologists. The initial part of the analysis determines whether or not a fracture is present. The second step involves categorizing the fracture into one of three types and determining where the break is located. When it comes to treating bone fractures, the kind of fracture that was sustained is quite crucial. The environment of MATLAB provides support for the methodologies used in this research through its variety of tools for processing images that may be used to accomplish the desired outcomes.

3.3.1. *Acquiring the X-ray Image for Input*

The X-ray images that were acquired from the patient under study were captured at 53 kV and 4 mA, and they were digitized from the X-ray mechanism at 7 bits per pixel. Even though there are also additional types of medical imaging, such as CT, ultrasonography, and MRI, X-ray imaging is most often used in the diagnosis of fractures [31]. They allow for simple and quick diagnosis by medical professionals to identify both joint and bone deformities. These images were taken at the Yangon Orthopedic Hospital, which is located in Myanmar, with the consent of the appropriate authorities. The X-ray images were gathered in the JPG format [79–85]. The dimensions of the processed images are often indicated as having a resolution of 400 × 400 pixels. The MATLAB programming software's "imread" function is used to read these images, and the "imshow" command is used to display them. Figures 3.6(a) and 3.6(b) show these images. The supplied images are then preprocessed in order to improve them in preparation for further processing to sharpen them, as depicted in Figures 3.6(c) and 3.6(d).

3.3.2. *Preprocessing*

Considering that the input picture is in color space, this part of the process begins by converting the X-ray image into a grayscale image. This is done because the conversion requires additional calculation time. The command "rgb2gray" available in MATLAB allows for the successful conversion of an RGB picture to a grayscale one. Due to the fact that the borderline region of the bone in the X-ray image is composed of higher-frequency output, the grayscale image is consequently treated by sharpening the picture using the USM tool. This is done in order to emphasize or boost the image's high-frequency information. "Imsharpen" is the action that is used for this stage in order to complete this activity [32]. Figures 3.7(a) and 3.7(b) show the effects of applying the sharpening filter to the picture. The resulting images may be viewed to ensure that the bone border area has been sharpened. Masking that is not sharp is particularly helpful for sharpening around the edge. Due to the fact that the USM technique already includes a blurring effect, the sharpening procedure is able to eliminate noise from input photos. Because USM is so efficient not just at filtering but also at edge identification, it has been selected for use in the preprocessing the different views of an image, as depicted in Figures 3.7(c) and 3.7(d).

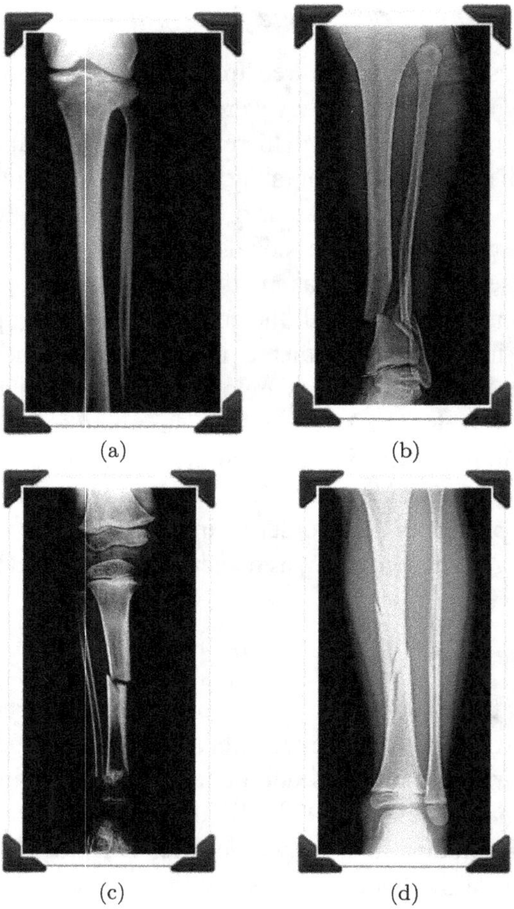

Figure 3.6. Input Original X-ray Scans: (a) Left Flexion View, (b) Right Flexion View, (c) Anteroposterior View, (d) Posteroanterior View

3.3.3. *Feature Extraction*

In the X-ray image, the areas of bone stand out more clearly than the surroundings. Because of this, fragments of shattered bones will show up according to the configuration of the lighter edges and darker areas of the scan picture when they are present. A broken bone fragment appears at the intersection of two edges, which triggers an abrupt shift in the image's luminance [33]. Another definition of a corner describes it as the point where two edges meet. As a result, the

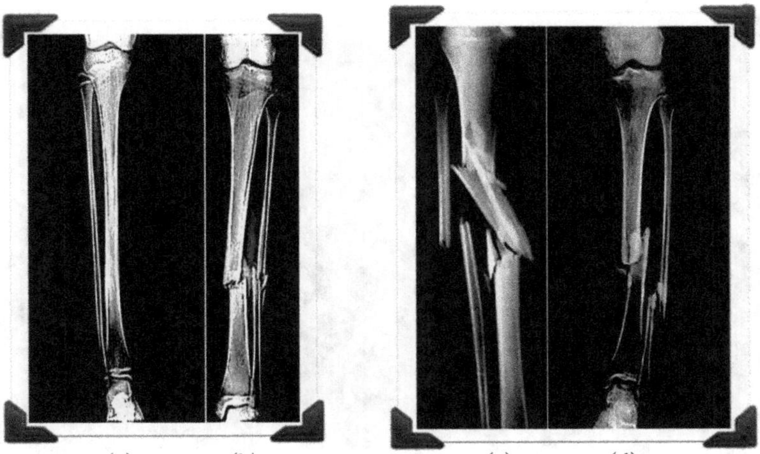

(a) (b) (c) (d)

Figure 3.7. Resulting Images from Image Sharpening: (a) Posteroanterior Sharpened View, (b) Anteroposterior Sharpened View, (c) Left Flexion Sharpened View, (d) Right Flexion Sharpened View

Harris corner identification technique was used to identify the fracture spots in the X-ray images of the leg bone. Because the suggested system is looking to acquire major fractured bone points, the Harris algorithm was modified such that it can create 10 corner points at its maximum. Figures 3.8(a)–3.8(c) display the outcome of using the corner detection algorithm.

The Harris algorithm identified seven corner points indicating that the bone was broken, as depicted in the figure. Because of the fractured nature of the bone, it is possible to discern eight corner points, as shown in Figures 3.8(a)–3.8(c). The next figure shows how the algorithm creates nine feature points using shattered bits of bone. However, the Harris algorithm will not output the corner points if the bone picture does not have a break point. The last stage is classification, which involves making use of the features that were extracted from the image. This classification will include standard bone, fractured bone, and fracture varieties, such as transversal, orthogonal, and confluent fractures, as well as fracture sites.

3.3.4. *Segmentation*

The bone defect detecting system employs two separate models of classification during the classification stage. Each of these models is

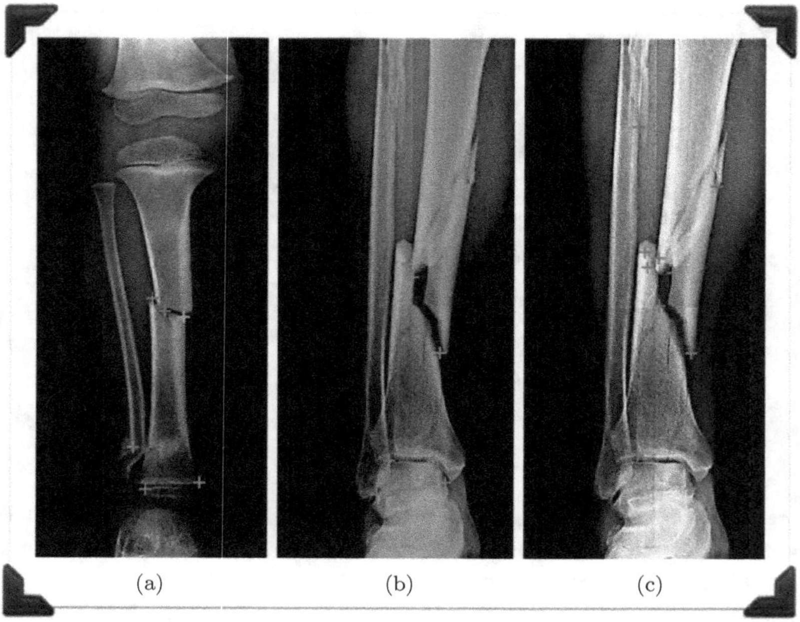

Figure 3.8. Harris Corner Detection: (a) Bone with Seven Corners, (b) Bone with Eight Corners, (c) Bone with Nine Corners

carried out for one of two basic reasons. The DT testing is used in conjunction with the KNN algorithm to ascertain whether or not a bone fracture has taken place, and then the KNN algorithm is used to classify the kind of fracture that has taken place. In DT, the points of features produced as a consequence of feature extraction are regarded as attributes of the root node, whereas broken bone and normal bone are regarded as classes of the tree. When using the KNN segregates, both the training and testing operations will be evaluated based on their respective Euclidean distances [34]. The test feature is then categorized using the numerous categories that were made available once this step had been completed. Because of this, training and examinations are necessarily included in the KNN classification system. The train and test features are both applied to the pattern feature. The status of the fracture, or whether it is normal, was already determined by the DT procedure, which is the sole reason why the fracture images from DT

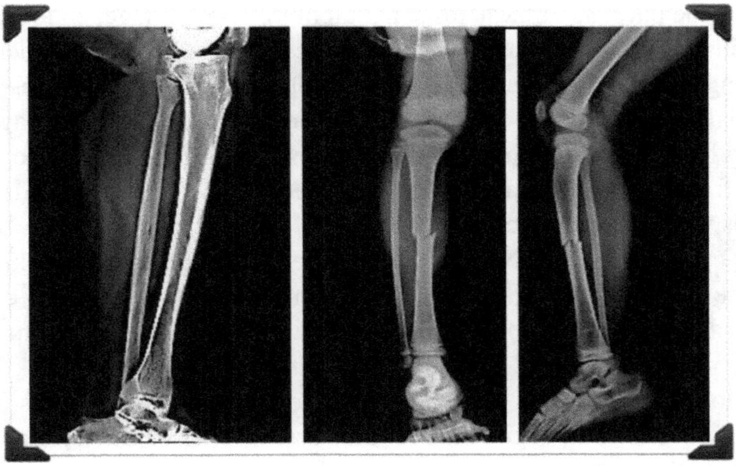

Figure 3.9. Training Normalized Feature Patterns

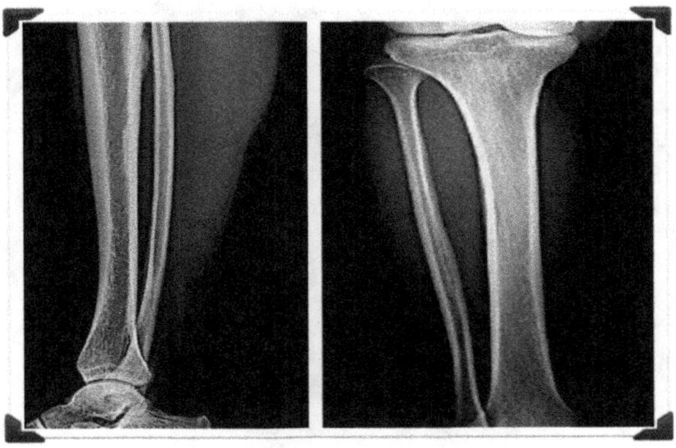

Figure 3.10. Testing Normalized Feature Patterns

are processed. In order to apply the fracture image, it must be normalized through ranging filtering, universal thresholding, the eradication of very minuscule objects, and the ramping of the image to 100 × 100 size. The procedures for training and evaluating candidates have been described in great depth in the previous chapter. Figures 3.9 and 3.10 provide an illustration of the feature patterns that should be used for training and testing, respectively.

For the purpose of these tests, a total of 40 X-ray images of leg bones are utilized as training images. These images include 10 standard pictures, 10 longitudinal fracture images, 10 elongated fracture images, and 10 tibial plateau fracture images. Then, there are also 12 X-ray images of the leg bone, including three regular images, three images with a longitudinal fracture, three images with a broken transverse bone, and three images with a lateral fracture.

In conclusion, the KNN method may be used to determine the kind of fracture that occurred after carrying out this process. The Harris corner points are responsible for determining the fractured location. Since the corner points of the leg bone coincide with the fracture or break sites, these locations serve as the indicative fracture locations. Figures 3.11–3.14 provide the output findings for the various types of fractures and the locations of the fractures.

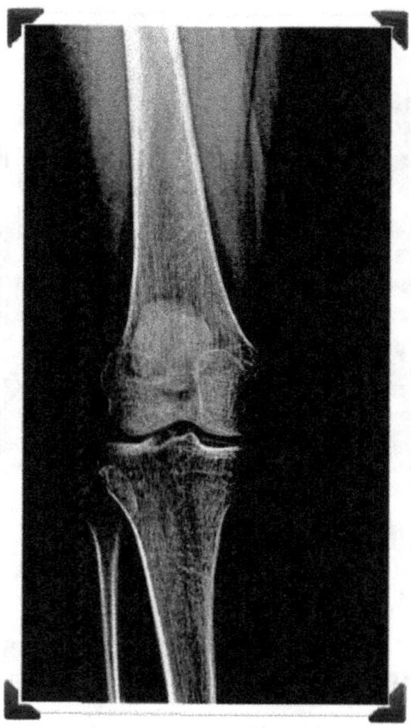

Figure 3.11. Normal Bone Image Outcome

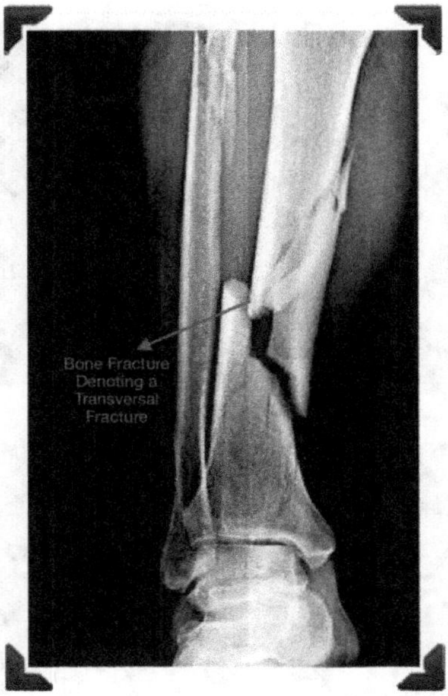

Figure 3.12. Transverse-Type Fracture

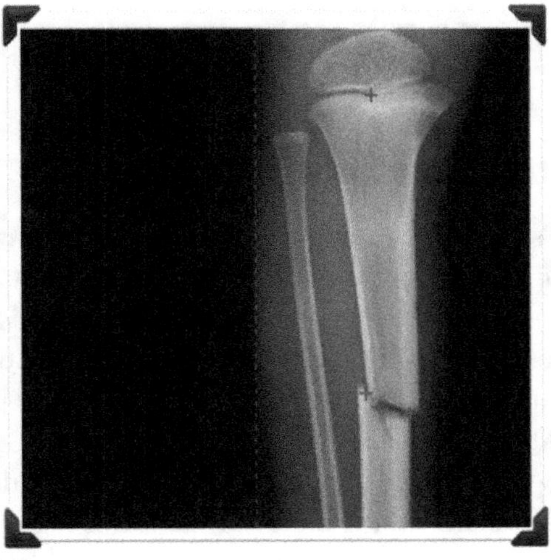

Figure 3.13. Bone Fracture Location

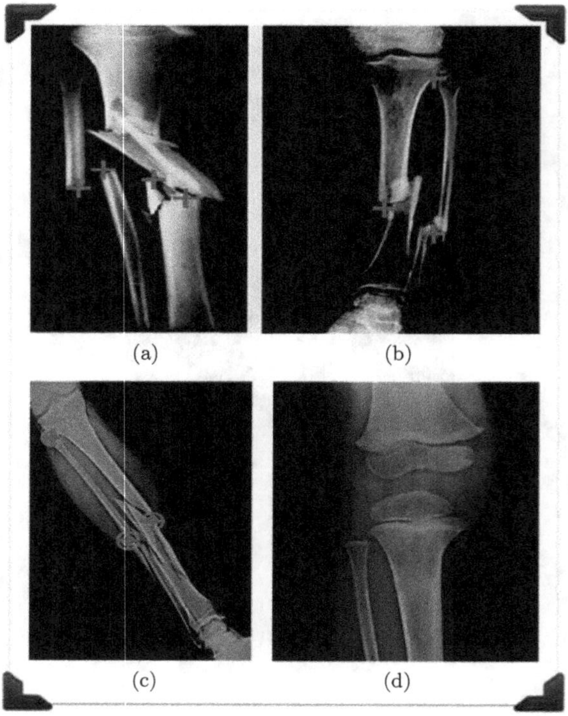

Figure 3.14. Outcome of Damaged Bone: (a) Horizontal Fracture Type and (b) Fracture Location; (c) Comminuted Fracture Type and (d) Fracture Location

3.4. Implication and Future Scope

The goal of this study was to demonstrate a computer-controlled model for the detection of broken bones implemented in a CAD system. To achieve this, an automatic leg bone fracture detection and localization algorithm was designed and developed in this work [86–92]. The technique relies on computer vision algorithms. This work is part of the larger project titled "Automatic Bone Fracture Diagnosis Using Computer-Aided Design." A completely automated method of identification and categorization of fractures in the bones of the lower leg is of great significance but difficult to achieve from the perspectives of both an orthopedic surgeon and a radiologist. Several different methods of image processing were used in the efforts to achieve this goal. In particular, image improvement and the sharpening of boundary conditions in photographs may both benefit from the use of USM [93–98].

The Harris corner detection system is a useful instrument for locating fractured places in the leg bone. A straightforward method of categorizing fractures and non-fractures is provided by a DT. The Harris corner points may be used to pinpoint the position of the fracture. KNN is useful for the identification of patterns and can aid in the classification of transverse, oblique, and comminated fractures, which are three distinct forms of fractures. The system is capable of working with a variety of image sizes below 400 × 400 pixels. In order to address the limitations of the previous studies, the method classifies several kinds of fractures. Examples include transverse, oblique, comminated, and normal. In this work, we calculated the performance accuracy with respect to fractures and non-fractures, and additionally, we evaluated the accuracy using the kappa coefficient for different types of fractures in order to ensure that the results are as accurate as possible. The kappa accuracy evaluation was employed because this technique takes into account the incorrect findings while evaluating the performance and identifying the different forms of fracture [99–103].

Though the proposed system identifies the different forms of fractures, a cracked or tattered state may not be acceptable or identified as a fracture. In order to provide trustworthy output results, the luminance and saturation of the input photographs need to be adjusted accordingly and remain consistent throughout the process. A system that includes a large training dataset would inspire more confidence in its ability to reach the desired degree of accuracy [104–110]. The section of the tibia bone known as the epiphysis, which is located between the ankle and the knee, is of particular significance to the system. In addition, the location of the fracture can be described in terms of centimeters or inches. Since X-rays are capable of producing highly exact measurements of bone structure, their utilization is widespread. This is due to the fact that there is no universally accepted definition of these factors for bone size and height. It is necessary to generate the production of medicinal actions on a real-time basis, and this may be done by directly linking the X-ray equipment with the computer that is processing the source code [110–115]. Despite its limitations, the framework setup efficiently generates output that is accurate and dependable while taking less time to process since it is based on the proposed approaches [116–119].

References

1. Jensen, J. R. and Lulla, K. Introductory digital image processing: A remote sensing perspective. Published online: 17 September 2008, Pearson Book.

2. Myint, S., Khaing, A. S., and Tun, H. M. (2016) Detecting leg bone fracture in X-ray images. *International Journal of Scientific & Research, 5*, 140–144.
3. Lehmann, T. M., Gonner, C., and Spitzer, K. (1999). Survey: Interpolation methods in medical image processing. *IEEE Transactions on Medical Imaging, 18*(11), 31–39.
4. Myint, W. W., Tun, K., Tun, H. M., and Myint, H. (2018). Analysis on leg bone fracture detection and classification using X-ray images. *Machine Learning Research, 3*(3), 49–59.
5. McInerney, T. and Demetri, T. (1996). Deformable models in medical image analysis: A survey. *Medical Image Analysis, 1*(2), 91–108.
6. Rinisha, B., Wadhwani, S., and Wadhwani, A. K. (2019). Automatic location of fracture region in a bone x-ray using MATLAB. In *IEEE national conference on advance trends and techniques in wireless communication and soft computing, February 2019*, vol. 6.
7. Hazra, T. K., Dutta, S. (2016). A new approach to identify the fracture zone and detection of bone diseases of X-ray image. *International Journal of Science and Research (IJSR), 5*(9), 1640–1646.
8. Anu, T. C., and Raman, R. (2015). Detection of bone fracture using image processing methods. *Int J Comput Appl*, 975, 8887.
9. Mahendran, S. K. and Santhosh Baboo, S. (2011). Automatic fracture detection using classifiers — A review. *International Journal of Computer Science Issues (IJCSI), 8*(6), 340.
10. Jiang, J., Trundle, P., and Ren, J. (2010). Medical image analysis with artificial neural networks. *Computerized Medical Imaging and Graphics, 34*(8), 617–631.
11. Bhardwaj, A., Kumar, P., Raj, B., and Anand, S. (2022). Design and performance optimization of doping-less vertical nanowire TFET using gate stack technique. *Journal of Electronic Materials (JEMS), 41*(7), 4005–4013.
12. Singh, J. and Raj, B. (2019). Tunnel current model of asymmetric MIM structure levying various image forces to analyze the characteristics of filamentary memristor. *Applied Physics A, 125*(3), 203.1 to 203.11.
13. Goyal, C., Ubhi, J. S., and Raj, B. (2019). Low leakage zero ground noise nanoscale full adder using source biasing technique. *Journal of Nanoelectronics and Optoelectronics, 14*, 360–370.
14. Singh, G., Sarin, R. K., and Raj, B. (2016). A novel robust exclusive-OR function implementation in QCA nanotechnology with energy dissipation analysis. *Journal of Computational Electronics, 15*(2), 455–465.
15. Wadhera, T., Kakkar, D., Wadhwa, G., and Raj, B. (2019). Recent advances and progress in development of the field effect transistor biosensor: A review. *Journal of Electronic Materials, 48*(12), 7635–7646.

16. Wadhwa, G., Kamboj, P., and Raj, B. (2019). Design optimisation of junctionless TFET biosensor for high sensitivity. *Advances in Natural Sciences: Nanoscience and Nanotechnology*, *10*(7), 045001.
17. Bansal, P. and Raj, B. (2016). Memristor modeling and analysis for linear dopant drift kinetics. *Journal of Nanoengineering and Nanomanufacturing*, *6*, 1–7.
18. Singh, A., Khosla, M., and Raj, B. (2016). Circuit compatible model for electrostatic doped Schottky barrier CNTFET. *Journal of Electronic Materials*, *45*(12), 4825–4835.
19. Rai, A., Vaithiyanathan, D., and Raj, B. (2021). Performance analysis of charge plasma induced graded channel Si nanotube. *Journal of Engineering Research*, *9*, 146–154.
20. Tomar, A. S., Magraiya, V. K., and Raj, B. (2013). Scaling of access and data transistor for high performance DRAM cell design. *Quantum Matter*, *2*, 412–416.
21. Jain, N. and Raj, B. (2018). Parasitic capacitance and resistance model development and optimization of raised source/drain SOI FinFET structure for analog circuit applications. *Journal of Nanoelectronics and Optoelectronins*, *13*, 531–539.
22. Singh, S., Vishvakarma, S. K., and Raj, B. (2018). Analytical modeling of split-gate junction-less transistor for a biosensor application. *Sensing and Bio-sensing*, 18, 31–36.
23. Gopal, M. and Raj, B. (2013). Low power 8T SRAM cell design for high stability video applications. *ITSI Transaction on Electrical and Electronics Engineering*, *1*(5), 91–97.
24. Raj, B., Mitra, J., Bihani, D. K., Rangharajan, V., Saxena, A. K., and Dasgupta, S. (2010). Analysis of noise margin, power and process variation for 32 nm FinFET based 6T SRAM cell. *Journal of Computer (JCP)*, *5*(6), 1–8.
25. Sharma, D., Mehra, R., and Raj, B. (2021). Comparative analysis of photovoltaic technologies for high efficiency solar cell design. *Superlattices and Microstructures*, *153*, 106861.
26. Kaur, P., Buttar, A. S., and Raj, B. (2021). A comprehensive analysis of nanoscale transistor based biosensor: A review. *Indian Journal of Pure and Applied Physics*, *59*, 304–318.
27. Yadav, D., Raj, B., and Raj, B. (2020). Design and simulation of low power microcontroller for IoT applications. *Journal of Sensor Letters*, *18*, 401–409.
28. Singh, S. and Raj, B. (2020). A 2-D analytical surface potential and drain current modeling of double-gate vertical *T*-shaped tunnel FET. *Journal of Computational Electronics*, *19*, 1154–1163.
29. Singh, J. and Raj, B. (2019). An accurate and generic window function for non-linear memristor model. *Journal of Computational Electronics*, *18*(2), 640–647.

30. Kaur, M., Gupta, N., Kumar, S., Raj, B., and Singh, A. K. (2021). Comparative RF and crosstalk analysis of carbon based nano interconnects. *IET Circuits, Devices & Systems*, *15*(6), 493–503.
31. Kandasamy, N., Ahmad, F., Ajitha, D., Raj, B., and Telagam, N. (2020). Quantum dot cellular automata based scan flip flop and boundary scan register. *IETE Journal of Research*, *66*, 535–548.
32. Sharma, S. K., Raj, B., Khosla, M. (2019). Enhanced photosensivity of highly spectrum selective cylindrical gate $In_{1-x}Ga_xAs$ nanowire MOSFET photodetector. *Modern Physics Letter-B*, *33*(12), 1950144 .
33. Singh, J. and Raj, B. (2019). Design and investigation of 7T2M NVSARM with enhanced stability and temperature impact on store/restore energy. *IEEE Transactions on Very Large Scale Integration Systems*, *27*(6), 1322–1328.
34. Bhardwaj, A. K., Gupta, S., Raj, B., Singh, A. (2019). Impact of double gate geometry on the performance of carbon nanotube field effect transistor structures for low power digital design. *Computational and Theoretical Nanoscience*, *16*, 1813–1820.
35. Jain, N. and Raj, B. (2019). Thermal stability analysis and performance exploration of asymmetrical dual-k underlap spacer (ADKUS) SOI FinFET for security and privacy applications. *Indian Journal of Pure & Applied Physics (IJPAP)*, *57*, 352–360.
36. Singh, A., Khosla, M., and Raj, B. (2019). Design and analysis of dynamically configurable electrostatic doped carbon nanotube tunnel FET. *Microelectronics Journal*, *85*, 17–24.
37. Jain, N. and Raj, B. (2019). Dual-k spacer region variation at the drain side of asymmetric SOI FinFET structure: Performance analysis towards the analog/RF design applications. *Journal of Nanoelectronics and Optoelectronics*, *14*, 349–359.
38. Singh, J., Sharma, S., Raj, B., and Khosla, M. (2018). Analysis of barrier layer thickness on performance of $In_{1-x}Ga_xAs$ based gate stack cylindrical gate nanowire MOSFET. *JNO*, *13*, 1473–1477.
39. Jain, N. and Raj, B. (2018). Analysis and performance exploration of high-k SOI FinFETs over the conventional low-k SOI FinFET toward analog/RF design. *Journal of Semiconductors (JoS)*, *39*(12), 124002-1-7.
40. Goyal, C., Ubhi, J. S., and Raj, B. (2018). A reliable leakage reduction technique for approximate full adder with reduced ground bounce noise. *Journal of Mathematical Problems in Engineering*, *2018*, 16 pages. Article ID 3501041.
41. Anuradha, J. S., Raj, B., and Khosla, M. (2018) Design and performance analysis of nano-scale memristor-based nonvolatile SRAM. *Journal of Sensor Letter*, *16*, 798–805.

42. Wadhwa, G. and Raj, B. (2018). Parametric variation analysis of charge-plasma-based dielectric modulated JLTFET for biosensor application. *IEEE Sensor Journal*, *18*(15), 6070–6077.
43. Singh, J. and Raj, B. (2018). Comparative analysis of memristor models for memories design. *JoS*, *39*(7), 074006-1-12.
44. Yadav, D., Chouhan, S. S., Vishvakarma, S. K., and Raj, B. (2018). Application specific microcontroller design for IoT based WSN. *Sensor Letter*, *16*, 374–385.
45. Singh, G., Sarin, R. K., and Raj, B. (2018). Fault-tolerant design and analysis of quantum-dot cellular automata based circuits. *IEEE/IET Circuits, Devices & Systems*, *12*, 638–64.
46. Singh, J. and Raj, B. (2018). Modeling of mean barrier height levying various image forces of metal insulator metal structure to enhance the performance of conductive filament based memristor model. *IEEE Nanotechnology*, *17*(2), 268–267.
47. Jain, A., Sharma, S., and Raj, B. (2018). Analysis of triple metal surrounding gate (TM-SG) III-V nanowire MOSFET for photosensing application. *Opto-electronics Journal*, Elsevier, *26*(2), 141–148.
48. Jain, A., Sharma, S., and Raj, B. (2016). Design and analysis of high sensitivity photosensor using cylindrical surrounding gate MOSFET for low power sensor applications. *Engineering Science and Technology*, *19*(4), 1864–1870.
49. Singh, A., Khosla, M., and Raj, B. (2017). Analysis of electrostatic doped Schottky barrier carbon nanotube FET for low power applications. *Journal of Materials Science: Materials in Electronics*, *28*, 1762–1768.
50. Kumar, G. S., Singh, A., and Raj, B. (2018). Design and analysis of gate all around CNTFET based SRAM cell design. *Journal of Computational Electronics*, *17*(1), 138–145.
51. Singh, G. P., Sohi, B. S., and Raj, B. (2017). Material properties analysis of graphene base transistor (GBT) for VLSI analog circuits. *Indian Journal of Pure & Applied Physics (IJPAP)*, *55*, 896–902.
52. Singh, A., Khosla, M., and Raj, B. (2016). Comparative analysis of carbon nanotube field effect transistor and nanowire transistor for low power circuit design. *Journal of Nanoelectronics and Optoelectronics*, *11*, 388–393.
53. Kumar, S. and Raj, B. (2017). Estimation of stability and performance metric for inward access transistor based 6T SRAM cell design using *n*-type/*p*-type DMDG-GDOV TFET. *IEEE VLSI Circuits and Systems Letter*, *3*(2), 25–39.
54. Sharma, S., Kumar, A., Pattanaik, M., and Raj, B. (2013). Forward body biased multimode multi-threshold CMOS technique for ground bounce

noise reduction in static CMOS adders" *International Journal of Information and Electronics Engineering*, *3*(3), 567–572.

55. Singh, H., Kumar, P., and Raj, B. (2013). Performance analysis of majority gate SET based 1-bit full adder. *International Journal of Computer and Communication Engineering* (IJCCE), *2*(4), 1–8.
56. Bhardwaj, A. K., Gupta, S., and Raj, B. (2020). Investigation of parameters for Schottky barrier (SB) height for Schottky barrier based carbon nanotube field effect transistor device. *Journal of Nanoelectronics and Optoelectronics*, *15*, 783–791.
57. Bansal, P. and Raj, B. (2017). Memristor: A versatile nonlinear model for dopant drift and boundary issues. *JCTN*, *14*(5), 2319–2325.
58. Jain, N. and Raj, B. (2016). An analog and digital design perspective comprehensive approach on Fin-FET (Fin-Field Effect transistor) technology — A review. *Reviews in Advanced Sciences and Engineering (RASE)*, *5*, 1–14.
59. Sharma, S., Raj, B., and Khosla, M. (2017). "Subthreshold performance of $In_{1-x}Ga_xAs$ based dual metal with gate stack cylindrical/surrounding gate nanowire MOSFET for low power analog applications. *Journal of Nanoelectronics and Optoelectronics*, *12*, 171–176.
60. Raj, B., Saxena, A. K., and Dasgupta, S. (2009). Analytical modeling for the estimation of leakage current and subthreshold swing factor of nanoscale double gate FinFET device. *Microelectronics International*, *26*, 53–63.
61. Soniya, S. S., Wadhwa, G., and Raj, B. (2019). An analytical modeling for dual source vertical tunnel field effect transistor. *International Journal of Recent Technology and Engineering (IJRTE)*, *8*(2), 603–608.
62. Singh, S. and Raj, B. (2019). Design and analysis of hetrojunction vertical *T*-shaped tunnel field effect transistor. *Journal of Electronics Material*, *48*(10), 6253–6260.
63. Goyal, C., Ubhi, J. S., and Raj, B. (2019). A low leakage CNTFET based inexact full adder for low power image processing applications. *International Journal of Circuit Theory and Applications*, *47*(9), 1446–1458.
64. Raj, B., Saxena, A. K., and Dasgupta, S. (2008). A compact drain current and threshold voltage quantum mechanical analytical modeling for FinFETs. *Journal of Nanoelectronics and Optoelectronics (JNO)*, *3*(2), 163–170.
65. Wadhwa, G. and Raj, B. (2020). An analytical modeling of charge plasma based tunnel field effect transistor with impacts of gate underlap region. *Superlattices and Microstructures*, *142*, 106512.
66. Singh, S. and Raj, B. (2020). Modeling and simulation analysis of SiGe hetrojunction double gatevertical *T*-shaped tunnel FET. *Superlattices and Microstructures*, *142*, 106496.

67. Singh, A., Saini, D. K., Agarwal, D., Aggarwal, S., Khosla, M., and Raj, B. (2016). Modeling and simulation of carbon nanotube field effect transistor and its circuit application. *Journal of Semiconductors (JoS)*, *37*, 074001-6.
68. Jain, N. and Raj, B. (2016). Device and circuit co-design perspective comprehensive approach on FinFET technology — A review. *Journal of Electron Devices*, *23*(1), 1890–1901.
69. Kumar, S. and Raj, B. (2016). Analysis of I_{ON} and ambipolar current for dual-material gate-drain overlapped DG-TFET. *Journal of Nanoelectronics and Optoelectronics*, *11*, 323–333.
70. Anjum, N., Bali, T., and Raj, B. (2013). Design and simulation of handwritten multiscript character recognition. *International Journal of Advanced Research in Computer and Communication Engineering*, *2*(7), 2544–2549.
71. Sharma, S., Raj, B., and Khosla, M. (2016). A Gaussian approach for analytical subthreshold current model of cylindrical nanowire FET with quantum mechanical effects. *Microelectronics Journal*, *53*, 65–72.
72. Singh, K. and Raj, B. (2015). Performance and analysis of temperature dependent multi-walled carbon nanotubes as global interconnects at different technology nodes. *Journal of Computational Electronics*, *14*(2), 469–476.
73. Kumar, S. and Raj, B. (2015). Compact channel potential analytical modeling of DG-TFET based on evanescent–mode approach. *Journal of Computational Electronics*, *14*(2), 820–827.
74. Singh, K. and Raj, B. (2015). Temperature dependent modeling and performance evaluation of multi-walled CNT and single-walled CNT as global interconnects. *Journal of Electronic Materials*, *44*(12), 4825–4835.
75. Sharma, V. K., Pattanaik, M., and Raj, B. (2014). INDEP approach for leakage reduction in nanoscale CMOS circuits. *International Journal of Electronics*, *102*(2), 200–215.
76. Singh, K. and Raj, B. (2015). Influence of temperature on MWCNT bundle, SWCNT bundle and copper interconnects for nanoscaled technology nodes. *Journal of Materials Science: Materials in Electronics*, *26*(8), 6134–6142.
77. Anjum, N., Bali, T., and Raj, B. (2013). Design and simulation of handwritten gurumukhi and devanagri numerical recognition. *International Journal of Computer Applications*, *73*(12), 16–21.
78. Khandelwal, S., Gupta, V., Raj, B., and Gupta, R. D. (2015). Process variability aware low leakage reliable nano scale DG-FinFET SRAM cell design technique. *Journal of Nanoelectronics and Optoelectronics*, *10*(6), 810–817.

79. Sharma, V. K., Pattanaik, M., and Raj, B. (2014). ONOFIC approach: Low power high speed nanoscale VLSI circuits design. *International Journal of Electronics*, *101*(1), 61–73.
80. Khandelwal, S., Raj, B., and Gupta, R. D. (2015). FinFET based 6T SRAM cell design: Analysis of performance metric, process variation and temperature effect. *Journal of Computational and Theoretical Nanoscience*, *12*, 2500–2506.
81. Singh, S., Yadav, S., Rahul, J., Srivastava, A., and Raj, B. (2015). Impact of HfO_2 in graded channel dual insulator double gate MOSFET. *Journal of Computational and Theoretical Nanoscience*, *12*(6), 950–953.
82. Sharma, V. K., Pattanaik, M., and Raj, B. (2014). PVT variations aware low leakage INDEP approach for nanoscale CMOS. *Circuits, Microelectronics Reliability*, *54*, 90–99.
83. Raj, B., Saxena, A. K., and Dasgupta, S. (2013). Quantum mechanical analytical modeling of nanoscale DG FinFET: Evaluation of potential, threshold voltage and source/drain resistance. *Journal of Material Science in Semiconductor Processing*, *16*(4), 1131–1137.
84. Gopal, M., Prasad, S. S. D., and Raj, B. (2013). 8T SRAM cell design for dynamic and leakage power reduction. *International Journal of Computer Applications*, *71*(9), 43–48.
85. Pattanaik, M., Raj, B., Sharma, S., and Kumar, A. (2012). Diode based trimode multi-threshold CMOS technique for ground bounce noise reduction in static CMOS adders. *Advanced Materials Research*, *548*, 885–889.
86. Raj, B., Saxena, A. K., and Dasgupta, S. (2011). Nanoscale FinFET based SRAM cell design: Analysis of performance metric, process variation, underlapped FinFET and temperature effect. *IEEE Circuits and System Magazine*, *11*(2), 38–50.
87. Sharma, V. K., Pattanaik, M., and Raj, B. (2013). Leakage current ONOFIC approach for deep submicron VLSI circuit design. *International Journal of Electrical, Computer, Electronics and Communication Engineering*, *7*(4), 239–244.
88. Chawla, T., Khosla, M., and Raj, B. (2021). Design and simulation of triple metal double-gate germanium on insulator vertical tunnel field effect transistor. *Microelectronics Journal*, *114*, 105125.
89. Kaur, P., Gill, S. S., and Raj, B. (2022). Comparative analysis of OFETs materials and devices for sensor applications. *Journal of Silicon*, *14*, 4463–4471.
90. Sharma, S. K., Kumar, P., Raj, B., and Raj, B. (2022). $In_{1-x}Ga_xAs$ double metal gate-stacking cylindrical nanowire MOSFET for highly sensitive photo detector. *Journal of Silicon*, *14*, 3535–3541.
91. Raj, B., Saxena, A. K., and Dasgupta, S. (2009). Analytical modeling of quasi planar nanoscale double gate FinFET with source/drain resistance

and field dependent carrier mobility: A quantum mechanical study. *Journal of Computer (JCP)*, *4*(9), 1–8.

92. Bhushan, S., Khandelwal, S., and Raj, B. (2012). Analyzing different mode FinFET based memory cell at different power supply for leakage reduction. In *Proceedings of Seventh International Conference on Bio-Inspired Computing: Theories and Applications (BIC-TA 2012)*, Springer India, *2*, pp. 89–100.
93. Singh, J. and Raj, B. (2018). Temperature dependent analytical modeling and simulations of nanoscale memristor. *Engineering Science and Technology*, *21*, 862–868.
94. Singh, S., Bala, S., Raj, B., and Raj, B. (2020). Improved sensitivity of dielectric modulated junctionless transistor for nanoscale biosensor design. *Sensor Letter*, *18*, 328–333.
95. Kumar, V., Vishvakarma, S. K., and Raj, B. (2020). Design and performance analysis of ASIC for IoT applications. *Sensor Letter*, *18*, 31–38.
96. Jaiswal, A., Sarin, R. K., and Raj, B., and Sukhija, S. (2018). A novel circular slotted microstrip-fed patch antenna with three triangle shape defected ground structure for multiband applications. *Advanced Electromagnetic (AEM)*, *7*(3), 56–63.
97. Wadhwa, G. and Raj, B. (2018). Label free detection of biomolecules using charge-plasma-based gate underlap dielectric modulated junctionless TFET. *Journal of Electronic Materials (JEMS)*, *47*(8), 4683–4693.
98. Singh, G., Sarin, R. K., and Raj, B. (2017). Design and performance analysis of a new efficient coplanar quantum-dot cellular automata adder. *Indian Journal of Pure & Applied Physics (IJPAP)*, *55*, 97–103.
99. Singh, A., Khosla, M., and Raj, B. (2017). Design and analysis of electrostatic doped Schottky barrier CNTFET based low power SRAM. *International Journal of Electronics and Communications*, *80*, 67–72.
100. Kaur, P., Pandey, V., and Raj, B. (2020). Comparative study of efficient design, control and monitoring of solar power using IoT. *Sensor Letter*, *18*, 419–426.
101. Bhardwaj, A. K., Gupta, S., and Raj, B. (2020). Development & analysis of compact model for double gate Schottky barrier CNTFET. *Journal of Nanoelectronics and Optoelectronics*, *15*, 1199–1208.
102. Wadhwa, G., Kamboj, P., Singh, J., and Raj, B. (2020). Design and investigation of junctionless DGTFET for biological molecule recognition. *Transactions on Electrical and Electronic Materials*, *22*(3), 1–6.
103. Chawla, T., Khosla, M., and Raj, B. (2020). Optimization of double-gate dual material GeOI-vertical TFET for VLSI circuit design. *IEEE VLSI Circuits and Systems Letter*, *6*(2), 13–25.
104. Verma, S. K., Singh, S., Wadhwa, G., and Raj, B. (2020). Detection of biomolecules using charge-plasma based gate underlap dielectric

modulated dopingless TFET. *Transactions on Electrical and Electronic Materials (TEEM)*, *21*, 528–535.

105. Jain, N. and Raj, B. (2017). Impact of underlap spacer region variation on Electrostatic and analog/RF performance of symmetrical high-k SOI FinFET at 20 nm channel length. *Journal of Semiconductors (JoS)*, *38*(12), 122002.
106. Singh, S. and Raj, B. (2020). Analytical modeling and simulation analysis of *T*-shaped III-V heterojunction vertical *T*-FET. *Superlattices and Microstructures*, *147*, 106717.
107. Girish, W., Kamboj, P., Singh, J., and Raj, B. (2021). Design and investigation of junctionless DGTFET for biological molecule recognition. *Transactions on Electrical and Electronic Materials*, *22*, 282–289.
108. Singh, A., Khosla, M., and Raj, B. (2016). Compact model for ballistic single wall CNTFET under quantum capacitance limit. *Journal of Semiconductors (JoS)*, *37*, 104001-8.
109. Singh, S., Khosla, M., Wadhwa, G., and Raj, B. (2021). Design and analysis of double-gate junctionless vertical TFET for gas sensing applications. *Applied Physics A*, *127*(16), 1–8.
110. Singh, I., Raj, B., Khosla, M., and Kaushik, B. K. (2020). Potential MRAM technologies for low power SoCs. *SPIN*, *10*(4), 2050027.
111. Singh, S. and Raj, B. (2020). Parametric variation analysis on hetero-junction vertical *T*-shape TFET for Supressing ambipolar conduction. *Indian Journal of Pure and Applied Physics*, *58*, 478–485.
112. Soniya, S. S., Wadhwa, G., and Raj, B. (2019). Design and analysis of dual source vertical tunnel field effect transistor for high performance. *Transactions on Electrical and Electronics Materials*, *21*, 74–82.
113. Kaur, M., Gupta, N., Kumar, S., Raj, B., and Singh, A. K. (2020). RF performance analysis of intercalated graphene nanoribbon based global level interconnects. *Journal of Computational Electronics*, *19*, 1002–1013.
114. Wadhwa, G. and Raj, B. (2019) Design and performance analysis of junctionless TFET biosensor for high sensitivity. *IEEE Nanotechnology*, *18*, 567–574.
115. Singh, J. and Raj, B. (2019). Enhanced nonlinear memristor model encapsulating stochastic dopant drift. *JNO*, *14*, 958–963.
116. Masud, M., Hossain, M. S., Alhumyani, H., Alshamrani, S. S., Cheikhrouhou, O., Ibrahim, S., Muhammad, G., Eldin Rashed, A. E., and Gupta, B. B. (2021). Pre-trained convolutional neural networks for breast cancer detection using ultrasound images. *ACM Transactions on Internet Technology (TOIT)*, *21*(4), 1–17.
117. Chui, K. T., Gupta, B. B., Jhaveri, R. H., Chi, H. R., Arya, V., Almomani, A., and Nauman, A. (2023). Multiround transfer learning and modified generative adversarial network for lung cancer detection. *International Journal of Intelligent Systems*, *2023*, 1–14.

118. Chui, K. T., Gupta, B. B., Chi, H. R., Arya, V., Alhalabi, W., Ruiz, M. T., and Shen, C. W. (2022). Transfer learning-based multi-scale denoising convolutional neural network for prostate cancer detection. *Cancers*, *14*(15), 3687.
119. Pathoee, K., Rawat, D., Mishra, A., Arya, V., Rafsanjani, M. K., and Gupta, A. K. (2022). A cloud-based predictive model for the detection of breast cancer. *International Journal of Cloud Applications and Computing (IJCAC)*, *12*(1), 1–12.

https://doi.org/10.1142/9789811279539_0004

Chapter 4

Urban Smart Connectivity Using Intelligent Technologies

Chirag Sethi[†,§], **Shailendra Singh**[*,†,¶], **and Balwant Raj**[‡,‖]

[†]*Department of Electronics and Communication Engineering, Pranveer Singh Institute of Technology, Kanpur, India*

[‡]*UIET PU Hoshiarpur Campus, Hoshiarpur, India*

[§]*sethichirag646@gmail.com*

[¶]*shailendras.nitj@gmail.com*

[‖]*b.raj255@gmail.com*

Abstract

In light of the rapid urbanization throughout the world, it is crucial to combine intelligent and environmentally friendly city design in terms of the carriage of passengers and commodities. This will play an important role in shaping how future civilizations handle transportation for decades to come. Intelligent transportation is a concept which is developed over time, contemporaneous with the increase in interest in the so-called "Intelligent Cities," in line with the United Nations' emphasis on the sustainability elements of development over the long term. However, other factors, such as promoting active and inclusive transportation, supporting the use of alternative ecologically friendly fuels, and engaging

[*]Corresponding author.

with individuals, must also be taken into account. Some of the crucial components of smart mobility include a decline in traffic congestion and new route optimizations with a reduced ecological footprint. Internet-of-Everything, artificial intelligence, blockchain, and Big Data technologies can become key points of initiation as well as core supports that encourage the growth of innovation, which will revolutionize the present conceptual framework for cities and their inhabitants. These technologies will also play a key role in the advancement of revolutionary new ideas that will alter every aspect of our lives. The development of current urban areas into intelligent cities will bring about a slew of modifications over the next several decades at least. Some of the impacts on services and applications include optimization of logistics, heavy traffic improvement, and driverless vehicles, and these changes will incorporate a variety of shifts. This study offers a comprehensive analysis of the most emerging advancements and solutions that have been offered within the realm of intelligent connectivity and the supporting advancements in science that are necessary for its success. The purpose of this chapter is to convey a breakdown of how intelligent mobility contributes to the emergence of smart cities by defining specific aspects and primary merits of using intelligent transportation within the context of a smart city ecosystem. This chapter discusses a variety of other potential advantages and constraints associated with smart mobility. In addition, the primary services and applications that are anticipated to emerge in the next few years in the realm of smart connectivity are probed, along with future potential trends and the scope of the industry.

Keywords: Intelligent connectivity, sustainability, intelligent cities, smart services

4.1. Introduction

Taking into consideration the 17 sustainable development objectives that have been established, the term "intelligent transportation" is a fairly new concept that is becoming more correlated with the term "stable and prosperous expansion." This notion was proposed by the United Nations for the year 2030. The concept of portability, in addition to the way in which it will be incorporated into the spatial organization of towns and the standard operating procedures for carrying both products and persons, would go through a period of profound transformation in the years to come. The expansion of the world's population continues to follow an exponential path; in 1999, there were roughly 6 billion people on the planet; by 2020, there were likely to be approximately 7.7 billion; and by 2040, the population is projected to reach approximately 9 billion. This expansion that we are seeing results in urbanization around the world, which inevitably means new algorithms are needed to optimize routes for automobiles and people, traffic management to prevent delays, and more performance improvement involved in logistics, among other things [1]. The notion of smart mobility, on the other hand, goes far beyond than just finding answers to these difficulties since it is anticipated that future contributions will represent differentiated and really new solutions. In the next few years, greater attention should be paid to the lifetimes of the proposed solutions, movement-based modes of transport, how we can burn hydrocarbons without harming the environment, and greater involvement with residents. These are all characteristics that should be included in smart mobility. Because of this, the aspects that will be affected will be diverse and will encompass economic viability, environmental balance, and quality-of-life conditions, all of which will have a direct influence on residents as well as on government institutions [2].

In the subsequent years, smart mobility will develop on the foundation of two major ideas: the conveyance of commodities, and the paradigm shift in the mobility of people, which will transition to connectivity services. The future holds great promise for smart transportation. When it comes to the transportation of commodities, we can already see various robot prototypes across the globe that carry essential items to the homes of individuals; this will require a change in the way in which companies work (grocery stores, shipping companies, and more). The Internet of Everything, Big Data, and artificial intelligence (AI) will each play an important role in the development of innovative solutions [3–4]. It is important to emphasize the fact that the changes that will occur in the years to come will inevitably lead to changes in the types of employment that will be available in the future. When it comes to the transportation of people, current trends are moving in the direction of a paradigm known as "mobility as a service." This will involve a significant decrease in the number of single-occupant cars that will be phased out in favor of multi-occupant, electric, lightweight "micro-transportation" and more compact automobiles also capable of autonomous driving [4–10].

This study presents a comprehensive literature review of many solutions spanning several areas interconnected with intelligent transportation in smart cities. The solutions are discussed in the context of smart cities. In addition to this, the chapter presents an outline of what smart mobility is, as well as the potential and problems associated with it [5]. In addition, the study sheds light on the enabling technologies that are being implemented to serve the public with insightful support from them. In the last part of this analysis, prospective trends and conclusions are discussed.

The global population density is rising at a very fast rate within metropolitan areas, which is compatible with the demand to assist the needs of citizens living in these localities [6]. The concept of a "Smart City" may be implemented successfully in complex metropolitan environments, including complex systems such as infrastructure, human behavior, technological advancements, sociopolitical institutions, and the economy [11–16]. A smart city provides an intelligent method for managing systems such as transportation, healthcare, educational institutions, energy, residences and buildings, and the environment. A feature of the smart city is that it is mostly supported by information and communication technologies. It provides assistance to the global community in the development, dissemination, and promotion of sustainable development techniques with the purpose of addressing the issues posed by growing urbanization [7]. These data are produced by using wireless sensor networks to connect the various components.

For the smart city vision to become a reality, billions of Internet of Things (IoT) devices need to be able to communicate with one another. Ningbo, Shanghai, Harbin, and Dalian are just some of the cities in China that have already begun building on their versions of the smart city. A city is considered smart if it utilizes information and communication technologies, services, monitoring, and management in a manner that is more aware, interactive, and efficient [17–23]. The rapidly developing IoT is the technical engine that propels and enables a city's smartness. The term "Internet of Things" refers to the process of linking devices, including electronics, computer software, and sensing capabilities, to the internet in order to provide such devices with the ability to gather and share data. The information gathered by sensors and actuators and sent via the gateway to the server is the foundation of the IoT [8–9]. The IoT allows for a variety of interactions, including device to device, object to device, and device to thought. Because every company, organization, and user is different, there is not a single agreement on the design of the IoT that is generally agreed upon. An IoT design can be split into three basic levels. The first layer is called the "Perception" layer, and it has a collection of nodes that may be used for monitoring, sensing, collecting data, and exchanging information with other devices through the internet [24–30].

This information may be sent on to the second layer of communication, which is known as the "Network" layer. This layer is in charge of sending the data that was gathered from the sensing layer to the application layer [10]. The primary objective of responsibility is to ensure that the application itself is the one to give this information to the end user or, more generally, to the platform. The third layer is the services of the "Application" layer, and its purpose is to receive and process the sensing data that has been sent up from the lower levels depending on the kind of application in order to fulfill the needs of the application. The primary emphasis of this chapter is the implementation of two different facets of the smart city that are based on the IoT [31–37].

4.1.1. *Necessity and Significance of Smart Mobility*

Residents are given the opportunity to travel and move around inside the intelligent cities' vicinity thanks to smart mobility.

It is possible for smart mobility to support better traffic control, backup routes in the case of accidents or emergencies, and specialized directions for emergency vehicles, government vehicles, and other crucial services [11]. It is necessary to have such mobility services in order to provide residents and administrative personnel with choices that are sustainable, favorable to the environment, and free of congestion. As depicted in Figure 4.1, the implementation of intelligent mobility solutions may result in benefits not just for the general public but also for administrative bodies [37–45].

4.1.2. *Detailed Organization Framework for Systematic Study*

The implementation of our desired goal is structured in seven distinct phases. In the subsequent part, new advancements in the intelligent transportation domains are presented, along with the data shortage and outstanding challenges [12].

In a part of this chapter, we discuss "smart mobility" and the function that it plays in smart cities. Following an introduction, we discuss the primary advantages and disadvantages associated with the implementation that will usher in a new era of intelligent mobility in the years to come. Many smart mobility services and applications are included in the last section. Connectivity services, existing traffic optimization, and supply chain management are all examples [46–53].

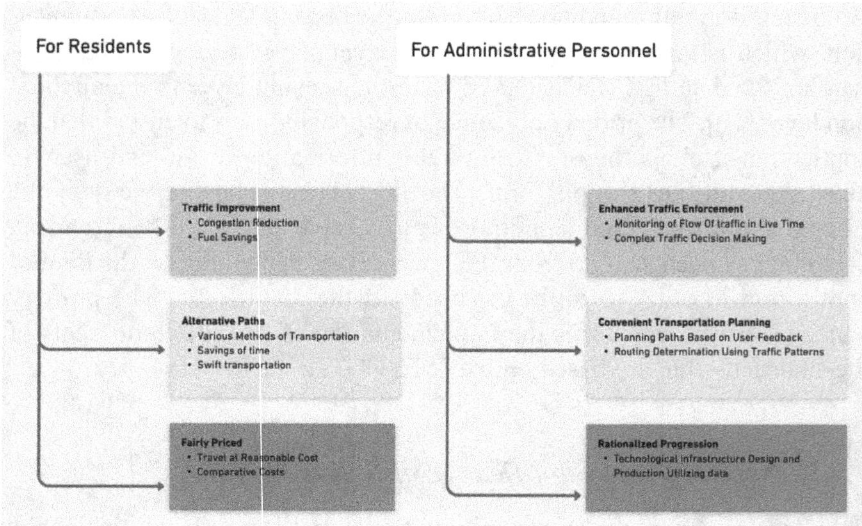

Figure 4.1. Need for Smart Mobility

Apart from navigational technologies and advanced driver assistance systems, we discuss the technologies that make transportation possible and how they contribute to the implementation of these strategies in the subsequent sections. The Internet of Everything, Big Data, and AI are all part of these innovations. Further, we discuss potential future trends and provide our primary findings. Figure 4.2 offers an overview of the structured framework for demonstration purposes.

4.1.3. *Crucial Accomplishments*

A thorough analysis of the intelligent transportation outcomes and associated support offered in recent years is one of this chapter's significant contributions. The explanations for the major advantages and characteristics that incorporate the concept of transportation models within the larger context of intelligent cities are discussed [54–61]. The possibilities for intelligent transportation to materialize in the near future are discussed, as are the main obstacles and hurdles to its fulfillment. Insightful observations are presented on the notion of connectivity services in the future. Moreover, the fundamentals of transportation management techniques and different perspectives, including the roles of AI, the IoT, blockchain, geospatial technologies, and Big Data are provided.

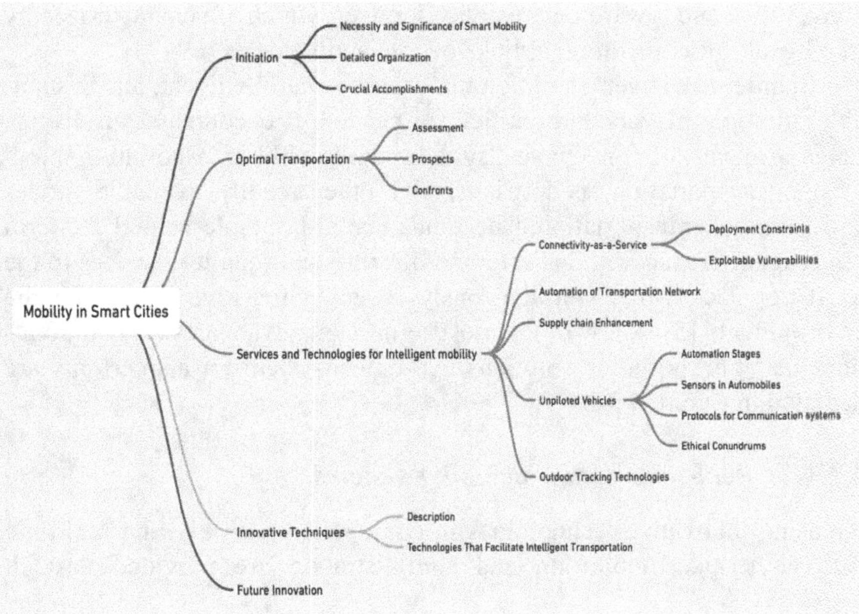

Figure 4.2. Structured Framework of Study

4.2. Optimal Transportation

4.2.1. *Assessment*

Mobility in the city is one of the most important aspects of a smart city and performs a vital function in intelligent and environmentally friendly development. Today, governments all over the world are making significant investments in transportation infrastructure as a strategy to guarantee that populations have access to efficient and cost-effective modes of transportation [13–14]. Intelligent mobility is one of the most important distinguishing characteristics of a smart city. The need for intelligent mobility is growing at an alarming rate in the majority of developing countries [62–68]. In these countries, rising urbanization is driving up the need for efficient and environmentally friendly means of transportation. The most important mobility problem in nations like these requires the widespread implementation of public transportation networks. Travel that is environmentally friendly, risk-free, and time- and money-saving is made possible thanks to the core characteristics of the smart mobility concept. The idea of "smart mobility" refers to the transition away from regular means of conveyance and toward connectivity services, which are characterized by intelligent infrastructure that links people and vehicles [69–74].

In order to deliver a solution that is effective, intelligent, and sustainable, multiple players and entities are needed. It encompasses a diverse range of means of conveyance, such as rapid public transportation, bicycles, e-transportation, pedestrians, and other readily available modes and forms of transportation. The fundamental principle behind the term "Intelligent Transportation" refers to offering high-quality services to the local populace while simultaneously reducing negative effects on the environment [15]. A few of the most significant advantages of incorporating smart transportation solutions into an ecosystem for a smart city are depicted in Figure 4.3.

4.2.1.1. *Real-time data processing systems*

By using innovative technologies in conjunction with existing real-time data gathering, monitoring and administration are provided through

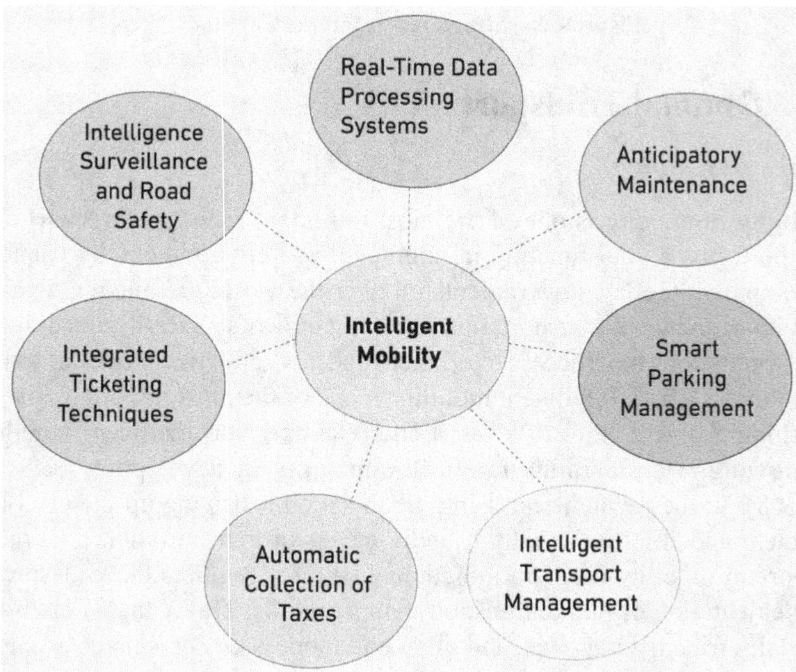

Figure 4.3. Main Perks of Utilizing Intelligent Mobility in a Smart City

tracking systems and the development of smart mobile devices. Information systems that are effective and intelligent are created using the data gathered from numerous linked organizations.

4.2.1.2. *Anticipatory maintenance*

The ideas of AI and machine learning may also be used for proactive maintenance of different systems. Continuous data collection and monitoring make this possible.

4.2.1.3. *Smart parking management*

When examined, the data gathered in real-time from sensors and other connected devices may give crucial inputs regarding the availability of parking places in a variety of locations.

4.2.1.4. *Intelligent transport management*

The use of effective data analytics may be of assistance in the monitoring and management of traffic in real-time, which can help prevent traffic jams and congestion. In addition, messages may be provided to linked cars in real-time on the status of parking, routes, and other relevant information [75–82].

4.2.1.5. *Automatic collection of taxes*

Automatic tax collection is a simplified method in which taxes are collected and processed electronically without the need for human involvement. This automated approach provides revenue collection efficiency, precision, and timeliness, eliminating administrative hassles for both taxpayers and government agencies.

4.2.1.6. *Integrated ticketing techniques*

The idea behind connectivity services allows for incorporating a number of different local services, which in turn makes it possible to implement a smart ticketing system that enables residents to access convenient and multimodal services [83–88].

4.2.1.7. *Intelligence surveillance and road safety*

Monitoring the flow of traffic and ensuring the safety of drivers are both made easier thanks to cameras and other surveillance equipment that are integrated into the linked network.

The development of technologies that are linked to one another has resulted in the rise of new opportunities in the fields of urban studies and administration [17]. Vehicle collisions are very expensive, both economically and in terms of human lives lost. Modern transportation management systems have made feasible the concept of intelligent transportation, which has improved the safety and efficiency of driving and commuting. Connection models for are shown in Figure 4.4. Transportation and other connected organizations may communicate more efficiently thanks to transportation models:

- **Vehicle-to-Infrastructure (V2I):** Automobiles are connected to traffic infrastructure, such as traffic lights, toll booths, parking lots, and crosswalks, to exchange real-time traffic data and predict traffic congestion and delays.

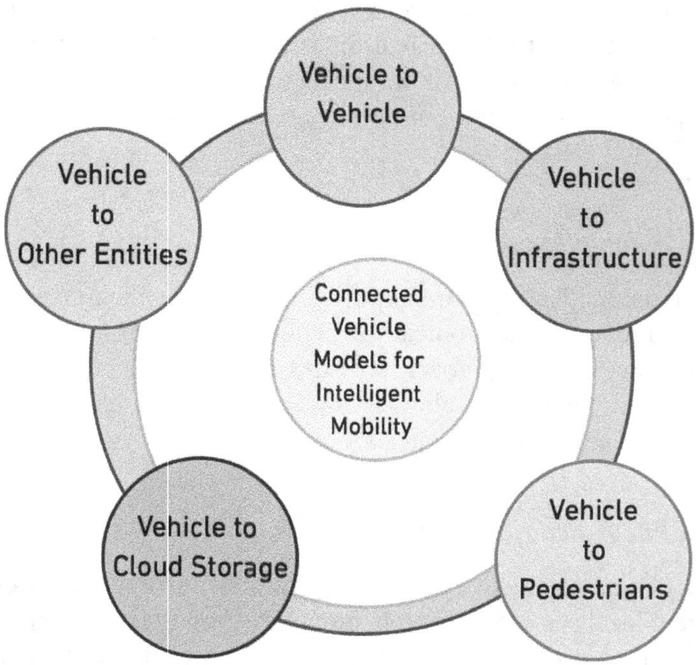

Figure 4.4. The Key Models of Smart Vehicles for Intelligent Mobility

- **Vehicle-to-Vehicle (V2V):** Automobiles are networked with one another to increase their overall level of safety on the road. The information that is sent between cars is an additional aid in combating congestion and managing parking spaces.
- **Vehicle-to-Cloud Storage:** Connectivity between vehicles and cloud storage facilitates the exchange of data in real-time, as well as its accumulation and analysis [89–94].
- **Vehicle-to-Pedestrians:** This connection framework is a concept that links cars to pedestrians via the use of intelligent devices, such as smartphones and smart wearables, in order to make walking safer and provide convenient, efficient transportation options and services in real-time.
- **Vehicles-to-Other-Entities:** By integrating transportation and intelligent infrastructure processes and components, this paradigm of full connection is ensured.

4.2.2. *Prospects*

Innovative solutions for urban transportation have the potential to provide previously unimaginable advantages to the ecosystem of smart cities. With the support of intelligent mobility solutions, things such as intelligent land use planning and management as well as intelligent traffic management are conceivable [18]. City planners may make use of intelligent mobility methods to develop designs that are not only effective and efficient but also sustainable and friendly to the environment. This is possible through the use of intelligent mobility techniques. There are many possibilities available for urban designers, administrators, citizens, and many other residents of intelligent cities that are related to smart mobility initiatives. These opportunities may be found in intelligent cities. More information on traffic patterns and municipal streets might be gleaned from the Big Data generated by the Internet of Everything that is being planned and developed as part of strategic route planning and development [19]. This information may be used further for the creation and planning of important routes. In order to maximize exposure to passing customers, businesses may benefit from strategically placing digital sign boards and hoardings at high-traffic intersections and other high-volume areas. This enables businesses to maximize the effectiveness of their marketing campaigns. The more traditional forms of advertising, such as posters and banners, may be upgraded to their digital equivalents, which can then be used to generate revenue. This offers advantages in two different ways. First, it cuts down on the amount of paper that is wasted, and second, since it can be dynamically updated via the use of basic coding or programming, it requires less time to set up and maintain. Lawmakers are able to make real-time choices for traffic management that are data-based and dynamic [20]. This information may be sent to commuters so that alternate routes can be suggested to them in the event that there is a potential for traffic congestion. Law-abiding persons who assist the government in obeying and executing these rules may be rewarded by the government for their efforts. Creators, enterprises, and producers have the ability to come up with creative solutions for the provision of environmentally friendly and sustainable alternatives to traditional methods [95–98]. This is an opportunity for companies that are focused on a certain sector or set of procedures to develop their company. Smart mobility solutions provide individuals with the opportunity to enjoy a higher quality of service (QoS) with respect to transportation costs, accessibility, and convenience [21, 22]. Citizens also benefit from a higher quality of service because of more and better transit choices. Techniques from the field of data analytics may be used in order to provide assistance to the general populace living in intelligent cities. Because what determines these norms and options is based on data, there is a good chance that they will increase the QoS [99–103].

4.2.3. *Confronts*

Contemporary smart mobility obstacles include the following:

Transportation System: To alleviate strain on most of the world's inadequate transportation networks, smart city systems need extensive infrastructure. In order to function, autonomous vehicles and electric vehicles need high-speed data connections and charging infrastructure. Smart mobility requires infrastructure to actualize smart system principles.

Last Mile Connection: Public transit systems struggle with connection at a cheap cost at the last mile. Solutions for smart transportation must include door-to-door connection, regardless of the means of travel.

Confidentiality and Safety: Connected devices access large volumes of personally identifiable information, raising privacy concerns. Sharing device and user data might compromise privacy and security. So, connected gadgets make the network open to outside assault and breaches.

Management: Smart mobility includes digital businesses and service providers in traditional transportation networks [23]. New actors necessitate smart mobility policy changes, which may leave smart mobility rules behind. Smart transportation systems cannot be widely used without standards. Smart mobility solutions need flawless traffic law integration.

Initial Adoption: Smart transportation's biggest challenge is educating prospective consumers. Most mobility solutions are still under development; therefore, their aims are unclear.

Digital Navigation and Automobile Connectivity: Dynamic routing systems analyze user travel needs and optimize resources to provide efficient solutions. This demands advanced software and technology.

Network Management: Millions of linked things make smart ecosystem network maintenance and monitoring difficult and expensive.

Procurement and Incorporation: Information is collected from a wide variety of sources, each of which may have different privacy and data protection policies in place [24]. It's challenging to handle data in real-time from connected devices.

Conflicts: Intelligent transportation requires a well-established legal framework and supportive policies for all stakeholders (payment processors, governments, local administrations, public–private transportation providers, customers, etc.).

4.3. Assistance & Technologies for Intelligent Mobility

4.3.1. *Connectivity-as-a-Service*

In smart cities, the Connectivity-as-a-Service (CaaS) concept will develop. Door-to-door travel does not meet sustainable development goals and is more expensive and slower than CaaS systems [25].

CaaS systems provide an alternative to private mobility, reducing traffic congestion and the capability of urban transportation. Within the next several years, a mainstream market will emerge that unifies many transportation systems, offers services on demand, and provides real-time and predictive information to enable multimodal routes. In addition, creating a global CaaS platform that integrates numerous geographically scattered and diverse systems may lead to uniform standards. Caas systems have basically two primary traits. Firstly, they must be user-centric, meaning the platform adapts to individual users' demands and preferences, ideally collected implicitly and transparently from their mobility history and behaviors. Secondly, these systems are designed to gather and utilize data from users' mobility history and behaviors to provide seamless and personalized connectivity services, ensuring a smooth and tailored experience for each individual user. Some of the problems that a CaaS will aid in resolving in a smart city are the inadequate location of specific places for bicycles, scooters, and parking links, as well as their absence of integration with the transportation infrastructure; the low use of active modes of transportation; the high number of private cars within a city; the low appeal of public transport; and the lack of unscrupulous connectedness and addressing the requirements of residents. The advantages offered by CaaS systems include enhancing consumer ease, efficiency, satisfaction, and convenience; expanding revenue for transportation service companies; improving the transportation network's efficacy, which reduces expenses for citizens; and reducing vehicular congestion and carbon impact.

4.3.2. *Deployment Constraints*

The introduction of the notion as a result of the rise of the "Intelligent city" has triggered a sea change in the way cities handle their public transit [26]. The transportation sector may once again be at the forefront of revolutionary change because of the proliferation of cutting-edge technological advancements, such as the Internet of Everything, wireless sensing systems, analytical data processing, and Big Data. Transport has evolved into a complicated system with services integrated into a single, technologically controlled platform. CaaS refers to a transport network that links various modes of transportation, and related services give consumers a seamless experience on demand. An effective CaaS system has the ability to facilitate business possibilities in important ways. Problems exist for most cities nowadays in terms of the accessibility of their public transit systems and amenities, including parking and traffic. Frameworks that provide CaaS have great potential to help solve transportation network problems and improve the environment by fostering the use of renewable energy. However, there are still many issues and concerns regarding the correct integration of smart city ecosystems, despite this ideal picture of on-demand, seamless mobility. CaaS platforms focus on an intelligent city's primary concerns and obstacles, as well as their possible solutions.

4.3.3. *Exploitable Vulnerabilities*

Autonomous cars employ technologies such as Wi-Fi sensor networks, AI, and ML to connect devices and collect data in the field, as well as gather, analyze, and exchange information in order to ultimately make intelligent judgments [27]. These autonomous cars are part of a bigger network consisting of tens of thousands of linked units. These sorts of linked ecosystems comprising heterogeneous devices rely heavily on the security standards of each individual device. This leaves autonomous cars exposed to a variety of security risks and assaults. These assaults may occur at several levels, including network, device, and software. Using driverless cars raises grave security and privacy issues due to the vulnerabilities at several levels. Some of the most pressing concerns in the world today include the safety of our data, networks, transportation, and finances. Potential threats to security include the following:

Information Vandalism: A seamless user experience is achieved via a network of linked entities, of which autonomous driving and self-driving cars are components [28]. Data from users, comprising payment information, facts, contacts, and traveling preferences and background are included in the data gathered by different businesses. There are security holes that might be used by hackers to get this information. The notion of autonomous cars is predicated on the handing over of the driving reins to the cars themselves. In these situations, cybercriminals and hackers may easily attack linked automobiles. These automobiles are susceptible to identity theft, in which hackers access and abuse vehicle identifying information.

Vehicle Hijacking: It is one of the most significant hazards facing autonomous vehicles. The attackers are able to modify the algorithms used for remote control and take command of the vehicle's system. Once hackers get access to a vehicle, they may wreck havoc with essential systems, including the NCU, engine, brakes, heating, communication, camera, and vision systems.

Attempt to Disrupt Service: Assailants may obtain control over the whole network and overload the linked entities by exploiting the weaknesses of the network's least secure devices. These sorts of assaults may be conducted against a single node, a connection between two vehicles, or between vehicles and infrastructure, causing disruption of the communication system.

Violation of Privacy: Autonomous cars create data continually, such as the vehicle's position. These data give personally identifiable information about the user's trip history and navigation, which may be used to monitor the user's present position and previous locations [29]. In nations, such as India, RFID readers in vehicles are used to automatically deduct toll fees, raising grave concerns about the monetary safety of the associated bank accounts. Unsecured devices may lead to financial fraud and other associated assaults, such as ransomware.

4.3.4. *Automation of Transportation Network*

Many cities' traffic problems stem from decades-old planning that didn't anticipate today's congestion. Population expansion globally has increased the number of cars on the roadways, yet infrastructures cannot adapt fast enough to prevent traffic congestion. The increasing number of cars that caused traffic congestion also increased fuel consumption, gas emissions, and accidents. These factors affect financial, ecological, and residential circumstances. The United States Department of Transportation identifies three primary factors that contribute to traffic jams: incidents that disrupt traffic flow, such as construction zones or inclement weather; variations in demand for transportation services; and preexisting infrastructure, such as traffic lights and roadblocks. Congestion is alleviated and productivity is boosted by traffic management systems. They achieve this by collecting information from vehicles, traffic lights, and sensors built into the road system [30]. Some of the challenges of traffic flow optimization include the best placement of charging stations for electric vehicles, integration of different domains (air, land, and sea), predictive traffic analysis, data analysis that enables real-time traffic planning and control, and optimization of road infrastructure (including e-scooters, e-bikes, carsharing, and carpooling). Future traffic management, driving, and parking conditions, as predicted, will benefit from a network of intelligent sensors, location-based applications, and an intelligent infrastructure. Solutions in technology will allow private vehicles to connect to smart roads, traffic lights, and transportation management systems (TMSs), as well as receive collision warnings, maintain lanes, and avoid accidents. There will be an increase in security and ease of use for all parties involved as a result of this integration. When vehicles, roadways, traffic lights, and command and control systems are all interconnected, real-time data processing will make traffic monitoring and preventive measures possible. By altering their driving speed or being aware of neighboring parks and parking spaces, users may be able to optimize their journeys in terms of time and reduce road congestion and accidents. Safer roads, fewer accidents, more efficient driving and route planning, less congestion, and less greenhouse gas emissions are all possible thanks to this integrated, real-time information and driver communication.

4.3.5. *Supply Chain Enhancement*

Urban transportation and its issues go beyond people and cars. Sustainable and eco-friendly transportation must also consider commodities mobility, commercial transport, and urban logistics. Commercial fleets contribute to traffic congestion, on which we previously focused on individual automobiles. E-commerce is increasingly used worldwide for its convenience. However, deliveries affect city planning and vehicular jams. Residence and eatery delivery have grown with Uber-Eats and others [31]. Thus, commercial mobility demands equal attention to private mobility. Logistics activities are fragmented and inefficient throughout the supply chain. Given the rise of e-commerce, home meal delivery, and other delivery services, these issues will likely grow. As with private cars, more deliveries and logistical operations will increase the use of gas, electricity, and raw materials, whose optimization is necessary.

Traffic flow optimization problems include inefficient route optimization for delivery, deliveries conducted using fossil-fuel-powered vehicles, inefficient coordination of water, atmosphere, and ground transport, and citizens utilizing personal transportation to shop instead of public transit. Delivery delays, a lack of understanding of the carbon footprint consequences of such a large number of deliveries, and a lack of collaboration amongst transport companies to build an integrated strategy including several fleets are all issues plaguing the delivery industry that coincide with city traffic congestion. Smart logistics integrates trucks and goods, and minimize system waste by loading units in such a way that the most efficient path and load are taken. Systems that include vehicles, roads, and loading units will enhance the adaptability and efficiency of freight transportation by land, air, and sea thanks to advances in data and sensor processing technology. Systematic fleet management platforms and optimum path computation are projected to improve logistics system operationalization and planning, decreasing costs, accidents, and product damage. Sensor technology can follow items and fleets, allowing real-time route modifications to avoid congestion and save time.

4.3.6. *Unpiloted Vehicles*

Driverless automobiles are the most ambitious automation application. Self-driving cars are ambitious because they engage humans in potentially life-threatening circumstances. Self-driving cars are rising in popularity, and in a few years, they will be fully automated. Today's city technologies enable us to predict that, in the following years, we will see key advances in vehicle automation for regular commutes. This will completely change urban transportation. This new reality will need major adjustments to city infrastructures. 5G, a hundred times faster than 4G, will be crucial [32]. The Netherlands leads the automotive vehicle readiness index in infrastructure readiness for future driverless vehicles, followed by Singapore and the UK. Changing city infrastructure requires three essential traits:

Lane markings: These must be improved to be machine-readable by modern automobiles. Roadside sensors should be on sidewalks, curbs, and lanes. Thus, driverless vehicles can be aware of their surroundings and avert harm.

Smart signage: Image recognition reads traffic signs. In the future, signals will be able to convey machine-readable signals to autonomous cars.

It is believed that the adoption of automated driving would make transportation safer and more dependable in urban areas.

4.3.7. *Automation Stages*

The U.S. Federal Highway Administration used the terminology developed by the Society of Skilled Machinists for the six levels of driverless vehicles. As can be seen in Figure 4.5, the transition from manually controlled to fully automated vehicles involves various stages.

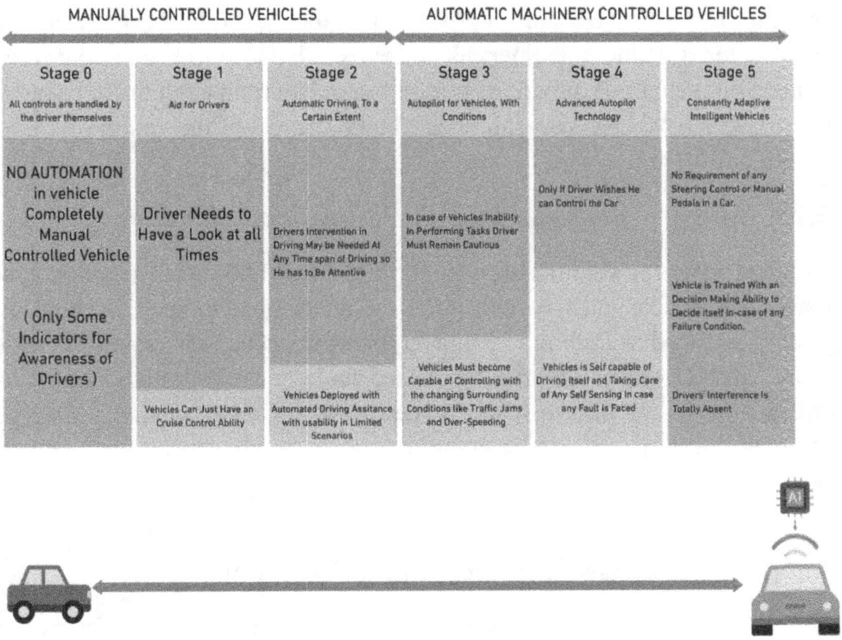

Figure 4.5. Vehicle Transformation Stages Toward Automation

4.3.7.1. *Stage 0 (all controls are handled by the driver themselves)*

The majority of cars on the road today are at this level, which means the driver is in complete control at all times. Assistance systems for the driver, such as an automatic braking system and a warning system, do not fall under the category of automation by definition.

4.3.7.2. *Stage 1 (aid for drivers)*

In this tier, the vehicle performs an autonomous action, such as speeding up or slowing down (cruise control), but the driver is still required to take over in the vast majority of driving situations.

4.3.7.3. *Stage 2 (automatic driving, to a certain extent)*

The presence of advanced driving aids (ADAs) that control the vehicle's speed and braking under certain circumstances defines this tier. This level is sometimes referred to as "partial driver automation" [33] because the degree of involvement of the machinery is currently relatively low, and continued vigilance on the part of the driver is required on the road at all times because they never know when they will need to take control of the vehicle.

4.3.7.4. *Stage 3* (*autopilot for vehicles, with conditions*)

This stage marks a significant step forward in technological development because the vehicle must be able to adapt to its surroundings, such as by adjusting its speed according to the behavior of the vehicle in front of it. This indicates that level 3 is the most advanced form of driving automation currently available. Despite this, the driver must always maintain a state of heightened awareness, as the automobile won't be able to perform certain actions.

4.3.7.5. *Stage 4* (*advanced autopilot technology*)

The primary function that is inducted at this stage is the capability of the vehicle to take over practically all driving autonomously and intervene if anything goes wrong or there is a malfunction. This is a highly advanced level of driving automation. Even if the driver's participation is not required in the vast majority of situations, they are still free to look after the vehicle if they so want.

4.3.7.6. *Stage 5* (*constantly adaptive intelligent vehicles*)

Stage 5 denotes when a car is fully automated and the steering wheel and pedals aren't needed anymore. This level is the highest. The car will have the ability to navigate anywhere and make all of the appropriate judgments on its own.

4.3.8. *Sensors in Automobiles*

The reliability of automated vehicles and the people's faith in these cars and the level of protection they can provide will both be significantly impacted by the function that sensors play. Sensors play a significant role in the communication between vehicles and infrastructure, known as V2V and V2I , and they provide a professional autopilot mode with the information necessary to formulate their recommendations for those who drive [34]. These suggestions can take the form of warnings to maintain a central position on the road, to avoid becoming drowsy or distracted while operating the cruise control, or, in more complex situations, to combine data on the external environment with data on the driver's state of mind (whether calm or agitated) and even the driver's heart rate. The purpose of this monitoring is to collect as much data as possible. In order to gather the information that was discussed before, it is important for the cars to be equipped with a number of sensors. These sensors must be environment sensors, which are in charge of gathering details about the vehicle's immediate surroundings. Cameras and remote radars are installed at the front of the vehicle as part of the environmental sensors. When traveling at high speeds, there are also long-range radars and lidar laser scanners available to help see obstacles in the path of the vehicle. When considering the effects of duration, detection angle, and maximum distance at which an item may be detected, it is typical for each of these sensors to be located in distinct locations on the vehicle. These factors all play a role in the identification process. The primary sensors, their locations, and the functions they serve are summarized in Table 4.1.

Table 4.1. The Function, Context, and Placing of Vehicle Sensors

Sensors Used	Placing	Functions
Long-range radar	Positioned precisely in the middle of the vehicle's front end	For use in an emergency pedestrian detection and crash-avoidance technology adaptative traction control
Light detection and ranging	Front, rear, and each of the four corners (diagonals)	Environmental scanning
Camera	The front, the rear, and both sides	Cautions for leaving your side road and observing traffic signs, smart panoramic view, side- and rear-view mirrors with 360 vision
Mid-range radar	Corners at the front, back, and rear of the phrase	Alerting for oncoming traffic, rear-end collision alert

4.3.9. *Protocols for Communication Systems*

Vehicular *ad hoc* networks are an essential component of today's modern transportation systems. In these networks, the creation of a routing protocol for efficient vehicular communications poses a number of issues, some of which are discussed in the following [35]:

- **Secure Routing:** The protection of the confidentiality of communications is a sensitive matter that is of utmost significance in all of the systems.
- **Communication that is Reliable:** Due to various VAN features, such as high mobility, sporadic connection, or city obstructions, link ruptures might occur quickly, and it can be difficult to know what to do in the event of packet losses.
- Choosing the best route while accounting for traffic density from a source to a destination and the latency of the network.
- Managing unauthorized message tampering, which has the potential to have catastrophic effects and potentially have an influence on life-or-death circumstances.

One of the most promising new protocols is position-based routing that is most effective method for use in vehicular *ad hoc* networks. It may be used not only on roads but also in urban environments, which provide the greatest difficulty due to the presence of obstructions. Numerous ideas have been made to implement this kind of communication standard V2V and V2I that can be found in the literature. These proposals take into account the difficulties associated with routing in an urban environment and include various categories of protocols.

4.3.10. *Ethical Conundrums*

Automakers from all over the globe have been interested in autonomous driving as a result of the recent surge in popularity of self-driving cars. Autonomous cars are vehicles that can go long distances with little or no human intervention. The mobility that these automobiles offer provides possible solutions to the problems with road safety that are caused by human mistakes. The shift from a vehicle that is only driven by humans to one that is entirely automated is not even close to being a possibility yet. However, a large number of automotive manufacturers are already providing some degree of automation in the form of emergency braking, parking assistance, and other similar features. The deployment of these cars on public roads has a number of significant obstacles relating to the mechanics, the equipment, the law, and, most crucially, the moral questions. This is despite the fact that they have a number of advantages and that their popularity is growing.

4.3.11. *Outdoor Tracking Technologies*

One of the services that may be provided to residents as part of smart mobility is one that involves outdoor navigation, which is an essential component of that service. In this part, we go through some of the approaches being taken to reach this objective and lay the groundwork for delivering apps and services to citizens in an effort to streamline one of the most routine processes people go through on a daily basis. This entails the situation in which we are in an unfamiliar location, as well as the situation in which we are discussing individuals with sensory processing disorders, vision impairments, mobility impairments, or autism who need special accommodations in order to navigate their environment. The global positioning system (GPS) was one of the primary and pioneering technologies that were developed for outdoor navigation. It is the foundation of the guiding systems that arose in automobiles, and its primary shortcoming is that it only provides approximate results. The GPS technology was initially made available with a predetermined error. This was referred to as selected availability (SA), which means that the technology could result in an error of up to 100 m for civilian use. This severely restricted the number of applications that required an accurate positioning. The GPS relies on signals that are received from a number of satellites that are not geostationary. The accuracy is directly impacted by the fact that the number of satellites, in addition to their locations, might change at any time. Even if the SA constraint was lifted in the year 2000, the level of accuracy that can be obtained by relying only on GPS to solve problems such as limitations on high-precision farming, pedestrian navigation in densely populated places, etc. To overcome these limitations, many alternatives have been developed, such as those based on differential corrections and allowing for technology such as RTK. Nonetheless, depending on the distance and quality of transmission, the signal may still be degraded or experience disruptions. Dual-gateway positioning system (DGPS) is a real-time positioning technology that uses at least two GPS receivers, one of which is fixed at a location with known coordinates (called a base or reference station), and the other is in motion along a specific route whose coordinates must be determined. To put it simply, differential global positioning in real-time is what DGPS is all about. To improve the accuracy of the mobile device's location, the base station computes the differential corrections and sends them to the mobile receiver. A second option is to utilize active beacons, which use radio, sonar, or laser to triangulate a signal and determine a target's or a person's location based on that triangulation. It is inaccurate because of the time lag in receiving the signal position that is acquired, and there are also substantial expenses connected with the related infrastructure since the presence of a large number of beacons is required.

4.4. Innovative Techniques

4.4.1. *Description*

The development of a number of technologies and concepts that, when combined, allow us to conceive of solutions that will effectively make a difference in the future. These differences will be perceptible in terms of new goods and new data processing algorithms and methods that will deliver valuable information to people in order to provide more autonomy and greater well-being in their everyday lives. This has contributed to the rise of a trend known as "smart mobility." This trend is anticipated to keep gaining followers in the years to come. There are several different enabling technologies that have the potential to make the implementation of smart mobility more straightforward. The enabling technologies are depicted in Figure 4.6 and discussed in the following:

Blockchain: The technology behind blockchain has the potential to provide an architecture for mobility services that is secure, transparent, and trustworthy for residents. It is possible to develop an architecture for the Internet of Vehicles (IoV) that is based on blockchain technology. This would allow for enhanced V2V contact and communication, as well as improved traffic monitoring and management in smart cities. The digital revolution and intelligent sensors can effectively sense and gather real-time traffic and mobility data of cars and people, allowing for efficient mobility management. These data can then be used to provide effective

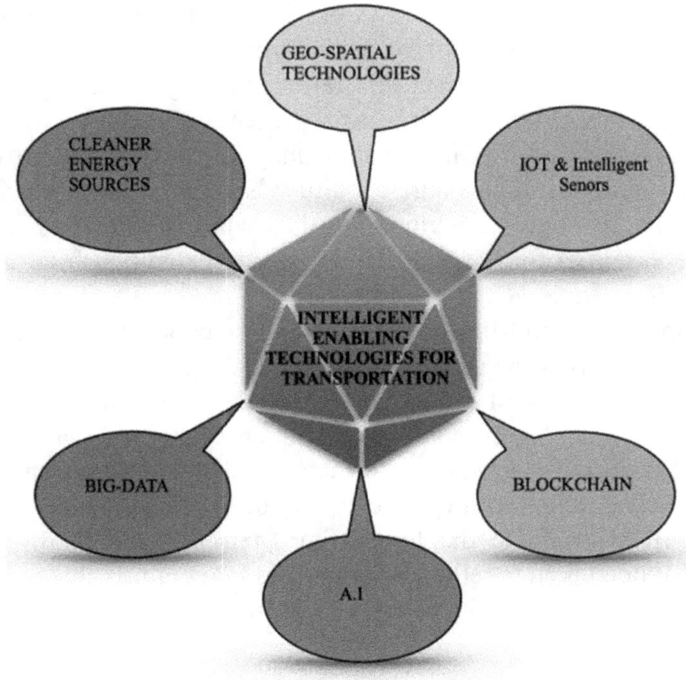

Figure 4.6. Technologies that Open Doors for Intelligent Connectivity

mobility solutions. In order to assist the running of traffic more smoothly, light poles and traffic signals are equipped with intelligent sensors able to independently make intelligent judgments depending on the circumstances they are confronted with in real-time [104–108].

Intelligent Systems: Algorithms based on AI may be built to make autonomous judgements and make predictions about a wide range of mobility-related services and entities based on data acquired from a variety of sources. Factors such as road maintenance, foot traffic, and lighting are some examples of these types of predictions. A further use for data-centric AI algorithms is customer-centric target marketing, which companies may utilize to better serve their clientele. Intelligent geolocation techniques provide trustworthy data to track the paths of people and vehicles. However, geospatial technology may be used to locate people. In the event of an emergency or natural catastrophe, the administration of alternate routes may also be facilitated by these technologies.

Big Data: The smart city ecosystem is held together by Big Data, which is critical to the health of the environment. The vast amounts of information generated by sensors and IoT gadgets may give useful insights on the object of study and the environment in which it is located. The use of this information may result in the discovery of important information. Smart mobility has to be accompanied by cleaner energy sources, including solar, wind, hydro, and biomass, which is essential for keeping the smart city ecosystem running with zero-emission fuels.

Figure 4.7 exemplifies how some of these concepts may contribute to smart mobility services and infrastructures, starting with the hardware-level gathering of data based on an IoT layer made of several devices at various levels, continuing with an aggregation layer, including Big Data and the production of datasets with enormous amounts of data, and concluding with the processing of that information using AI, which will also enable the forecasting of trends and help decision-making. When taken together, these factors make it possible to provide end-to-end services to city residents and hence boosting mobility in smart cities [109–112].

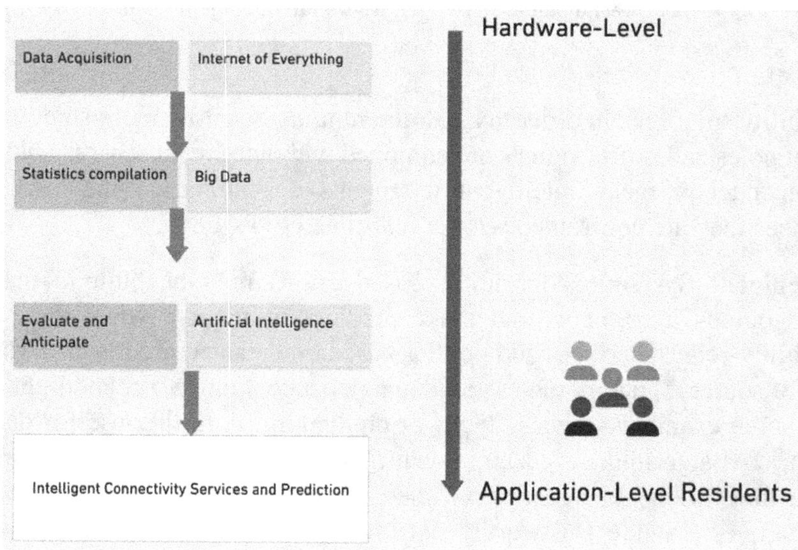

Figure 4.7. Stages of Implications Made by Intelligent Enabling Technologies

4.4.2. *Technologies That Facilitate Intelligent Transportation*

In this section, we examine how these enabling technologies are now being implemented in some of the most vital services and applications for transportation models that came before this one:

Function Within the Context of Mobility-as-a-Service (MaaS): When it comes to CaaS, having a central operator that is accessible via an intermediary layer of the system is absolutely necessary in order to handle all of the communication that takes place between the various transport providers, passengers, and other stakeholders of the system. Solutions based on blockchain technology do away with this intermediary layer, thus increasing confidence and transparency among all parties involved and getting rid of Business contracts are necessary between Cloud-based deployment providers. The use of AI and smart contracts enabled by blockchain technology makes transactions simple, speedy, and trustworthy. The use of geospatial technology in MaaS systems has a wide range of potential applications and brings additional value to the overall system as well as the services that are linked with it. One example is the surveillance of people who use public transportation.

Utilize Multiple-Access Edge Computing: We may devise a system for CaaS that collects and analyzes data pertaining to passenger mobility in order to prescribe optimal routes that account for the needs and preferences of individual commuters. This framework makes use of MEC, messaging protocols for the IoT, and virtualization (also known as digital twins). All of the above-mentioned tools collect and analyze information from commuters and are examples of CaaS systems that make use of Big Data. Commuters' data are collected and processed by these solutions.

In this chapter, a particular Big Data architecture is described, with the special needs of mobility analytics serving as the primary motivation. In relation to the generation of renewable energy, the CaaS approach as a whole advocates for a gradual shift away from private modes of transportation and toward public and shared modes of transportation, such as carpooling and carsharing. This change in perspective will bring with it a corresponding drop in the number of cars as well as the amount of carbon dioxide released into the environment. At this time, research is being done to determine when the transformation from sensors to MaaS may occur on a larger scale and what effect it will have on the options available for types of mobility [113–118].

4.4.3. *The Role in Warehousing and Transportation Enhancement*

When it comes to mobility, one of the most pressing issues is the optimization of traffic. A method based on blockchain technology is centered on the optimization of air traffic, in which blockchain technology is utilized to guarantee a platform that is safe, transparent, and decentralized. The contribution of the transportation sector to CO_2 emissions is 24%, with gridlock responsible for 35% of this statistic. As a result, it is vital to optimize the flow of traffic (both privately and also from the perspective of logistics) in order to minimize CO_2 emissions. It is very necessary to do traffic analysis in real-time in order to effectively manage traffic in metropolitan areas. Monitoring and analysis of traffic via the use of remotely piloted vehicles uses AI technology to analyze live footage in order to recognize and track motion information. A system for collecting and analyzing patterns of traffic using cameras placed at junctions aims to reduce the amount of time spent waiting in traffic. Processing at the network's edge, rather than in the cloud, is what the authors suggest should be done to alleviate the strain that congestion puts on networks. The processing is carried out with the assistance of deep learning strategies and algorithms. It is essential to emphasize the fact that these methods may also contribute to the solution of issues associated with transportation and supply chain. Utilizing sensing equipment and an IoT group for the purpose of optimizing and monitoring traffic is the product of many different contributions found in the relevant literature. It has been suggested that ultrasonic sensors that are connected to the road's power cables and running on a Raspberry Pi might be used as a solution to improve the flow of traffic at junctions. This would take into account the amount of traffic that is present. The significance of utilizing the IoT to drive lies in ensuring more intelligent and secure gadgets on the road. Transparency is required for the performance of the many players in supply chains, including transport logistics, in order for the system to be linked, intelligent, and effective. Blockchain technology can provide verifiable performance monitoring in the logistics sector. However, when you consider the fact that there is no way to track the logistical transaction and that there are doubts about the veracity and privacy of the information, an adaptive blockchain-based tracking protocol applicable to a wide range of supply-chain operations becomes necessary [119–122].

4.4.4. *Function Within Self-Driving Vehicles*

There are now a number of ideas in the published research that look at using blockchain technology in autonomous cars. It can help a system keep track of things that have happened inside the autonomous system car. This information may be helpful in the case of an accident involving many vehicles or a person. In these kinds of circumstances, determining who is responsible requires the use of a system that is designed to make it feasible to comprehend the many different events that have transpired within the autonomous car. A further use of blockchain technology in autonomous vehicles involves ride-hailing services and the mutual trust that must exist between participants. The concept of CaaS is also intrinsically linked to this particular use case. Designers and builders of the infrastructure of driver assistance systems that are more vulnerable to assaults are turning to blockchain technology and smart contracts in order to authenticate and protect the integrity of the process by which they update the vehicles' software. In order to provide a wide variety of mobility services, geospatial technologies are designed to monitor and record the locations of moving vehicles and humans. Using a concept called frequency modulated continuous wave (FMCW) radar, a system is described here that can detect and keep track of an autonomous car in real-time. It is model that addresses both the precision of tracking and the stability of the vehicle. The purpose of the model is to carry out the ability to foretell what will happen, which plans to keep an eye out for any shifts in the trajectory to ensure the vehicle stays upright, even at incredible velocities. The selection of the most effective path is one of the important considerations involved with respect to autonomous vehicles. This specific option uses techniques such as information extraction and evaluation of authentic crash data from everyday life, in addition to data collected in real-time from linked automobiles and an automated computing-based, real-time accident detection system. A research acquired information about Istanbul's traffic flow in 2015 using a wide array of sensors through Big Data processing techniques. These approaches are also integrally tied to the IoT and how it affects self-driving automobiles. In the long-term, it is anticipated that all cars will be linked to the internet and be electric. The future will be represented by autonomous automobiles. This development is inextricably tied to climate and environmental sustainability, especially when one considers a European Union mandate that calls for a 40% decrease in CO_2 emissions by the year 2030 in comparison to 1990 levels.

The shift toward electric cars will be substantially responsible for the decrease in carbon dioxide emissions caused by automobiles.

In autonomous cars, vision-based AI enables a collection of characteristics that may include the identification of impediments as well as methods to avoid them. The computation integrated device architecture, in tandem with the believe propagation algorithm, was used to accomplish this. The employment of artificial intelligence to in-car the development of vision systems to power improved driving assistance systems is presently in progress as a strategy for boosting the amount of computer power available in order to figure out how far away the obstacles are from the desired location [123–125].

4.4.5. *Future Innovation in Urban Smart Connectivity*

A paradigm change is being sparked within the field of transportation planning of intelligent communities as a direct result of the implementation of smart mobility technologies. To advance the current technologically advanced in intelligent connectivity, a number of novel strategies are currently being developed. There are a few essential aspects, such as the ones in the following, which must be taken into account: accessibility for those with vision impairments thanks to electric automobiles; accessibility enhancements in the event of catastrophes of a geological nature; residents' freedom to move with varying degrees of physical ability; and inclusive, environmentally friendly transportation that is both ecofriendly and functional. With the help of the IoT, traffic may be managed dynamically, with more transparency and decentralization and with greater protection for people, gadgets, and automobiles.

When it comes to cutting-edge innovations, AI technology is right up there and has been used in recent years to provide solutions for mobility that involve state-of-the-art technology. It is possible to increase visibility in settings where eyesight is impaired by using object recognition and identification methods that are based on deep learning. Some examples of these conditions are locations with low light, surroundings that are foggy, windy, or wet. The Internet of Everything keeps tabs on the state of the roads, the surroundings, and the neighborhood around them, which might provide useful insights that have never been seen before. This information can be utilized to anticipate future circumstances, which can help in determining the appropriate steps to take. Blockchain, a distributed ledger technology, may offer a transparent and immutable ledger.

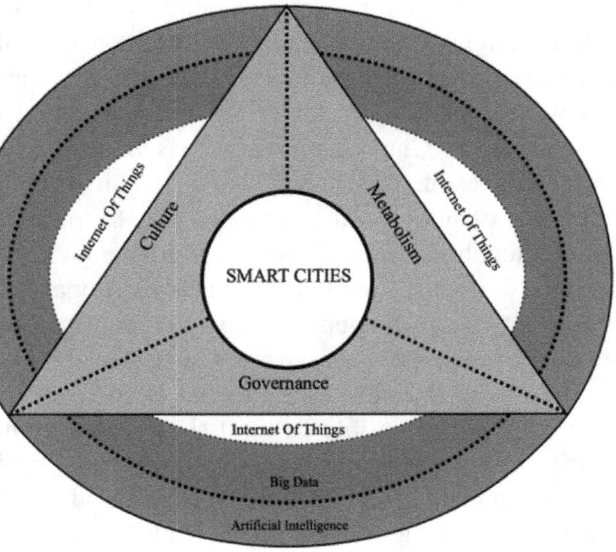

Figure 4.8. AI and Big Data Integration in Smart Cities for Livability

Figure 4.8 depicts AI and Big Data integration in smart cities for Vehicle-to-Everything networks that enable efficient, secure, and confidential administration of the smart mobility ecosystem [126–131]. The development of technology that produces decreased carbon dioxide emissions, more availability of sustainable energy, greater efficiency in electric cars, and more powerful batteries all contribute to the conservation of natural resources and progress toward environmentally beneficial sustainable development objectives, which in turn contributes to an enhanced standard of living . In order to maintain connection in the face of natural catastrophes and calamities of any kind, the area of communications has been the subject of much study, and research is being done on comparatively inexpensive communication solutions based on satellites.

4.5. Conclusion

With regard to transportation in particular, the past few years have brought about a number of advancements and established the groundwork for the building of smart cities of the future. The increasing population of the world will have far-reaching consequences for urban transportation planning. This is due in no small part to the fact that urban areas were designed tens, if not hundreds, of years ago, when populations were considerably lower and infrastructural needs were substantially different. Changing urban areas to fit modern needs is challenging. As a result, people will have to quit driving alone and instead make use of public transportation, which might cut down on pollution and greenhouse gas emissions. On the other hand, smart transportation and urbanization will not need additional road construction or parking lots. People's dependence on personal automobiles will instead have to be phased out as part of a larger change in cultural norms. In addition to the obvious technical challenge brought on by this shift, an equally perceptible cultural shift is at play here as well. As an example, consider the rationale that there exists a distinct level of ease and convenience associated with the use of our own transportation, and it won't be straightforward to switch to a different mode of transportation overnight. Citizens need to be made aware of the negative effects that maintaining the current state will have on the environment. It is difficult to regulate the proliferation of automobiles in the world's most densely populated cities. This cultural and cognitive change will be beneficial for all of us, but it will take work to get the word out about it. Sharing of vehicles, carpooling, and carsharing will usher in a new era as a result of the exponential development of technology, which is forecast to incline sharply in the future. This facilitates a shift from the intelligent transportation paradigm to the CaaS framework, which will allow for the establishment of a comprehensive smart transportation tier. The IoT and the interconnection that is proven respectively in vehicles, traffic lights, and commuters, among other things, will authorize the progression to sophisticated transportation infrastructure with significant cuts and even the number of crashes, which is the predominant line of argument that Stage 5 driverless vehicles are aiming to achieve. Big Data and AI will be able to make a significant contribution toward this goal. Together, they will deal with the humongous amounts of information gathered by the IoT. As a result, residents will have access to services that tremendously enhance their overall quality of life. As a consequence, citizens will be able to make more informed decisions about the flow of automobile, fleet, and foot traffic.

References

1. Zhang, C., Sun, Z., Xing, Q., Sun, J., Xia, T., and Yu, H. (2021). Localizing Indicators of SDG11 for an integrated assessment of urban sustainability — A case study of Hainan province. *Sustainability*, *13*(19), 11092.
2. Leridon, H. (2020). World population outlook: Explosion or implosion? *Population Societies 573*(1), 1–4.
3. Gouveia, J. P., Seixas, J., and Giannakidis, G. (2016). Smart city energy planning. In *Proceedings of the 25th International Conference Companion on World Wide Web — WWW '16 Companion*, Montréal, QC, Canada, 11–16 April 2016, pp. 345–350.
4. Ruhlandt, R. W. S. (2018). The governance of smart cities: A systematic literature review. *Cities*, *81*, 1–23.
5. Amoretti, M., Belli, L., and Zanichelli, F. (2017). UTravel: Smart mobility with a novel user profiling and recommendation approach. *Pervasive and Mobile Computing*, *38*, 474–489.
6. Belbachir, A., El Fallah-Seghrouchni, A., Casals, A., and Pasin, M. (2019). Smart mobility using multi-agent system. *Procedia Computer Science*, *151*, 447–454.
7. Backhouse, J. (2020). A taxonomy of measures for smart cities. In *Proceedings of the 13th International Conference on Theory and Practice of Electronic Governance*, Athens, Greece, 23–25 September 2020.
8. Mirri, S., Prandi, C., Salomoni, P., Callegati, F., Melis, A., and Prandini, M. (2016). A service-oriented approach to crowdsensing for accessiblesmart mobility scenarios. *Mobile Information Systems*, *2016*, 2821680.
9. Lopez, D. and Farooq, B. (2020). A multi-layered blockchain framework for smart mobility data-markets. *Transportation Research Part C: Emerging Technologies*, *111*, 588–615.
10. Bravo, Y., Ferrer, J., Luque, G., and Alba, E. (2016). Smart mobility by optimizing the traffic lights: A new tool for traffic control centers. In *International Conference on Smart Cities*, Springer: Cham, Switzerland, pp. 147–156.
11. Aletà, N. B., Alonso, C. M., and Ruiz, R. M. A. (2017). Smart mobility and smart environment in the Spanish cities. *Transportation Research Procedia*, *24*, 163–170.
12. Camero, A., Toutouh, J., Stolfi, D. H., and Alba, E. (2018). Evolutionary deep learning for car park occupancy prediction in smart cities. In *International Conference on Learning and Intelligent Optimization*, Springer: Cham, Switzerland, pp. 386–401.
13. Alonso, J. M. and Mencar, C. (2017). Building cognitive cities with explainable artificial intelligent systems. In *Proceedings of the CEX@ AI* IA Conference*, Bari, Italy, 14–17 November 2017.

14. Kaplan, C. J. and Hegarty, E. D. (2006). *Understanding GPS: Principles and Applications*. ArtechHouse: Worwood, MA, USA.
15. Panzieri, S., Pascucci, F., and Ulivi, G. (2002). An outdoor navigation system using GPS and inertial platform. *IEEE/ASME Transactions on Mechatronics, 7*, 134–142.
16. Parkinson, B. W. and Spilker, J. J. (1996). *Global Positioning System: Theory and Applications*. American Institute of Aeronautics and Astronautics: Reston, VA, USA.
17. Wellenhof, B. H., Lichtenegger, H., and Collins, J. (1992). *Global Positioning System* (5th edn.). Springer: Berlin/Heidelberg, Germany.
18. Garau, C., Masala, F., and Pinna, F. (2015). Benchmarking smart urban mobility: A study on Italian cities. In *International Conference on Computational Science and Its Applications*, Springer: Cham, Switzerland, pp. 612–623.
19. Jeekel, H. (2017). Social sustainability and smart mobility: Exploring the relationship. *Transportation Research Procedia, 25*, 4296–4310.
20. Kronsell, A. and Mukhtar-Landgren, D. (2020). Experimental governance of smart mobility: Some normative implications. In *Shaping Smart Mobility Futures: Governance and Policy Instruments in times of Sustainability Transitions*. Emerald Publishing Ltd.: Bingley, UK.
21. Pereira, J., Ricardo, L., Luís, M., Senna, C., and Sargento, S. (2019). Assessing the reliability of fog computing for smart mobility applications in VANETs. *Future Generation Computer Systems, 94*, 317–332.
22. Faria, R., Brito, L., Baras, K., and Silva, J. (2017). Smart mobility: A survey. In *Proceedings of the 2017 International Conference on Internet of Things for the Global Community (IoTGC)*, Funchal, Portugal, 10–13 July 2017, pp. 1–8.
23. Dowling, R. (2018). Smart mobility: Disrupting transport governance? In *Governance of the Smart Mobility Transition*. Emerald Publishing Ltd.: Bingley, UK.
24. El-Din, D. M., Hassanien, A. E., and Hassanien, E. E. (2020). Information integrity for multi-sensors data fusion in smart mobility. In *Toward Social Internet of Things (SIoT): Enabling Technologies, Architectures and Applications*. Springer: Cham, Switzerland, pp. 99–121.
25. Jin, X., Feng, C., Ponnamma, D., Yi, Z., Parameswaranpillai, J., Thomas, S., and Salim, N. (2020). Review on exploration of graphene in the design and engineering of smart sensors, actuators and soft robotics. *Chemical Engineering Journal Advances, 4*, 100034.
26. De Souza, A. M., Brennand, C. A., Yokoyama, R. S., Donato, E. A., Madeira, E. R., and Villas, L. A. (2017). Traffic management systems: A classification, review, challenges, and future perspectives. *International Journal of Distributed Sensor Networks, 13*(4), 1550147716683612.

27. Hank, P., Müller, S., Vermesan, O., and Van Den Keybus, J. (2013). Automotive ethernet: In-vehicle networking and smart mobility. In *Proceedings of the 2013 Design, Automation & Test in Europe Conference & Exhibition (DATE)*, Grenoble, France, 18–22 March 2013, pp. 1735–1739.
28. Bittencourt, L. F., Diaz-Montes, J., Buyya, R., Rana, O. F., and Parashar, M. (2017). Mobility-aware application scheduling in Fog computing. *IEEE Cloud Computing*, 4, 26–35.
29. Davies, E. R. (2017). *Computer Vision: Principles, Algorithms, Applications, Learning* (5th edn.). Academic Press: Cambridge, MA, USA.
30. Mauri, A., Khemmar, R., Decoux, B., Ragot, N., Rossi, R., Trabelsi, R., Boutteau, R., Ertaud, J. Y., and Savatier, X. (2020). Deep learning for real-time 3D multi-object detection, localisation, and tracking: Application to smart mobility. *Sensors*, *20*, 532.
31. Allam, Z. and Dhunny, Z. A. (2019). On big data, artificial intelligence and smart cities. *Cities*, *89*, 80–91.
32. Ullah, Z., Al-Turjman, F., Mostarda, L., and Gagliardi, R. (2020). Applications of artificial intelligence and machine learning in smart cities. *Computer Communications*, *154*, 313–323.
33. Batty, M. (2018). Artificial intelligence and smart cities. *Environment and Planning B: Urban Analytics and City Science*, *45*(1), 3–6.
34. Srivastava, S., Bisht, A., and Narayan, N. (2017). Safety and security in smart cities using artificial intelligence — A review. In *2017 7th International Conference on Cloud Computing, Data Science & Engineering-Confluence*, pp. 130–133. IEEE.
35. Batool, T., Abbas, S., Alhwaiti, Y., Saleem, M., Ahmad, M., Asif, M., and Elmitwal, N. S. (2021). Intelligent model of ecosystem for smart cities using artificial neural networks. *Intelligent Automation & Soft Computing*, *30*(2), 513–525.
36. Anuradha, J., Singh, R., B., and Khosla, M. (2018). Design and performance analysis of nano-scale memristor-based nonvolatile SRAM. *Journal of Sensor Letter*, *16*, 798–805.
37. Wadhwa, G. and Raj, B. (2018). Parametric variation analysis of charge-plasma-based dielectric modulated JLTFET for biosensor application. *IEEE Sensor Journal*, *18*(15), 6070–6077.
38. Singh, J. and Raj, B. (2018). Comparative analysis of memristor models for memories design. *JoS*, *39*(7), 074006-1-12.
39. Yadav, D., Chouhan, S. S., Vishvakarma, S. K., and Raj, B. (2018). Application specific microcontroller design for IoT based WSN. *Sensor Letter*, *16*, 374–385.
40. Singh, G., Sarin, R. K., and Raj, B. (2018). Fault-tolerant design and analysis of quantum-dot cellular automata based circuits. *IEEE/IET Circuits, Devices & Systems*, *12*, 638–664.

41. Singh, J. and Raj, B. (2018). Modeling of Mean barrier height levying various image forces of metal–insulator–metal structure to enhance the performance of conductive filament based memristor model. *IEEE Transactions on Nanotechnology*, *17*(2), 268–275.
42. Jain, A., Sharma, S., and Raj, B. (2018). Analysis of triple metal surrounding gate (TM-SG) III-V nanowire MOSFET for photosensing application. *Opto-Electronics Review*, *26*(2), 141–148.
43. Jain, A., Sharma, S., and Raj, B. (2016). Design and analysis of high sensitivity photosensor using cylindrical surrounding gate MOSFET for low power sensor applications. *Engineering Science and Technology, an International Journal*, *19*(4), 1864–1870.
44. Singh, A., Khosla, M., and Raj, B. (2017). Analysis of electrostatic doped Schottky barrier carbon nanotube FET for low power applications. *Journal of Materials Science: Materials in Electronics*, *28*, 1762–1768.
45. Kumar, G. Saiphani, Singh, A., and Raj, B. (2018). Design and analysis of gate all around CNTFET based SRAM cell design. *Journal of Computational Electronics*, *17*(1), 138–145.
46. Singh, G. P., Sohi, B. S., and Raj, B. (2017). Material properties analysis of graphene base transistor (GBT) for VLSI analog circuits. *Indian Journal of Pure & Applied Physics (IJPAP)*, *55*, 896–902.
47. Singh, A., Khosla, M., and Raj, B. (2016). Comparative analysis of carbon nanotube field effect transistor and nanowire transistor for low power circuit design. *Journal of Nanoelectronics and Optoelectronics*, *11*, 388–393.
48. Kumar, S. and Raj, B. (2017). Estimation of stability and performance metric for inward access transistor based 6T SRAM cell design using *n*-type/*p*-type DMDG-GDOV TFET IEEE VLSI. *Circuits and Systems Letter*, *3*(2), 25–39.
49. Sharma, S., Kumar, A., Pattanaik, M., and Raj, B. (2013) Forward body biased multimode multi-threshold CMOS technique for ground bounce noise reduction in static CMOS adders. *International Journal of Information and Electronics Engineering*, *3*(3), 567–572.
50. Singh, H., Kumar, P., and Raj, B. (2013). Performance analysis of majority gate SET based 1-bit full adder. *International Journal of Computer and Communication Engineering (IJCCE)*, *2*(4), 2933–2940.
51. Bhardwaj, A. K., Gupta, S., and Raj, B. Investigation of parameters for Schottky barrier (SB) height for Schottky barrier based carbon nanotube field effect transistor device. *Journal of Nanoelectronics and Optoelectronics*, *15*, 783–791.
52. Bansal, P. and Raj, B. (2017). Memristor: A versatile nonlinear model for dopant drift and boundary issues. *JCTN*, *14*(5), 2319–2325.
53. Jain, N. and Raj, B. (2016). An analog and digital design perspective comprehensive approach on fin-FET (fin-field effect transistor) technology — A review. *Reviews in Advanced Sciences and Engineering (RASE)*, *5*, 1–14.

54. Sharma, S., Raj, B., and Khosla, M. (2017). Subthreshold performance of $In_{1-x}Ga_xAs$ based dual metal with gate stack cylindrical/surrounding gate nanowire MOSFET for low power analog applications. *Journal of Nanoelectronics and Optoelectronics*, *12*, 171–176.
55. Raj, B., Saxena, A. K., and Dasgupta, S. (2009). Analytical modeling for the estimation of leakage current and subthreshold swing factor of nanoscale double gate FinFET device. *Microelectronics International*, *26*, 53–63.
56. Soniya, S. S., Wadhwa, G., and Raj, B. (2019). An analytical modeling for dual source vertical tunnel field effect transistor. *International Journal of Recent Technology and Engineering (IJRTE)* *8*(2), 603–608.
57. Singh, S. and Raj, B. (2019). Design and analysis of hetrojunction vertical T-shaped tunnel field effect transistor. *Journal of Electronics Material*, *48*(10), 6253–6260.
58. Goyal, C., Ubhi, J. S., and Raj, B. (2019). A low leakage CNTFET based inexact full adder for low power image processing applications. *International Journal of Circuit Theory and Applications*, *47*(9), 1446–1458.
59. Raj, B., Saxena, A. K., and Dasgupta, S. (2008). A compact drain current and threshold voltage quantum mechanical analytical modeling for FinFETs. *Journal of Nanoelectronics and Optoelectronics (JNO)*, *3*(2), 163–170.
60. Wadhwa, G. and Raj, B. (2020). An analytical modeling of charge plasma based tunnel field effect transistor with impacts of gate underlap region. *Superlattices and Microstructures*, *142*, 106512.
61. Singh, S. and Raj, B. (2020). Modeling and Simulation analysis of SiGe hetrojunction double gatevertical *T*-shaped tunnel FET. *Superlattices and Microstructures*, *142*, 106496.
62. Singh, A., Saini, D. K., Agarwal, D., Aggarwal, S., Khosla, M., and Raj, B. (2016). Modeling and simulation of carbon nanotube field effect transistor and its circuit application. *Journal of Semiconductors (JoS)*, *37*, 074001-6.
63. Jain, N. and Raj, B. (2016). Device and circuit co-design perspective comprehensive approach on FinFET technology — A review. *Journal of Electron Devices*, *23*(1), 1890–1901.
64. Kumar, S. and Raj, B. (2016). Analysis of I_{ON} and ambipolar current for dual-material gate-drain overlapped DG-TFET. *Journal of Nanoelectronics and Optoelectronics*, *11*, 323–333.
65. Anjum, N., Bali, T., and Raj, B. (2013). Design and simulation of handwritten multiscript character recognition. *International Journal of Advanced Research in Computer and Communication Engineering*, *2*(7), 2544–2549.
66. Sharma, S., Raj, B., and Khosla, M. (2016). A Gaussian approach for analytical subthreshold current model of cylindrical nanowire FET with quantum mechanical effects. *Microelectronics Journal*, *53*, 65–72.

67. Singh, K. and Raj, B. (2015). Performance and analysis of temperature dependent multi-walled carbon nanotubes as global interconnects at different technology nodes. *Journal of Computational Electronics*, *14*(2), 469–476.
68. Kumar, S. and Raj, B. (2015). Compact channel potential analytical modeling of DG-TFET based on evanescent–mode approach. *Journal of Computational Electronics*, *14*(2), 820–827.
69. Singh, K. and Raj, B. (2015). Temperature dependent modeling and performance evaluation of multi-walled CNT and single-walled CNT as global interconnects. *Journal of Electronic Materials*, *44*(12), 4825–4835.
70. Sharma, V. K., Pattanaik, M., and Raj, B. (2014). INDEP approach for leakage reduction in nanoscale CMOS circuits. *International Journal of Electronics*, *102*(2), 200–215.
71. Singh, K. and Raj, B. (2015). Influence of temperature on MWCNT bundle, SWCNT bundle and copper interconnects for nanoscaled technology nodes. *Journal of Materials Science: Materials in Electronics*, *26*(8), 6134–6142.
72. Anjum, N., Bali, T., and Raj, B. (2013). Design and simulation of handwritten Gurumukhi and Devanagri numerical recognition. *International Journal of Computer Applications*, *73*(12), 16–21.
73. Khandelwal, S., Gupta, V., Raj, B., and Gupta, R. D. (2015). Process variability aware low leakage reliable nano scale DG-FinFET SRAM cell design technique. *Journal of Nanoelectronics and Optoelectronics*, *10*(6), 810–817.
74. Sharma, V. K., Pattanaik, M., and Raj, B. (2014). ONOFIC approach: Low power high speed nanoscale VLSI circuits design. *International Journal of Electronics*, *101*(1), 61–73.
75. Khandelwal, S., Raj, B., and Gupta, R. D. (2015). FinFET based 6T SRAM cell design: Analysis of performance metric, process variation and temperature effect. *Journal of Computational and Theoretical Nanoscience*, *12*, 2500–2506.
76. Singh, S., Yadav, S., Rahul, J., Srivastava, A., and Raj, B. (2015). Impact of HfO_2 in graded channel dual insulator double gate MOSFET. *Journal of Computational and Theoretical Nanoscience*, *12*, 950–953.
77. Sharma, V. K, Pattanaik, M., and Raj, B. (2014). PVT variations aware low leakage INDEP approach for nanoscale CMOS circuits. *Microelectronics Reliability*, *54*, 90–99.
78. Raj, B., Saxena, A. K., and Dasgupta, S. (2013). Quantum mechanical analytical modeling of nanoscale DG FinFET: Evaluation of potential, threshold voltage and source/drain resistance. *Journal of Material Science in Semiconductor Processing*, *16*(4), 1131–1137.
79. Gopal, M., Prasad, S. S. D., and Raj, B. (2013). 8T SRAM cell design for dynamic and leakage power reduction. *International Journal of Computer Applications*, *71*(9), 43–48.

80. Pattanaik, M., Raj, B., Sharma, S., and Kumar, A. (2012). Diode based tri-mode multi-threshold CMOS technique for ground bounce noise reduction in static CMOS adders. *Advanced Materials Research, 548*, 885–889.
81. Raj, B., Saxena, A. K., and Dasgupta, S. (2011). Nanoscale FinFET based SRAM cell design: Analysis of performance metric, process variation, underlapped FinFET and temperature effect. *IEEE Circuits and System Magazine, 11*(2), 38–50.
82. Sharma, V. K., Pattanaik, M., and Raj, Balwinder (2013). Leakage current ONOFIC approach for deep submicron VLSI circuit design. *International Journal of Electrical, Computer, Electronics and Communication Engineering, 7*(4), 239–244.
83. Chawla, T., Khosla, M., and Raj, B. (2021). Design and simulation of triple metal double-gate germanium on insulator vertical tunnel field effect transistor. *Microelectronics Journal, 114*, 105125.
84. Kaur, P., Gill, S. S., and Raj, B. (2022). Comparative analysis of OFETs materials and devices for sensor applications. *Journal of Silicon, 14*, 4463–4471.
85. Sharma, S. K., Kumar, P., Raj, B., and Raj, B. (2022). $In_{1-x}Ga_xAs$ double metal gate-stacking cylindrical nanowire MOSFET for highly sensitive photo detector. *Journal of Silicon, 14*, 3535–3541.
86. Raj, B., Saxena, A. K., and Dasgupta, S. (2009). Analytical modeling of quasi planar nanoscale double gate FinFET with source/drain resistance and field dependent carrier mobility: A quantum mechanical study. *Journal of Computer (JCP), 4*(9), 1–8.
87. Bhushan, S., Khandelwal, S., and Raj, B. (2012). Analyzing different mode FinFET based memory cell at different power supply for leakage reduction, seventh international conference on bio-inspired computing: Theories and application. *Advances in Intelligent Systems and Computing, 202*, 89–100.
88. Singh, J. and Raj, B. (2018). Temperature dependent analytical modeling and simulations of nanoscale memristor. *Engineering Science and Technology, an International Journal, 21*, 862–868.
89. Singh, S., Bala, S., Raj, B., and Raj, B. (2020). Improved sensitivity of dielectric modulated junctionless transistor for nanoscale biosensor design. *Sensor Letter, 18*, 328–333.
90. Kumar, V., Vishvakarma, S. K., and Raj, B. (2020). Design and performance analysis of ASIC for IoT applications. *Sensor Letter, 18*, 31–38.
91. Jaiswal, A., Sarin, R. K., Raj, B., and Sukhija, S. (2018). A novel circular slotted microstrip-fed patch antenna with three triangle shape defected ground structure for multiband applications. *Advanced Electromagnetic (AEM), 7*(3), 56–63.
92. Wadhwa, G. and Raj, B. (2018). Label free detection of biomolecules using charge-plasma-based gate underlap dielectric modulated junctionless TFET. *Journal of Electronic Materials (JEMS), 47*(8), 4683–4693.

93. Singh, G., Sarin, R. K., and Raj, B. (2017). Design and performance analysis of a new efficient coplanar quantum-dot cellular automata adder. *Indian Journal of Pure & Applied Physics (IJPAP)*, *55*, 97–103.
94. Singh, A., Khosla, M., Raj, B. (2017). Design and analysis of electrostatic doped Schottky barrier CNTFET based low power SRAM. *International Journal of Electronics and Communications*, *80*, 67–72.
95. Kaur, P., Pandey, V., and Raj, B. (2020). Comparative study of efficient design, control and monitoring of solar power using IoT. *Sensor Letter*, *18*, 419–426.
96. Bhardwaj, A. K., Gupta, S., and Raj, B. (2020). Development & analysis of compact model for double gate Schottky barrier CNTFET. *Journal of Nanoelectronics and Optoelectronics*, *15*, 1199–1208.
97. Wadhwa, G., Kamboj, P., Singh, J., and Raj, B. (2021). Design and investigation of junctionless DGTFET for biological molecule recognition. *Transactions on Electrical and Electronic Materials, 22,* 282–289.
98. Chawla, T., Khosla, M., and Raj, B. Optimization of double-gate dual material GeOI-vertical TFET for VLSI circuit design. *IEEE VLSI Circuits and Systems Letter*, *6*(2), 13–25.
99. Verma, K., Singh, S., Wadhwa, G., and Raj, B. (2020). Detection of biomolecules using charge-plasma based gate underlap dielectric modulated dopingless TFET. *Transactions on Electrical and Electronic Materials (TEEM)*, *21*, 528–535.
100. Jain, N. and Raj, B. (2017) Impact of underlap spacer region variation on Electrostatic and analog/RF performance of symmetrical high-k SOI FinFET at 20 nm channel length. *Journal of Semiconductors (JoS)*, *38*(12), 122002.
101. Singh, S. and Raj, B. (2020). Analytical modeling and Simulation analysis of T-shaped III-V heterojunction vertical T-FET. *Superlattices and Microstructures*, *147*, 106717.
102. Singh, G., Sarin, R. K., and Raj, B. (2017). Design and analysis of area efficient QCA based reversible logic gates. *Journal of Microprocessors and Microsystems*, *52*, 59–68.
103. Singh, A., Khosla, M., and Raj, B. (2016). Compact model for ballistic single wall CNTFET under quantum capacitance limit. *Journal of Semiconductors (JoS)*, *37*, 104001-8.
104. Singh, S., Khosla, M., Wadhwa, G., and Raj, B. (2021). Design and analysis of double-gate junctionless vertical TFET for gas sensing applications. *Applied Physics A*, *127*(16).
105. Singh, I., Raj, B., Khosla, M., and Kaushik, B. K. (2020). Potential MRAM Technologies for low power SoCs. *SPIN*, *10*(4), 2050027.
106. Singh, S. and Raj, B. (2020). Parametric variation analysis on hetero-junction vertical *T*-shape TFET for Supressing ambipolar conduction. *Indian Journal of Pure and Applied Physics*, *58*, 478–485.

107. Soniya, S. S., Wadhwa, G., and Raj, B. (2019). Design and analysis of dual source vertical tunnel field effect transistor for high performance. *Transactions on Electrical and Electronics Materials*, *21*, 74–82.
108. Kaur, M., Gupta, N., Kumar, S., Raj, B., and Singh, A. K. (2020). RF performance analysis of intercalated graphene nanoribbon based global level interconnects. *Journal of Computational Electronics*, *19*, 1002–1013.
109. Wadhwa, G. and Raj, B. (2019). Design and performance analysis of junctionless TFET biosensor for high sensitivity. *IEEE Nanotechnology*, *18*, 567–574.
110. Singh, J. and Raj, B. (2019). Enhanced nonlinear memristor model encapsulating stochastic dopant drift, *JNO*, *14*, 958–963.
111. Kaur, P., Buttar, A. S., and Raj, B. (2021). A comprehensive analysis of nanoscale transistor based biosensor: A review. *Indian Journal of Pure and Applied Physics*, *59*, 304–318.
112. Yadav, D., Raj, B., and Raj, B. (2020). Design and simulation of low power microcontroller for IoT applications. *Journal of Sensor Letters*, *18*, 401–409.
113. Singh, S. and Raj, B. (2020). A 2-D analytical surface potential and drain current modeling of double-gate vertical t-shaped tunnel FET. *Journal of Computational Electronics*, *19*, 1154–1163.
114. Singh, J. and Raj, B. (2019). An accurate and generic window function for non-linear memristor model. *Journal of Computational Electronics*, *18*(2), 640–647.
115. Kaur, M., Gupta, N., Kumar, S., Raj, B., and Singh, A. K. (2021). Comparative RF and crosstalk analysis of carbon based nano interconnects. *IET Circuits, Devices & Systems*, *15*(6), 493–503.
116. Kandasamy, N., Ahmad, F., Ajitha, D., Raj, B., and Telagam, Nagarjuna (2020). Quantum dot cellular automata based scan flip flop and boundary scan register. *IETE Journal of Research*, *66*, 535–548.
117. Sharma, S. K., Raj, B., and Khosla, M. (2019). Enhanced photosensivity of highly spectrum selective cylindrical gate in 1-xGaxAs nanowire MOSFET photodetector. *Modern Physics Letter B*, *33*(12), 1950144.
118. Singh, J. and Raj, B. (2019). Design and investigation of 7T2M NVSARM with enhanced stability and temperature impact on store/restore energy. *IEEE Transactions on Very Large Scale Integration Systems*, *27*(6), 1322–1328.
119. Bhardwaj, A. K., Gupta, S., Raj, B., and Singh, A. (2019). Impact of double gate geometry on the performance of carbon nanotube field effect transistor structures for low power digital design. *Computational and Theoretical Nanoscience*, *16*, 1813–1820.
120. Jain, N. and Raj, B. (2019). Thermal stability analysis and performance exploration of asymmetrical dual-k underlap spacer (ADKUS) SOI FinFET for security and privacy applications. *Indian Journal of Pure & Applied Physics (IJPAP)*, *57*, 352–360.

121. Singh, A., Khosla, M., and Raj, B. (2019). Design and analysis of dynamically configurable electrostatic doped carbon nanotube tunnel FET. *Microelectronics Journal*, *85*, 17–24.
122. Jain, N. and Raj, B. (2019). Dual-k spacer region variation at the drain side of asymmetric SOI FinFET structure: Performance analysis towards the analog/RF design applications. *Journal of Nanoelectronics and Optoelectronics*, *14*, 349–359.
123. Singh, J., Sharma, S., Raj, B., and Khosla, M. (2018). Analysis of barrier layer thickness on performance of $In_{1-x}Ga_xAs$ based gate stack cylindrical gate nanowire MOSFET. *JNO*, *13*, 1473–1477.
124. Jain, N. and Raj, B. (2018). Analysis and performance exploration of high-k SOI FinFETs over the conventional low-k SOI FinFET toward analog/RF design. *Journal of Semiconductors (JoS)*, *39*(12), 124002-1-7.
125. Goyal, C., Ubhi, J. S., and Raj, B. (2018). A reliable leakage reduction technique for approximate full adder with reduced ground bounce noise. *Journal of Mathematical Problems in Engineering*, *2018*, 16. Article ID 3501041.
126. Plageras, A. P., Psannis, K. E., Stergiou, C., Wang, H., and Gupta, B. B. (2018). Efficient IoT-based sensor BIG Data collection–processing and analysis in smart buildings. *Future Generation Computer Systems*, *82*, 349–357.
127. Memos, V. A., Psannis, K. E., Ishibashi, Y., Kim, B. G., and Gupta, B. B. (2018). An efficient algorithm for media-based surveillance system (EAMSuS) in IoT smart city framework. *Future Generation Computer Systems*, *83*, 619–628.
128. Esposito, C., Ficco, M., and Gupta, B. B. (2021). Blockchain-based authentication and authorization for smart city applications. *Information Processing & Management*, *58*(2), 102468.
129. Li, D., Deng, L., Gupta, B. B., Wang, H., and Choi, C. (2019). A novel CNN based security guaranteed image watermarking generation scenario for smart city applications. *Information Sciences*, *479*, 432–447.
130. Mishra, A., Hsu, C. H., Arya, V., Chaurasia, P., and Li, P. (2023). A hybrid approach for protection against rumours in a IoT enabled smart city environment. In *International Conference on Cyber Security, Privacy and Networking (ICSPN 2022)* (pp. 101–109). Springer International Publishing: Cham.
131. Hammad, M., Alkinani, M. H., Gupta, B. B., and Abd El-Latif, A. A. (2021). Myocardial infarction detection based on deep neural network on imbalanced data. *Multimedia Systems*, *28*, 1–13.

https://doi.org/10.1142/9789811279539_0005

Chapter 5

Artificial Neural Network for RF MEMS Circuits

S. Kanthamani[*,†], J. Joslin Percy[‡], and S. Mohamed Mansoor Roomi[§]

MEMS Design Centre, Thiagarajar College of Engineering, Madurai, India

[†]*skmece@tce.edu*

[‡]*joslin@student.tce.edu*

[§]*smmroomi@tce.edu*

Abstract

The evolution of communication technologies with high-frequency radio-frequency (RF) devices increased the demand for compact and efficient designs. Micro-electromechanical systems (MEMS) technology revolutionized microwave and RF applications because of its ability to be engineered into miniaturized devices that are highly linear and power efficient. It is more challenging to perform numerical analysis and optimization of such complex MEMS devices. Electromagnetic (EM) simulation-based optimization software using different methodologies employing coupled domains and time domain analysis of MEMS devices requires repeated simulation, which makes it computationally expensive. The artificial neural network (ANN) model is an alternative

[*]Corresponding author.

to these conventional simulation-based design methodologies to expedite the design process. ANN models for RF and microwave modeling are known to be effective, precise, and flexible. ANN is capable of producing accurate results with less computational time than sophisticated EM models. An overview of various RF MEMS components and an introduction to ANN are provided in this chapter. In addition, this chapter presents the concept of modeling an RF MEMS shunt switch using ANN as a case study.

Keywords: MEMS, Radio frequency, TCAD simulations, ANN, HFSS

5.1. Introduction

Communication devices are rapidly increasing and adapting to technological evolutionary changes, leading to various generations up to 8G [1]. Advanced consumer devices, such as smart phones, tablets, and other portable communication devices, are established at different base stations to provide uninterrupted access to data, with better quality, higher speeds, increased bandwidths, better compatibility, and reduced costs [2]. As the upcoming communication technologies are to be developed in higher frequency bands due to their benefits in terms of compactness and bandwidth, such networks must be developed for usage from microwave up to THz frequencies [3]. Hence, the challenges of spectrum efficiency, reconfigurability, and miniaturization can be met by low-cost, low-power, high-performance, and adaptive RF devices [4].

The radio-frequency (RF) transceiver components are designed and fabricated on a single planar substrate using semiconductors based on transistors and diodes. Though semiconductor technology continues to dominate the development of digital and analog circuits, it has become evident that this technology has certain disadvantages. Recently, research on micro-electromechanical systems (MEMS) and nano-electromechanical systems (NEMS) has proved the ability of these technologies in the development of devices and circuits to be used in RF and microwave applications [5]. RF MEMS is an evolving MEMS technology that can revolutionize microwave and RF applications. RF MEMS devices find applications in millimeter wave (MMW) sensors and commercial and military wireless communication. In addition, sensor and navigation systems can employ RF MEMS components, such as transmitters, switches, antennas, phase shifters, exciters, filters, and IF/RF receivers [6].

MEMS device structures are electromechanically coupled, geometrically complicated, and inherently 3D. It is more challenging to develop reliable, rapid, and efficient CAD tools for the analysis of MEMS devices than for conventional electrical or mechanical systems. A 3D electromechanical systems analysis employs two coupled domains, namely electrostatics and elastomechanics, which have been studied in the literature [7–9], and coupled domain analyses of MEMS circuits have been done using relaxation methods [10, 11], the surface Newton method [12, 13], the coupled Newton method [14, 15], and the multilevel Newton method [15].

To design the movable structure of MEMS devices with reference to time, a finite difference time domain (FDTD) formulation [16], a multi-resolution time domain technique, an adaptive extension of the FDTD technique which utilizes wavelets to decrease the computational complexity of the FDTD analysis [17], and a method of moments (MOMs) for beam structure analysis have been proposed for designing RF-MEMS switches [18].

The limitations of the above methods include the generation of surface meshing/2D meshing or 3D mesh elements/adaptive volume mesh elements on the MEMS components and the need for performing exterior electrostatic analysis based on boundary element analysis and finite-element-based elastic analysis [19]. The surface and volume meshes have to be interoperable with one another, and for better convergence, interpolation solutions have to be selected for coupled domain analysis. All these 2D and 3D models for RF analysis can be done using electromagnetic (EM) simulation-based optimization software, such as High-Frequency Structure Simulator (HFSS), Advanced Design System (ADS), CST Studio Suit, and Electromagnetic Professional (EMPro). The EM simulation model is computationally expensive as it involves repeated processing of the EM solver until the response of the RF MEMS devices and circuit converges to the desired goals of the RF design engineer [20]. These approaches are significantly based on time-consuming trial-and-error methods.

As an alternative to these traditional simulation-based design approaches, the artificial neural network (ANN) model is employed for various applications to hasten the design process [21]. ANN models have been recognized as quick, precise, and adaptable modeling tools for RF and microwaves [22]. Recent studies of RF MEMS device modeling have demonstrated that ANN is capable of providing accurate results much faster than complex EM models [23].

This chapter gives an overview of various RF MEMS components and an introduction to ANN. The chapter also proposes a formulation for modeling RF MEMS switches using ANNs as a case study presentation.

5.2. RF MEMS Components

5.2.1. *RF MEMS Switches*

The RF MEMS switches are widely used in transmitters and receivers [24], mode selection [25], antennas [26], and other microwave and RF communication equipment [27]. These switches are explicit micromechanical switches designed to function between 0.1 and 100 GHz. Compared to semiconductor switches, such as PIN diodes, FETs, and GaAs, RF MEMS switches provide a number of advantages, including less weight, microstructure, minimal insertion loss, high isolation, linearity, and less power requirement. RF MEMS switches also have the drawbacks of high actuation voltage and poor speed [28]. Micromachined RF MEMS switches use a mechanical component in a RF transmission line to act as a short circuit or open circuit. They are classified based on actuating mechanisms as electromagnetic [29], electrothermal [30], piezoelectric [31], and electrostatic [32]. In electrostatic switches, the static force actuates the switch with zero current usage but requires a high actuation voltage. On the contrary, electromagnetic actuation needs less actuation voltage but significant current [33]. The most preferred switches are electrostatic switches, which operate in two configurations: series [34] and shunt [35] as shown in Figures 5.1 and 5.2 respectively.

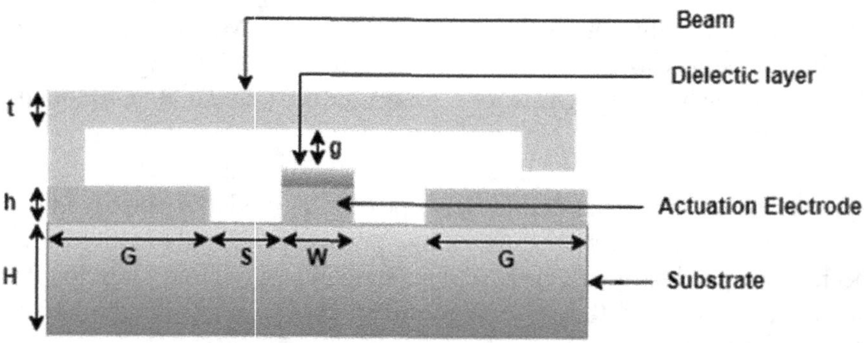

Figure 5.1. RF MEMS Series Switch

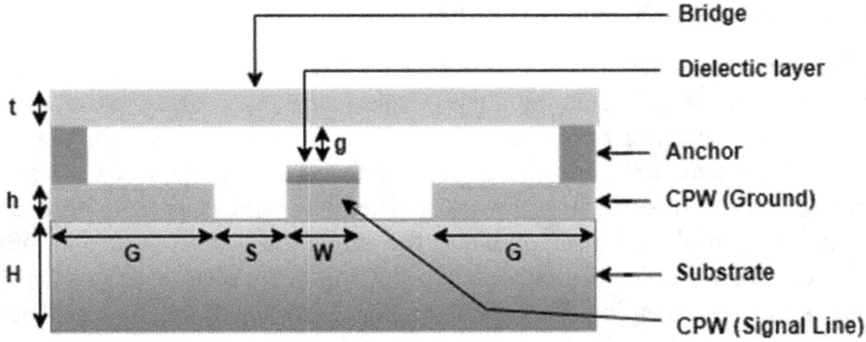

Figure 5.2. RF MEMS Shunt Switch

Shunt-type switches are used as capacitive switches, while resistive switches are employed as series switches. RF MEMS series switches are mostly cantilever structures consisting of a thin transmission line and a dielectric layer, which are attached at one end and remain movable at the other. Shunt switches consist of a thin membrane that is suspended over a central conductor and attached at both ends to the ground plane of a coplanar waveguide (CPW). The dielectric layer is placed above the central conductor to avoid stiction problems, and it forms a capacitance between the transmission line and the membrane. When the switch is in the OFF state, the CPW transmission line conducts the RF signal. On applying a DC voltage, an electric field is developed, which generates an electrostatic force. At a particular threshold voltage, the switch membrane is pulled down, the conductor establishes contact with both sides of the transmission line, and the switch is said to be in the ON state. The voltage at which the switch snaps down is called the pull down voltage [36]:

$$V_p = \sqrt{\frac{8k}{27 \epsilon_0 Ww} g_0^{\ 3}} . \tag{5.1}$$

RF MEMS switches play a significant role in 5G mobile communications [37–39]. The switches with low pull-down voltages that operate between 10 MHz and 60 GHz are the best option for MMW 5G communications [40]. The various applications of RF MEMS switches include variable oscillators, tunable filters, multi-band antennas, and adjustable amplifiers. They are utilized for mode/frequency band selection in antennas to receive and transmit the antenna signals [41].

5.2.2. *RF MEMS Phase Shifters*

5G communications necessitates advanced phased array antennas to satisfy the aims of better quality of service, high data rates, and increased capacity. To connect and improve the performance for many users at the same time, the phased array antenna is employed by generating concurrent directional beams [42]. The phase shifter is an important component in phased array antennas for automatic beam steering in the desired direction. The automatic beam steering in the phased array antennas is very challenging for object tracking. The beam steering can be performed mechanically or electronically. The position of the antenna is changed by mechanical beam steering using motors. Though this technique is quite easy, it slows down the system performance and generates noise. The most preferred beam steering method for faster applications is electrical beam steering. A phase shifter is implemented in electrical beam steering to change the beam position quickly [43]. The phase shifters have been modeled using GaAs FETs for the past decade as they have narrow-band frequency, low power consumption, and high speed. But they suffer from loss and are difficult to maintain isolation, which depends on circuit configuration [44]. The P–I–N diode phase shifter requires a higher DC power than FETs but provides lower loss. The ferrite phase shifters are used in radars due to their small size and low weight but are complex and expensive. These conventional phase shifters use more power, are less efficient at microwave frequencies, and cannot withstand high power [45]. When the power divider, conventional phase shifter, and antenna element are combined, it results in more losses and more cost. The RF MEMS-based phase shifter successfully miniaturizes and improves performance [46].

The RF MEMS phase shifters are used over a wide band because of their high linearity, high isolation, low DC power consumption, and low loss [47]. The RF MEMS phase shifters that utilize MEMS bridges, which provide better performance, were first proposed in Ref. [48], known as distributed MEMS transmission lines (DMTLs), as an evolution from the conventional method [49]. The DMTL true time delay (TTD) phase shifter using MIM capacitors is used to reduce the size and insertion loss [50, 51]. Two back-to-back SP4T switches [52], SP8T switches, and SP16T switches [53] were used to model a phase shifter from Ku to V band in order to give less loss and a minimum phase error when compared to other state-of-the-art phase shifters.

The conventional DMTL phase shifters are designed using CPW transmission lines. As frequency rises, the CPW transmission line experiences insertion loss due to a discontinuity in the air–dielectric medium. The width of the central conductor (W) and gap (G) play a major role in determining waveguide loss and bandwidth, which in turn determine the length of the MEMS membrane. The membrane length determines the speed of the switch, isolation, and insertion loss. The constraints on the central conductor width and gap have a great impact on the final design of the membrane and hence increase the pull down voltage [54]. Variants of phase shifters such as switched delay line phase shifters [55], distributed MEMS phase shifters [55], polymer-based phase shifters [56], ferroelectric phase shifters [56], interdigital capacitor phase shifters [57], ECPW-based phase shifters, fractal-based phase shifters [58], and three-bit DMTL phase shifters [59] are widely used in communication systems.

5.3. Introduction to Artificial Neural Networks

ANNs have recently proven to be an efficient tool for EM modeling and design optimization. An ANN model consists of three primary processing units, grouped to form the input layer, output layer, and hidden layer, as shown in Figure 5.3.

Each neuron is linked to other neurons in neighboring layers, but there is no connection between neurons in the same layer. Each connection is weighted, and a transfer function defines each neuron. Consider an ANN having I neurons in the input layer, O neurons in the output layer, and H neurons in the hidden layer. The output of the ANN model can be represented as in Equation (5.2):

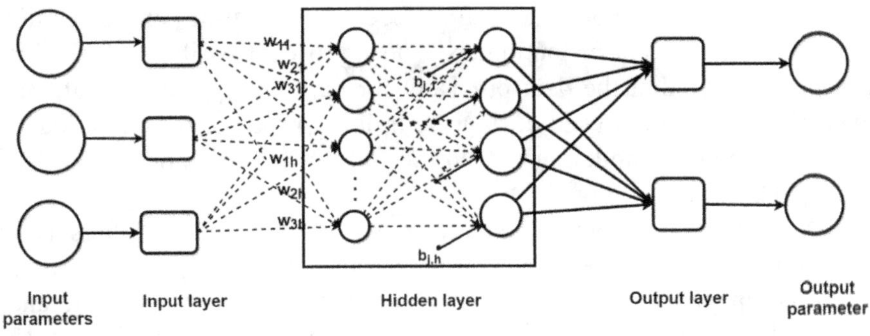

Figure 5.3. Artificial Neural Network Model

$$Y = f\left(\sum_{i=1}^{n} x_i w_i + b\right), \tag{5.2}$$

where Y is the ANN model output, f is the transfer function, x is the neuron input, w is the weight, b is bias, n is the number of inputs from the incoming layer, and i is the iteration number. ANN has the capability to learn and map the relationship between the input and output data [60]. Once the input data are given to the input neurons, the output is computed, and the ANN compares its output with the targeted output to determine the errors. Once all training datasets have been given to the ANN, error derivatives are calculated and added for each weight. The weights for the neurons are updated using the error derivatives. The training proceeds until the errors drop below predetermined levels.

Multi-layer perceptron (MLP), wavelet network, recurrent neural network (RNN), knowledge-based neural network (KBNN), and radial basis function (RBF) are some of the widely used neural network architectures. Training neural networks is an optimization process and is frequently carried out using training algorithms, such as quasi-Newton, gradient descent (GD), scaled conjugate gradient (SCG), gradient descent with momentum (GDM), Bayesian regularization (BR), one-step secant (OSS), variable learning rate gradient descent (GDX), conjugate gradient with Fletcher Reeves restarts (CGF), and the Levenberg–Marquardt (LM) algorithm. The ANN architecture and training algorithms need to be selected based on the application and problem.

The ANN model validation is performed by comparing the predicted outputs of the ANN model to the EM simulator toolbox output. The efficiency of the ANN model is measured by different parameters, such as cross-entropy, error histogram, receiver operating characteristic (ROC) curve, and confusion matrix. The significant attribute of ANN is its ability to predict the data precisely for various input combinations, even if they were not used during the training process, which validates ANN as a helpful modeling tool in the field of RF and microwave circuits [61].

A range of components and circuits, including waveguide filters, passive components, microstrip filters, interconnects, vias, spiral inductors, CPW components, and RF MEMS switches, has been modeled and designed using ANN [60]. ANN is used to minimize the design time of RF MEMS circuits by training the nonlinear relationship between geometrical parameters and the electrical, mechanical, and electromagnetic performances of RF MEMS devices and circuits.

5.3.1. *Types of ANN Model*

5.3.1.1. *Direct ANN model*

The direct approach to any ANN-based RF device and circuit modeling refers to the training of an ANN to predict the electrical, mechanical, and RF characteristics (outputs) from the considered geometrical or physical dimensions, such as width, length, height, thickness, and dielectric as inputs [23, 62–66], as shown in Figure 5.4.

The neural network consists of input neurons with respect to physical parameters and output neurons corresponding to electrical, mechanical, and RF parameters. In the direct method of ANN modeling, the following steps need to be followed:

(1) Calculate the electrical, mechanical, and RF performance parameters obtained from any EM simulator for several combinations of physical parameters.
(2) Divide the available data into two subsets: training data and test data.
(3) Train ANN with different numbers of hidden neurons to model electrical, mechanical, and RF performance parameters that are dependent on physical parameters.
(4) Validate ANN by comparing the ANN model response with the targeted values.
(5) Choose the ANN model that gives the least error and is the best fitted model.
(6) Express the final ANN model as **RF parameters = *f*(physical parameters)**.

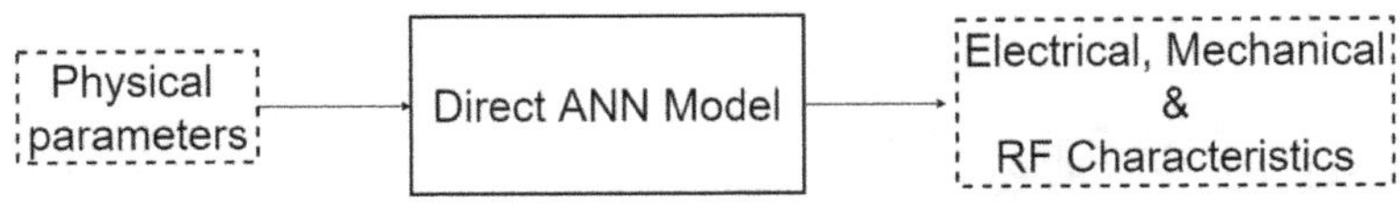

Figure 5.4. Block Diagram of Direct ANN Model

5.3.1.2. *Indirect ANN model*

The indirect approach to any ANN-based RF device and circuit modeling refers to the training of an ANN to predict the geometrical or physical dimensions, such as width, length, height, thickness, and dielectric (outputs), from the considered electrical, mechanical, and RF characteristics as inputs [67–70], as shown in Figure 5.5.

The neural network contains input neurons with respect to electrical, mechanical, and RF parameters and output neurons corresponding to geometrical parameters. In the indirect method of the ANN model, the following steps need to be followed:

(1) Calculate the electrical, mechanical, and RF performance parameters obtained from any EM simulator for several combinations of physical parameters.
(2) Divide the available data into two subsets: training data and test data.
(3) Train ANN with different numbers of hidden neurons to model any one physical parameter for given electrical, mechanical, and RF performance parameters and any other physical parameter.
(4) Validate ANN by comparing the ANN model response with the targeted values.
(5) Choose the ANN model that gives the least error and is the best fitted model.
(6) Express the final inverse ANN model as **any one physical parameter =** ***f*** **(RF parameters, any other physical parameters)**.

5.4. Case Study

5.4.1. *RF MEMS Switch Modeling Using Artificial Neural Network*

The RF MEMS shunt switch structure proposed for analysis using ANN is shown in Figure 5.6. It consists of CPW transmission lines on the

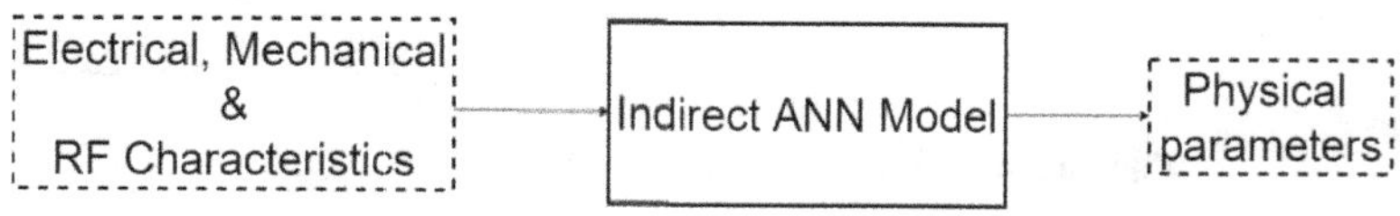

Figure 5.5. Block Diagram of Indirect ANN Model

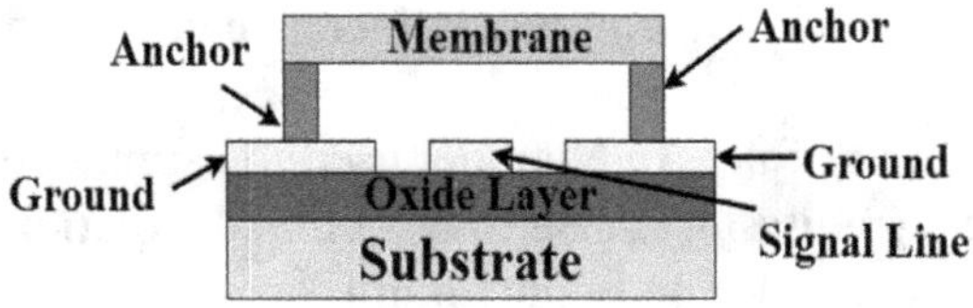

Figure 5.6. Schematic Diagram of RF MEMS Shunt Switch

Table 5.1. Physical Dimensions of RF MEMS Shunt Switch

Physical Parameter	Value
Length of the central conductor of CPW	100 μm
Length of the ground conductor of CPW	60 μm
Thickness of silicon substrate	3,000 Å
Length of the membrane	300 μm
Width of the anchors	40 μm
Height of the membrane from the CPW (upstate)	1.5 μm
Height of the membrane from the CPW (downstate)	0.5 μm
Thickness of membrane	2 μm

silicon substrate and a membrane suspended above the CPW. The membrane is supported by anchors on the ground conductor of the CPW. A dielectric layer is placed on the central conductor of the CPW to avoid a short circuit. The physical parameters of the RF MEMS shunt switch chosen are given in Table 5.1.

The membrane moves toward the CPW when the electrostatic force is generated by charges. This charge is developed as a result of the potential difference between the membrane and the ground. Consequently, the membrane loses its stability and snaps down on the central conductor of the CPW when the voltage difference reaches above the threshold. When the membrane comes into contact with the CPW, the switch is in the ON state. This threshold voltage is said to be the pull-down voltage. When the voltage is released, the switch is in an OFF state.

The optimization of the physical parameters of the switch needs to be performed to find better RF performance for the RF MEMS shunt switch. However, the optimization of microswitch structures in conventional EM simulators is time intensive. With the help of ANNs though, the anticipation of the input and output parameters can be done within a few minutes.

A block diagram of the ANN model used to estimate the output scattering parameters (return loss, insertion loss, and isolation loss) in terms of input parameters (W, w, f) is shown in Figure 5.7.

According to [28], the S parameters of an RF MEMS switch are a function of the following parameters: width of the membrane (w), width of the central conductor (W), frequency (F), dielectric thickness (t_d), and relative dielectric permittivity ε_r.

In this case study, the direct approach to modeling scattering parameter for RF MEMS shunt switch is carried out by changing the switch's physical parameters, such as w, W, and f, which greatly influence the RF performance of the switch. Figure 5.8 shows the development procedure of the ANN model used for the analysis of RF MEMS switches.

The physical parameters considered as input to the ANN-based RF MEMS shunt switch modeling to obtain the desired S parameters (desired outputs) are: 1 μm $\leq w \leq$ 100 μm, 1 μm $\leq W \leq$ 100 μm, and 1 GHz $\leq f \leq$ 40 GHz.

The training data samples are to be generated for both the upstate and downstate of the considered RF MEMS switch. Training data samples are generated using conventional EM simulation software, such as HFSS. Training data samples obtained from the EM simulator are in the form of relationships between the input (W, w, f) and output (S) parameters. Out of the samples obtained, 15% are utilized for testing, 15% for validation, and 70% are chosen for training. The multi-layer perceptron feedforward neural network model is selected and trained. Out of 70% of the training

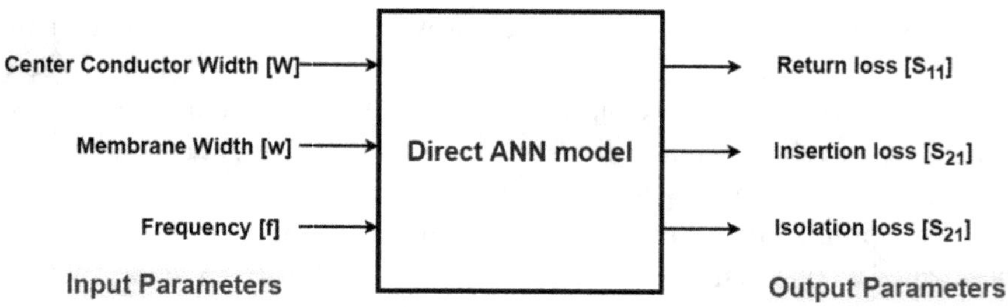

Figure 5.7. Block Diagram of Direct ANN Model

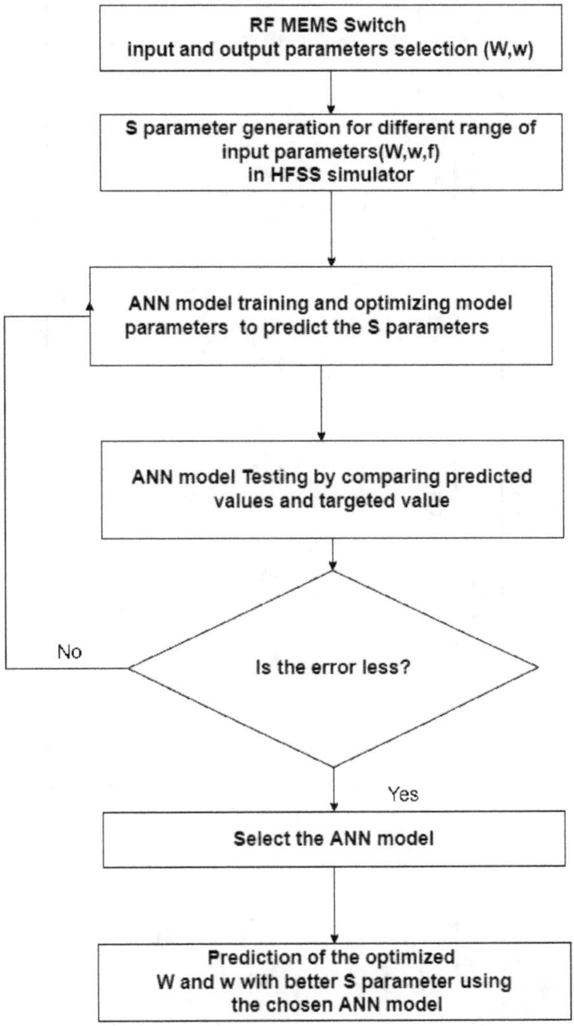

Figure 5.8. Flow Chart of Direct ANN Model Development

dataset, several neural network models are trained with various numbers of hidden layers.

Initially, as a trial-and-error process, the considered RF MEMS switch model was trained with 397 samples, which was found to be insufficient to train the ANN model, and the error was greater. Consequently, the samples and hidden layers were increased from 1,985 to 3,970 until the error was minimized, as shown in Table 5.2.

Table 5.2. Samples Taken for Optimization of RF MEMS Shunt Switch

Hidden	397 Samples			1,985 Samples			3,970 Samples		
Layers	Training	Validation	Testing	Training	Validation	Testing	Training	Validation	Testing
10	99.27	100	100	86.46	99.32	98.32	77.32	92.78	90.43
20	99.27	100	100	77.10	98.65	97.31	52.19	75.50	75.50
30	98.55	100	100	73.36	92.95	93.62	34.41	55.70	59.06
40	98.91	100	100	49.60	81.87	81.20	17.49	33.55	37.58
50	97.47	100	100	31.89	63.08	66.77	11.87	25.67	26.34
60	98.55	100	100	15.91	42.28	42.95	2.015	3.85	6.87
70	98.91	100	100	24.40	60.06	57.71	2.23	7.21	7.55
80	98.91	100	100	7.70	26.84	24.49	1.69	4.86	5.704
90	99.63	100	100	6.76	24.16	31.54	7.55	3.52	3.02
100	98.55	100	100	2.87	15.77	16.10	5.39	1.51	2.68

The considered RF MEMS switches were trained using various ANN models, and the corresponding error percentages are given in Table 5.3.

Table 5.3 shows that the neural network model based on the scaled conjugate gradient algorithm with 3,970 samples and 60 hidden layers provides a lower error percentage compared to the other ANN models. The modeling parameters considered in the SCG-based neural network model relevant to the development of RF MEMS switches are given in Table 5.4.

Table 5.3. Error Comparison of Various Training Algorithms

Training Algorithms	Error (%)
Resilient backpropagation	0.5466
Scaled Conjugate Gradient	**0.0076**
Conjugate gradient with Beale restarts	0.0856
Polak–Ribere conjugate gradient	0.0403
Gradient descent	0.99
Variable learning rate gradient descent	0.8791
One-step secant	0.9924
Gradient descent with momentum	0.995
Levenberg–Marquardt	—
Bayesian regularization	—
BFGS quasi-Newton	—

Table 5.4. Parameters Involved in Modeling RF MEMS Shunt Switch Using SCG

Transfer Function	Sigmoid
Hession indefiniteness regulating parameter	5.0×10^{-7}
Marquardt adjustment parameter	0.005
Network architecture	**3-60-40**
Maximum epochs	1,000
Training dataset (%)	**70**
Testing dataset (%)	**15**
Verifying dataset (%)	**15**
Maximum validation failures	6
Second-derivative approximation changing weight	5.0×10^{-5}
Minimum performance gradient	3.38×10^{-5}

Having selected the ANN model, the next step is to validate the performance of the considered RF MEMS switch for the following ranges of input parameters: 1 μm ≤ w ≤ 100 μm, 1 μm ≤ W ≤ 100 μm, and 1 GHz ≤ f ≤ 40 GHz and to determine the losses in terms of scattering parameters (output parameters).

The developed ANN model is validated using two approaches:

(i) validating the ANN model using data that **were used** during the training process;
(ii) validating the ANN model using data that **were not used** during the training process.

(1) **Discussions on validation of RF MEMS switch using SCG ANN model with data that were used for training process**

RF MEMS switch validation in both the ON and OFF states is performed using the predicted values obtained from the SCG-based direct ANN model and the HFSS-based EM simulator. Tables 5.5 and 5.6 display the

Table 5.5. Comparison of Return Loss (S_{11}) of RF MEMS Switch in OFF and ON States Using HFSS and ANN Model During the Training Process with Various Input Parameters

			OFF State			ON State		
W	w	f	HFSS	ANN	Error	HFSS	ANN	Error
1	1.2	1	–33.4125	–33.9945	0.582	–14.3493	–14.97298	0.62368
4	4.8	4	–22.9958	–22.9367	0.0591	–4.32197	–4.11289	0.20908
7	8.4	7	–18.2883	–18.3403	0.052	–1.97298	–1.43901	0.53397
10	12	10	–15.3172	–15.1762	0.141	–1.13145	–1.44356	0.31211
13	15.6	13	–13.1766	–13.1166	0.06	–0.76154	–0.64421	0.11733
16	19.2	16	–11.5725	–11.4215	0.151	–0.57684	–0.571	0.00584
19	22.8	19	–10.3515	–10.9920	0.6405	–0.47699	–0.2271	0.24989
22	26.4	22	–9.42781	–9.0182	0.40961	–0.42029	–0.621	0.20071
25	30	25	–8.75491	–8.20120	0.55371	–0.39866	–0.4261	0.02744
28	33.6	28	–8.3403	–8.3403	0	–0.38938	–0.31045	0.07893
31	37.2	31	–8.3027	–8.8120	0.5093	–0.3720	–0.3721	0.0001
34	40.8	34	–8.99163	–8.99480	0.00317	–0.3103	–0.42017	0.10987
37	44.4	37	–11.2671	–11.7701	0.503	–0.20047	–0.2115	0.01103
40	48	40	–17.812	–17.0943	0.7177	–0.20578	–0.20334	0.00244

Table 5.6. Comparison of Insertion and Isolation Loss (S_{21}) of RF MEMS Switch in OFF and ON States Using HFSS and ANN Model During the Training Process with Various Input Parameters

			OFF State			ON State		
W	*w*	*f*	HFSS	ANN	Error	HFSS	ANN	Error
1	1.2	1	–0.01593	–0.0672	0.05127	–0.20802	–0.2771	0.06908
4	4.8	4	–0.05603	–0.07723	0.0212	–2.14764	–2.42109	0.27345
7	8.4	7	–0.11325	–0.16278	0.04953	–4.69415	–4.8021	0.10795
10	12	10	–0.19098	–0.7296	0.53862	–7.01804	–7.7329	0.71486
13	15.6	13	–0.28375	–0.27519	0.00856	–9.01943	–9.33012	0.31069
16	19.2	16	–0.39468	–0.279	0.11568	–10.7547	–10.6221	0.1326
19	22.8	19	–0.51677	–0.482	0.03477	–12.2932	–12.8452	0.552
22	26.4	22	–0.63994	–0.622	0.01794	–13.6991	–13.61982	0.07928
25	30	25	–0.74971	–0.83105	0.08134	–15.0468	–15.39321	0.34641
28	33.6	28	–0.82579	–0.6445	0.18129	–16.4655	–16.90321	0.43771
31	37.2	31	–0.84058	–0.9331	0.09252	–18.2781	–18.55712	0.27902
34	40.8	34	–0.76094	–0.73392	0.02702	–21.6089	–21.9032	0.2943
37	44.4	37	–0.56621	–0.61982	0.05361	–34.1604	–34.1604	0
40	48	40	–0.31179	–0.38845	0.07666	–18.1289	–18.273	0.1441

direct ANN model validation in predicting S_{11} and S_{21} that are used during the training process. The developed ANN model is proven to be efficient as it predicts the scattering parameters of the RF MEMS with an error of less than 0.77177%.

The SCG-based ANN model was tested on 14 random samples that were used during the training process in both the OFF and ON states of the RF MEMS switch. Figure 5.9 displays the convergence of the proposed ANN model and HFSS EM simulator model in terms of the S parameters (S_{11}, S_{21}).

(2) **Discussions on validation of RF MEMS switch using SCG ANN model with data that were not used for training process**

RF MEMS switch validation in both the ON and OFF states is performed using the predicted value obtained from the SCG-based direct ANN model and the HFSS-based EM simulator. Tables 5.7 and 5.8 display the direct ANN model validation in predicting S_{11} and S_{21} that are not used

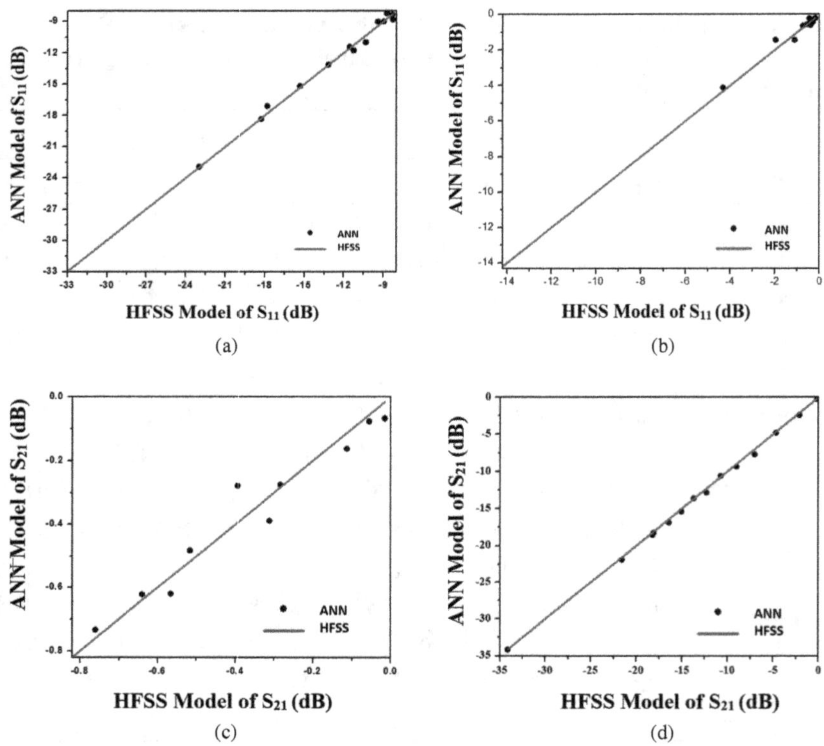

Figure 5.9. (a) Return Loss Comparison (S_{11}) of ANN and HFSS Models in OFF State. (b) Return Loss Comparison (S_{11}) of ANN and HFSS Models in ON State. (c) Insertion Loss Comparison (S_{21}) of ANN and HFSS Models in OFF State. (d) Isolation Loss Comparison (S_{21}) of ANN and HFSS Models in ON State

during the training process. The developed ANN model is proven to be efficient as it predicts the scattering parameters of the RF MEMS with an error of less than 1.6%.

The SCG-based ANN model was tested on 14 random samples that were used during the training process in both the OFF and ON states of the RF MEMS switch. Figure 5.10 displays the convergence of the proposed ANN model and the HFSS EM simulator model in terms of the S parameters (S_{11}, S_{21}).

To show the validity of training and testing the developed SCG-based neural network model for the considered RF MEMS shunt switch, precise

Table 5.7. Comparison of Return Loss (S_{11}) of RF MEMS Switch in OFF and ON States Using HFSS and ANN Model not Used During the Training Process with Various Input Parameters

			OFF State			ON State		
W	w	f	HFSS	ANN	Error	HFSS	ANN	Error
1.28	1.6	2	–33.1478	–34.3678	0.9553	–14.2358	–13.4769	0.8724
2.23	2.6	5	–23.4377	–23.8614	0.8656	–4.6794	–5.0113	0.6893
3.14	3.7	8	–18.1158	–18.1621	0.1262	–1.2014	–1.20578	0.7672
5.6	6.7	11	–15.6399	–16.9234	1.6062	–1.5463	–1.20047	0.0690
8.84	10.6	14	–13.2086	–13.0113	0.1653	–0.6442	–0.71632	0.0452
12.4	14.9	17	–11.9635	–11.2342	0.3383	–0.4084	–0.41977	0.1570
16.65	19.9	20	–10.7844	–10.6789	0.3274	–0.6914	–0.31009	0.1669
20.12	24.2	23	–9.3654	–9.1987	0.2291	–0.931	–0.99973	0.5794
24.12	28.9	26	–8.0145	–8.0012	0.7537	–0.5841	–0.0241	0.3745
29.32	35.2	29	–8.2598	–9.1356	0.7953	–0.2835	–0.39163	0.0022
35.16	42.1	32	–8.3321	–8.7184	0.4157	–0.193	–0.3546	0.0174
40.3	48.5	35	–8.863	–8.3403	0.6513	–0.1353	–0.3851	0.0748
43.34	52.0	38	–11.843	–11.3172	0.0501	–0.6841	–0.39163	0.1911
51.6	61.9	39	–17.7564	–17.9958	0.1838	–0.1404	–0.2096	0.0039

Table 5.8. Comparison of Insertion and Isolation Loss (S_{21}) of RF MEMS Switch in ON and OFF States Using HFSS and ANN Model not Used During the Training Process with Various Input Parameters

			OFF State			ON State		
W	w	f	HFSS	ANN	Error	HFSS	ANN	Error
1.28	1.6	2	–0.0234	–0.07621	0.06028	–0.3541	–0.21739	0.00937
2.23	2.6	5	–0.022	–0.09325	0.03722	–2.2136	–2.37729	0.22965
3.14	3.7	8	–0.2387	–0.2861	0.17285	–4.7415	–4.9621	0.26795
5.6	6.7	11	–0.2732	–0.83319	0.64221	–7.0257	–7.49921	0.48117
8.84	10.6	14	–0.3001	–0.20116	0.08259	–9.0543	–9.0721	0.05267
12.4	14.9	17	–0.442	–0.3218	0.07288	–10.8452	–10.8324	0.0777
16.65	19.9	20	–0.6245	–0.96321	0.44644	–12.3145	–12.501	0.2078
20.12	24.2	23	–0.7865	–0.74551	0.10557	–13.6984	–13.9835	0.2844
24.12	28.9	26	–0.8548	–0.78901	0.0393	–15.0123	–15.8432	0.7964

(*Continued*)

Table 5.8. (*Continued*)

			OFF State			ON State		
W	*w*	*f*	HFSS	ANN	Error	HFSS	ANN	Error
29.32	35.2	29	−0.9284	−0.83351	0.00772	−16.5841	−16.1053	0.3602
35.16	42.1	32	−0.8796	−0.79261	0.04797	−18.3084	−18.7210	0.4429
40.3	48.5	35	−0.8501	−0.79032	0.02938	−21.7412	−21.183	0.4259
43.34	52.0	38	−0.6987	−0.43121	0.135	−34.1546	−34.5921	0.4317
51.6	61.9	39	−0.421	−0.36512	0.05333	−18.1621	−18.5210	0.3921

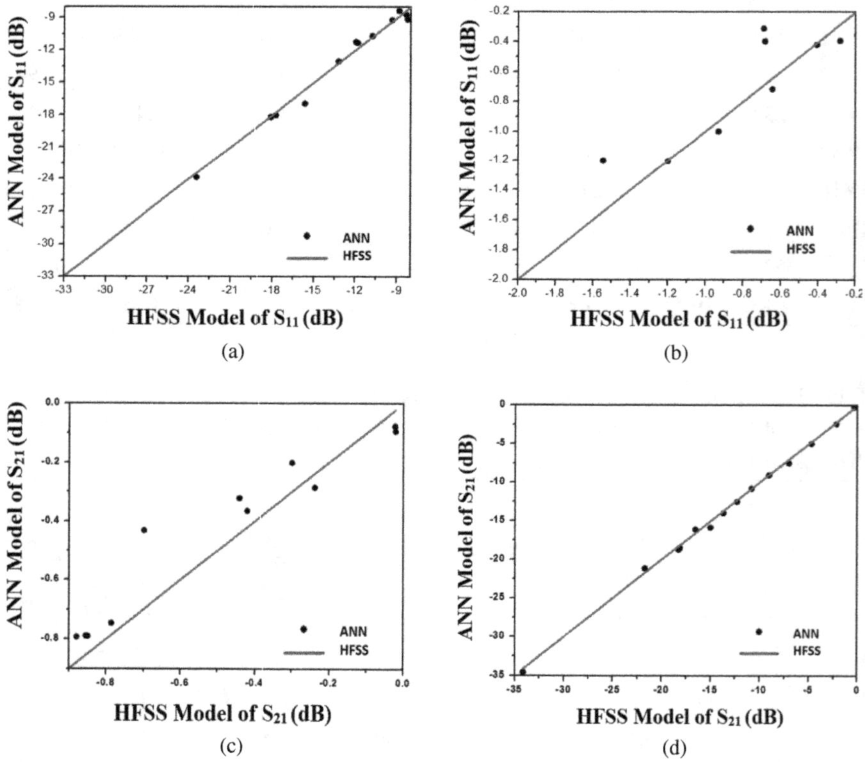

Figure 5.10. (a) Return Loss Comparison (S_{11}) of ANN and HFSS Models in OFF State. (b) Return Loss Comparison (S_{11}) of ANN and HFSS Models in ON State. (c) Insertion Loss Comparison (S_{21}) of ANN and HFSS Models in OFF State. (d) Isolation Loss Comparison (S_{21}) of ANN and HFSS Models in ON State

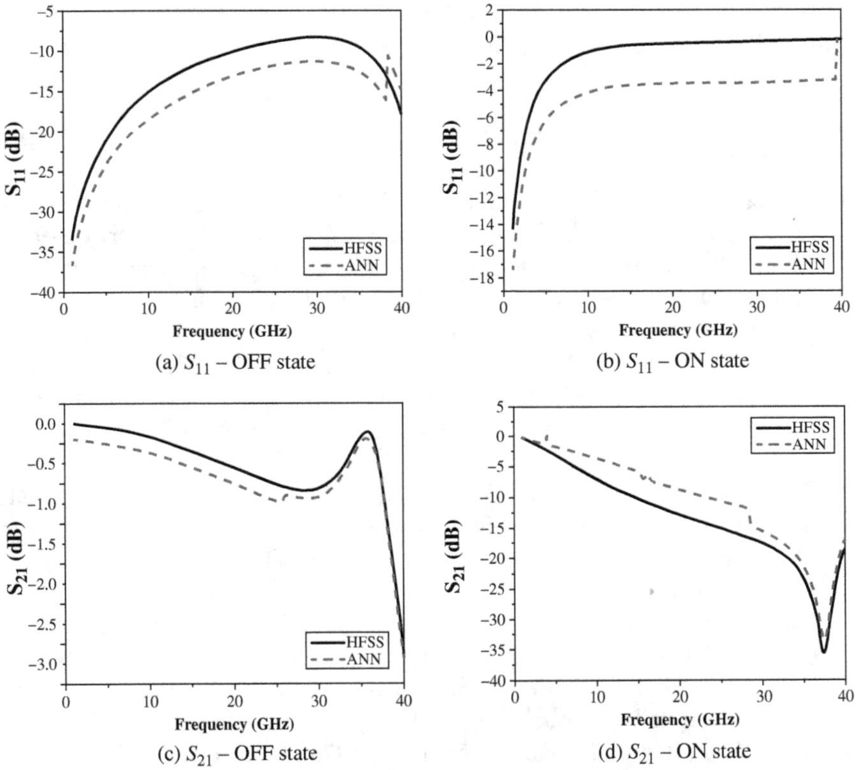

Figure 5.11. Scattering Parameters (S_{11}, S_{21}) for ANN Model and HFSS Model for Different Frequencies at $W = 100$ μm and $w = 80$ μm

parameters of a membrane width (w) of 80 μm and CPW central conductor width (W) of 100 μm at various frequencies (f) are chosen. Figure 5.11 compares the HFSS-simulated results with the ANN model.

5.5. Conclusion

The design, analysis, and optimization of RF MEMS circuits are quite challenging using EM CAD tools since they are a time-consuming process to attain a solution. To overcome the drawbacks present in optimization, ANN has been considered a suitable candidate. In this chapter, an SCG-based ANN model of RF MEMS switch modeling was considered and developed as a case study. It revealed that the ANN model suits well for miniaturized devices and circuits.

References

1. Mihret, E. T. and Haile, G. (2021). 4G, 5G, 6G, 7G and future mobile technologies. *American Journal of Computer Science Technology, 9*(February), 3–9.
2. Khutey, R., Rana, G., Dewangan, V., Tiwari, A., and Dewamngan, A. (2015). Future of wireless technology 6G & 7G. *International Journal of Electronics Engineering Research (IJEER), 3*(2), 583–585.
3. Li, L., Wang, D., Niu, X., Chai, Y., Chen, L., He, L., Wu, X., Zheng, F., Cui, T., and You, X. (2018). mmWave communications for 5G: Implementation challenges and advances. *Science China Information Sciences, 61*(2), 1–19.
4. Balteanu, F., Modi, H., Choi, Y., Lee, J., Drogi, S., and Khesbak, S. 5G RF front end module architectures for mobile applications. In *2019 49th European Microwave Conference EuMC 2019,* 1–3 October, Paris, France, pp. 252–255.
5. Iannacci, J. (2022). *RF-MEMS Technology for High-Performance Passives* (2nd edn.). IOP Publishing: Bristol, UK.
6. Rebeiz, G. M. (2003). *RF MEMS Theory Design and Technology*. John Wiley and Sons Ltd: Hoboken, New Jersey.
7. Bathe, K. J., Ramm, E., and Wilson, E. L. (1975). Finite element formulation for large deformation dynamic analyses. *International Journal for Numerical Methods in Engineering, 9*(2), 353–386.
8. Nabors, K. and White, J. (1991). FastCap: A multipole-accelerated 3-D capacitance extraction program. *IEEE Transactions on Computer-Aided Design, 10*, 1447–1459.
9. Phillips, J. R. and White, J. (1994). A precorrected — FFT method for capacitance extraction of complicated 3-D structures. In *Proceedings of International Conference on Computer Aided Design*, 6–10 November, San Jose California, USA, pp. 268–271.
10. Cai, X., Yie, H., Osterberg, P., Gilbert, J., Senturia, S., and White, J. (1993). A relaxation/multipole-accelerated scheme for self-consistent electromechanical analysis of complex 3-D microelectromechanical structures. In *Proceedings of 11th International Conference on Computer-Aided Design*, 7–11 November, Santa Clara, California, USA, pp. 283–287.
11. Gilbert, J. R., Legtenberg, R., and Senturia, S. D. (1995). 3D coupled electro mechanics for MEMS: Applications of CoSolve-EM. In *Proceedings of IEEE Micro Electro mechanical Systems*, 29 January–02 February, Amsterdam, Netherlands, pp. 122–127.

12. Bachtold, M., Korvink, J. G., Funk, J., and Baltes, H. (1995). New convergence scheme for self-consistent electromechanical analysis of iMEMS. In *Proceedings of the International Electron Devices Meeting (IEDM)*, 10–13 December, Washington, DC, USA, pp. 605–608.
13. Yie H., Cai X., and White J. (1994). Convergence properties of relaxation versus the surface-Newton generalized-conjugate residual algorithm for self-consistent electro mechanical analysis of 3-D micro-electromechanical structures. In *Proceedings of Numerical Process and Device Modeling*, 5–6 June, Honolulu, Hawaii, pp. 137–140.
14. Aluru, N. R. and White, J. (1996). Direct-Newton finite-element/boundary element technique for micro-electro-mechanical analysis. In *Technical Digest IEEE Solid-State Sensor and Actuator Workshop*, 2–6 June Hilton Head Island, South Carolina, pp. 54–57.
15. Aluru, N. R. and White, J. (1997). An efficient numerical technique for electromechanical simulation of complicated micro electro mechanical structures. *Sensors and Actuators A*, *58*, 1–11.
16. Tentzeris, E. M. (2002). Adaptive modeling and design of highly integrated 3D microwave-millimeter wave radio front ends. In *IEEE GaAs Integrated Circuit Symposium*, 20–23 October, Monterey, CA, USA, pp. 109–112.
17. Bushyager, N. A. and Tentzeris, M. M. (2001). Modeling and design of RF MEMS structures using computationally efficient numerical techniques. In *Proceedings of IEEE Electronic Components and Technology Conference*, 29 May–01 June, Orlando, FL, USA, pp. 327–330.
18. Wang, J., Ren, Z., and Nguyen, C. T. C. (2003). 1.14-GHz self-aligned vibrating micromechanical disk resonator. In *IEEE Radio Frequency Integrated Circuits (RFIC) Symposium*, 8–10 June, Philadelphia, Pennsylvania, pp. 335–338.
19. Ananthasuresh, G. K., Gupta, R. K., and Senturia, S. D. (1996). An approach to macro modeling of MEMS for nonlinear dynamic simulation. Microelectromechanical Systems (MEMS). *ASME Dynamic Systems and Control (DSC) Series*, *59*, 401–407.
20. Jargon, J. A., Gupta, K. C., and DeGroot, D. C. (2002). Applications of artificial neural networks to RF and microwave measurements. *International Journal RF Microwave Computer-Aided Engineering*, *12*(1), 3–24.
21. Kabir, H., Cao, Y., Cao, Y., and Zhang, Q. J. (2010). Advances of neural network modeling methods for RF/microwave applications. *Applied Computational Electromagnetics Society Journal*, *25*(5), 423–432.
22. Lee, Y. and Filipovic, D. S. (2005). ANN based electromagnetic models for the design of RF MEMS switches. *IEEE Microwe and Wireless Components Letters*, *15*(11), 823–825.
23. Ćirić, T., Marinković, Z., Pronić-Rančić, O., Marković, V., and Vietzorreck, L. (2017). ANN Approach for modeling of mechanical characteristics of

RF MEMS capacitive switches — An overview. *Microwe Review*, *23*(1), 25–34.

24. Chaloun, T., Tabarani, F., and Tolunay Wipf, S. (2018). A modular phased array transceiver with RF-MEMS SPDT switches in a 0.25 μm SiGe BiCMOS technology. *12th European Conference on Antennas and Propagation (EuCAP 2018)*, 9–13 April, London, UK, pp. 1–5.
25. Xu, Y., Tian, Y., Zhang, B., Duan, J., and Yan, L. (2018). A novel RF MEMS switch on frequency reconfigurable antenna application. *Microsystem Technologies*, *24*(9), 3833–3841.
26. Srinivasa Rao, K., Ashok Kumar, P., Guha, K., Sailaja, B. V. S., Vineetha, K. V., Baishnab, K. L., Girija Sravani, K. (2021). Design and simulation of fixed–fixed flexure type RF MEMS switch for reconfigurable antenna. *Microsystem Technologies*, *27*(2), 455–462.
27. Kurmendra and Kumar, R. (2021). A review on RF micro-electro-mechanical-systems (MEMS) switch for radio frequency applications. *Microsystem Technologies*, *27*(7), 2525–2542.
28. Muldavin, B. and Rebeiz, G. M. (2000). High-isolation CPW MEMS shunt switches-part 1: Modeling. *IEEE Transactions on Microwave Theory and Techniques*, *48*(6), 1038–1044.
29. Sharma, K. and Gupta, N. (2015). Electromagnetic modeling and parameter extraction of RF-MEMS switch. *Microsystem Technologies*, *21*(1), 181–185.
30. Yadav, D., Murthy, N. S., Palathingal, S., Shekhar, S., Giridhar, M. S., and Ananthasuresh, G. K. (2019). A two-terminal bistable electrothermally actuated microswitch. *Journal of Microelectromechanical Systems*, *28*(3), 540–549.
31. Benoit, R. R., Rudy, R. Q., Pulskamp, J. S., and Polcawich, R. G. (2020). Piezoelectric RF MEMS switches on Si-on-sapphire substrates. *Journal of Microelectromechanical Systems*, *29*(5), 1087–1090.
32. Kashani Ilkhechi, H., Mirzajani, E., Aghdam, N., and Badri Ghavifekr, H. (2017). A new electrostatically actuated rotary three-state DC-contact RF MEMS switch for antenna switch applications. *Microsystem Technologies*, *23*(1), 231–243.
33. Lysenko, E., Tkachenko, A. V., Sherova, E. V., and Nikitin, A. V. (2018). Analytical approach in the development of RF MEMS switches. *Electronics*, *7*(12), 1–23.
34. Khodadady, K. and Ganji, B. A. (2016). Design and modeling of a novel RF MEMS series switch with low actuation voltage. *Microsystem Technologies*, *22*(12), 2921–2929.
35. Naji, M. K., Farhood, A. D., and Ali, A. H. (2019). Novel design and analysis of RF MEMS shunt capacitive switch for radar and satellite communications. *Indonesian Journal of Electrical Engineering and Computer Science*, *15*(2), 971–978.

36. Rebeiz, G. M. and Muldavin, J. B. (2001). RF MEMS switches and switch circuits. *IEEE Microwave Magazine*, *2*(4), 59–71.
37. Iannacci, J. and Tschoban, C. (2017). RF-MEMS for future mobile applications: Experimental verification of a reconfigurable 8-bit power attenuator up to 110 GHz. *Journal of Micromechanics and Microengineering*, *27*. Article ID 04003.
38. Iannacci, J., Huhn, M., Tschoban, C., and Pötter, H. (2016). RF-MEMS technology for 5G: Series and shunt attenuator modules demonstrated up to 110 GHz. *IEEE Electron Device Letters*, *37*, 1336–1339.
39. Iannacci, J., Huhn, M., Tschoban, C., and Pötter, H. (2016). RF-MEMS technology for future mobile and highfrequency applications: Reconfigurable 8-bit power attenuator tested up to 110 GHz. *IEEE Electron Device Letters*, *37*, 1646–1649.
40. Shekhar, S., Vinoy, K. J., and Ananthasuresh, G. K. (2018). Low-voltage high-reliability MEMS switch for millimeter wave 5G applications. *Journal of Micromechanics and Microengineering*, *28*(7), 75012.
41. Cetiner, B. A., Qian, J. Y., Chang, H. P., Bachman, M., Li, G. P., and De Flaviis, F. (2003). Monolithic integration of RF MEMS switches with a diversity antenna on PCB substrate. *IEEE Transactions on Microwave Theory Techniques*, *51*(1), 332–335.
42. Hong, W., Jiang, Z. H., Yu, C., Zhou, J., Chen, P., Yu, Z., Zhang, H., Yang, B., Pang, X., Jiang, M., Cheng, Y., Taher, M. K., Al-Nuaimi, Y., Chen, J., and He, S. (2017). Multibeam antenna technologies for 5G wireless communications. *IEEE Transactions on Antennas and Propagation*, *65*(12), 6231–6249.
43. Real, S. M., Turjo, R. R., Sobhan, F. F., and Ashraf, N. (2018). Phased array antennas for mm wave frequency spectrum for 5G communication — Thesis, Brac University, Bangladesh.
44. Gaspari, R. A. and Lee, H. H. (1978). Microwave GaAs FET switching. *IEEE-MTT-S International Microwave Symposium Digest*, 27–29 June, Ottawa, ON, Canada, pp. 58–60.
45. Miao, M., Wu, G., Li, Z., Jin, Y., and Haol, Y. (2006). A bulk micromachined distributed digital microwave phase shifter with butterfly multilayer bridges and MAM capacitors. In *Proceedings of the 1st IEEE International Conference on Nano/Micro Engineered and Molecular Systems*, 18–21 January, Zhuhai, China, pp. 479–483.
46. Guclu, C., Cetintepe, C., Civi, O. A., Demir, S., and Akın, T. (2010). 35 GHz phased array antenna using DMTL phase shifters. In *10th Mediterranean Microwave Symposium*, 25–27 August, Northern Cyprus, pp. 114–117.
47. Rebeiz, G. M., Tan, G.-L., and Hayden, J. S. (2002). RF MEMS phase shifters: Design and applications. *IEEE Microwave Magazine*, *3*(2), pp. 72–81.

48. Barker, S. and Rebeiz, G. (1998). Distributed MEMS true-time delay phase shifters and wide-band switches. *IEEE Transactions on Microwave Theory Techniques*, *46*(II), 1881–1889.
49. Hayden, J. S. and Rebeiz, G. M. (2000). 2-Bit MEMS distributed X-band phase shifters. *IEEE Microwave Guided Wave Letters*, *10*(12), 540–542.
50. Campbell, C. F. and Brown, S. A. (2000). A compact 5-bit phase-shifter MMIC for K-band satellite communication systems. *IEEE Transactions on Microwave Theory and Techniques*, *48*, 2652–2656.
51. Wallace, J, Redd, H., and Furlow. R. (1999). Low cost MMIC DBS chip sets for phased array applications. In *1999 IEEE MTT-S Int. Microwave Symposium Digest*, Anaheim, CA, pp. 677–680.
52. Dey, S., Koul, S. K., Poddar, A. K., and Rohde, U. L. (2017). Ku to V-band 4-Bit MEMS phase shifter bank using high isolation SP4T switches and DMTL structures. *Journal of Micromechanics and Micro Engineering*, *27*(10), 1–10.
53. Koul, S. K. and Dey, S. (2018). K-band 4-bit phase shifter using two back to back MEMS SP16T switching networks. *Journal of Micro Electro Mechanical Systems*, *27*(4), 643–655.
54. Rock, J. and Hudson, T. (2004). Design fabrication and testing of coplanar waveguide transmission lines for use in RF MEMS switches. *IEEE Aerospace Conference Proceedings*, 6–13 March, USA, *2*, 1–8.
55. Puyal, V., Leti, C., Cnrs, L., and Titz, D. (2013). RF MEMS phase shifters for wireless applications. In *Handbook of Mems for Wireless and Mobile Applications*, Woodhead Publishing Limited, Cambridge, UK, pp. 136–175.
56. Shah, U., Decrossas, E., Jung-Kubiak, C., Reck, T., Chattopadhyay, G., Mehdi, I., and Oberhammer, J. (2016). Submillimeter-wave 3.3-bit RF MEMS phase shifter integrated in micromachined waveguide. *IEEE Transactions on Terahertz Science and Technology*, *6*(5), 706–715.
57. Nikfalazar, M., Sazegar, M., Mehmood, A., Wiens, A., Friederich, A., Maune, H., Binder, J. R., and Jakoby, R. (2017). Two-dimensional beam-steering phased-array antenna with compact tunable phase shifter based on BST thick films. *IEEE Antennas and Wireless Propagation Letters*, *16*, 585–588.
58. Chakraborty, A. and Gupta, B. (2017). Paradigm phase shift: F MMS phase shifters: An overview. *IEEE Microwave Magazine*, *18*(1), 22–41.
59. Sindhuja, N. M. M. and Kanthamani, S. (2019). Three three-bit DMTL phase shifter for phased array antennas. *Journal of Circuits, Systems and Computers*, *28*(7), 1950112.
60. Zhang, C., Zhang, L., and Zhang, Q.-J. (2013). Neural networks for radio frequency/microwave modeling. Simulation-Driven Design in Microwave Engineering, Imperial College Press, London, pp. 191–213.

61. Zhang, Q. J., Gupta, K. C., and Devabhaktuni, V. K. (2003). Artificial neural networks for RF and microwave design — From theory to practice. *IEEE Transactions on Microwave Theory and Techniques*, *51*(4 II), 1339–1350.
62. Kelley, M., Malshe, A. P., and Barlow, F. (2005). Modeling-based design optimization of wafer-level and chip-scale packaging for RF-MEMS devices. In *Electronic Components and Technology Conference (ECTC)*, 31 May–03 June, Lake Buena Vista, FL, USA, No. 479, pp. 1814–1818.
63. Ciric, T., Marinkovic, Z., Dhuri, R., Pronic-Rancic, O., and Markovic, V. (2020). Hybrid neural lumped element approach in inverse modeling of RF MEMS switches. *Facta Universitatis, Series: Electronics and Energetics*, *33*(1), 27–36.
64. Suganthi, S., Murugesan, K., and Raghavan, S. (2012). ANN model of RF MEMS lateral SPDT switches for millimeter wave applications. *Journal of Microwaves, Optoelectronics and Electromagnetic Applications*, *11*(1), 130–143.
65. Marinkovic, Z., Markovic, V., Ciric, T., Vietzorreck, L., and Pronic-Rancic, O. (2016). Artifical neural networks in RF MEMS switch modelling. *Facta Universitatis, Series: Electronics and Energetics*, *29*(2), 177–191.
66. Ćirić, T., Dhuri, R., Marinković, Z., Pronić-Rančić, O., Marković, V., and Vietzorreck, L. (2018). Neural based lumped element model of capacitive RF MEMS switches. *Frequenz*, *72*(11–12), 539–546.
67. Kim, T., Marinkovic, Z., Markovic, V., Milijic, M., Pronic-Rancic, O., and Vietzorreck, L. (2013). Efficient modelling of an RF MEMS capacitive shunt switch with artificial neural networks. *2013 International Symposium on Electromagnetic Theory, EMTS 2013 — Proceedings*, 20–24 May, Hiroshima, Japan, pp. 550–553.
68. Marinkovi, Z., Aleksi, A., Tomislav, Ć., Proni, O., Markovi, V., and Vietzorreck, L. (2015). Inverse electro-mechanical ANN model of RF MEMS capacitive switches — Applicability evaluation. *L INT. SC. Conference on Information, Communication and Energy Systems and Technologies*, 24–26 June, Sofia, Bulgaria, Vol. 1, pp. 24–26.
69. Marinkovis, Z., Kim, T., Markovis, V., Milijis, M., Pronis-Ranšis, O., Siris, T., and Vietzorreck, L. (2016). Artificial neural network based design of RF MEMS capacitive shunt switches. *Applied Computational Electromagnetics Society (ACES)*, *31*(7), 756–764.
70. Joslin Percy. J., Kanthamani. S., Sethuraman. S., Roomi, M. M., and Uma Maheswari, P. (2022). Artificial neural network approach to model sidewall metallization of silicon-based bistable lateral RF MEMS switch for redundancy applications. *Silicon*, *14*, 9175–9185.

https://doi.org/10.1142/9789811279539_0006

Chapter 6

Management of COVID-19 Patients Through IoT-Based Smart Ambu Bag

Ankit Jain[*,†,§], Shailendra Singh[†,¶], Puspraj Singh Chauhan[†,‖], and Anita Shukla[‡,]**

[†]*Department of Electronics and Communication Engineering, Pranveer Singh Institute of Technology, Kanpur, Uttar Pradesh, India*

[‡]*Department of Basic Science & Humanities, Pranveer Singh Institute of Technology, Kanpur, Uttar Pradesh, India*

[§]*ankit2483jain@gmail.com*

[¶]*shailendras.nitj@gmail.com*

[‖]*puspraj.chauhan@gmail.com*

[**]*shukla.anita27@gmail.com*

Abstract

Coronavirus disease 2019 (COVID-19), which causes severe acute respiratory syndrome (SARS-CoV-2), is a persistent global epidemic. The coronavirus was discovered for the first time on March 11, 2020, following an outbreak in the Chinese city of Wuhan in December 2019. This virus causes respiratory issues, fever, coughing, and shortness of breath.

[*]Corresponding author.

In the worst circumstances, infection might result in mortality due to pneumonia, severe acute respiratory syndrome, and other conditions. The second COVID-19 wave in India showed how ill-prepared the country's healthcare systems were to deal with the sudden rise in the number of daily positive cases. Infected patients were left gasping for air during the second wave of COVID-19 as a result of the oxygen and bed shortage issues, particularly in India. This chapter proposes a wireless, Internet of Things (IoT)-based healthcare system featuring an automatically operated Ambu bag that might be utilized to save patients' lives. The system can use NodeMCU with IoT technology to continually monitor a patient's oxygen saturation, pulse rate, and body and room temperatures using a DS18B20 temperature sensor from any location in the world.

Keywords: Ambu bag, COVID-19, IoT, NodeMCU, DS18B20 temperature sensor, pulse sensor

6.1. Introduction

The coronavirus disease 2019 (COVID-19) pandemic has caused an unimaginable number of fatalities since its outbreak. Due to the rising number of patients suffering from COVID-19, the need for ventilators has grown daily. Due to the insurmountable gap between demand and supply, particularly in India, the available resources and facilities are unable to cover it. As a result, many perished during the second wave because they were not even able to reach the hospitals. The severe acute respiratory syndrome coronavirus 2 (SARS-CoV-2), or the 2019 novel coronavirus (2019-nCoV), as it is now known, originated in Wuhan City of China's Hubei Province, from where it spread to the rest of the world [1]. This illness can spread from asymptomatic people even before symptoms appear [2]. It transmits through large droplets released by sick patients during coughing and sneezing. Research has shown that there is no difference in viral load between symptomatic and asymptomatic people and that viral loads are greater in the nasal cavity than in the throat [3]. Patients with COVID-19 may also exhibit fever, breathlessness, a decrease in oxygen saturation, dry cough, nausea, vomiting, sore throat, headache, loss of taste, and physical pain [4]. If a patient has a high fever, a low oxygen saturation, and an irregular pulse rate, they are regarded as a critical patient. Shortness of breath and low oxygen saturation levels are symptoms of hypoxemia and hypoxia, respectively. The likelihood of the patient surviving is decreased when hypoxemia and a low pulse rate are combined. Patients occasionally fail to notice hypoxemia and an accelerating pulse, and as a result, they pass away without receiving the appropriate care. Therefore, it is crucial that COVID-19 patients' health conditions, particularly their body temperature, heart rate, and oxygen saturation (SpO_2), are regularly assessed. In such circumstances, IoT-based solutions may be advantageous to people for routine health examinations and ongoing patient health parameter monitoring [5, 6]. IoT technology has emerged as a crucial invention with multiple uses. It specifically refers to any arrangement of physical objects which acquire and swap data across wireless networks without the involvement of a person [7, 8]. Nearly every country was unable to handle the situation during the second wave. For COVID-19 patients, regular body temperature and pulse rate monitoring is required. The number of pulses per minute is known as the pulse rate or beat rate. The average pulse rate for healthy people is determined to be 60–100 beats per minute (bpm), while it is between 70 and 75 bpm for adult males and females, respectively. Typically, the pulse rates of women above the age of 12 are higher than those of men [9, 10]. The pulse rate is observed to be abnormal in COVID-19 individuals, necessitating the assistance of an emergency medical assistant. When it comes to body temperature, it differs from person to person and is influenced by a number of variables, including gender and environmental temperature. In a healthy adult, the body temperature ranges from 97.8°F (36.5°C) to 99°F (37.2°C) [10–12]. In addition to these causes, diseases such as influenza and low-temperature hypothermia can also affect the body temperature. Hence, it becomes crucial to regularly check the body temperature of a patient with COVID-19. Now, in addition to heart rate and body temperature, oxygen saturation is a vital factor for the treatment of COVID-19 patients. The usual range of the human body's SpO_2 is 95–100%. A COVID-19 patient needs emergency medical attention if their SpO_2 level is less than 95%. Silent hypoxia is caused by the SARS-COV-2 coronavirus and is defined as SpO_2 90% without shortness of breath. Using a pulse oximeter to measure SpO_2, silent hypoxia can be identified [11, 12]. A patient with COVID-19 has a very low oxygen saturation risk of death. Monitoring of early symptoms, including fever, cough, heart rate, and SpO_2 levels, becomes crucial for managing this contagious and fatal illness. To measure these values, a variety of tools and gadgets are on the market. For instance, most countries, people have easy access to fingertip pulse oximeters, which monitor SpO_2 and pulse rate [13]. The deluxe handheld pulse oximeter, which costs around USD 299 [14] and determines SpO_2 and heart rate, is also an option. A wrist-worn pulse oximeter that measures SpO_2 and heart rate is available over the counter. Similar to the above-listed devices, this one lacks features for measuring body temperature. The wrist-worn pulse oximeter, at USD 179, is somewhat pricey. Analog and digital thermometers are also sold on the market for measuring body temperature [15], although the majority of them are pricey [13]. The disadvantage of the currently available traditional equipment is that they are not based on the Internet of Things (IoT). Therefore, when patients are arriving in droves, it becomes challenging for a doctor to gather updates from every patient at once. Patients with significant symptoms of COVID-19 are in need of immediate monitoring. Patients can obtain COVID-19 treatment utilizing their mobile phones at home thanks to technology [16]. The pulse rate of an individual is influenced by their age, physical stature, cardiovascular health, and emotional stability [17]. Since oxygen saturation and pulse rate are inversely correlated, when the oxygen level of a patient drops, his pulse rate rises. All of these problems are resolved by using IoT technology. The IoT-based smart healthcare system functions as a real-time patient monitoring system, which can considerably benefit the current healthcare sector [18]. This is the rationale behind the rise in research interest in IoT-based smart healthcare equipment. In a study, the examined literature outlines the creation of smart healthcare monitoring systems in an IoT setting [19]. In this investigation, a pulse monitoring system equipped with temperature, SpO_2, and heart rate sensors and based on the Android operating system was developed. The measured data from a different investigation, in which the SpO_2 measurement sensor was not employed, is available online [20]. An IoT-based lung function monitoring system that did not take temperature, SpO_2, or pulse rate into account could be helpful for asthma patients [21]. Systems for monitoring heart rate using an Arduino, Android, or microcontroller have been suggested in Refs. [22, 23]. The system [24] is based on cloud computing and the Arduino Uno, for which only a hardware prototype was created. There aren't any statistics from actual testing, though. A mobile app-based heart rate monitoring system was established in Ref. [25]. The device detected the patient's pulse rate using a pulse rate sensor, and an Arduino was used to interpret the results. The measured data were delivered to the Android application. The experiment employed only a small number of sensors [25]. Furthermore, several authors have suggested various IoT-based wireless health-monitoring systems using creative working theories, with research advancing to great heights in this direction. However, IoT-based smart solutions for estimating various health parameters for COVID-19 patients, including temperature, heart rate, and SpO_2, have apparently not yet been introduced.

In this study, a brand-new IoT-based smart health monitoring system is developed and used for COVID-19 patients to evaluate their various vital signs, such as body temperature, pulse, and SpO_2. Through a mobile application, the device may monitor and show measured values for human body temperature, oxygen saturation level, and pulse rate, enabling the patient to seek medical assistance even if a doctor is not physically present. Doctors must enquire about the patient's medical issues when caring for COVID-19 patients (including oxygen saturation level and pulse rate). Patients with COVID-19 and other illnesses, including asthma and chronic obstructive pulmonary disease (COPD), can also benefit from this gadget. In 2005, COPD contributed to 5% of all fatalities globally and is expected to continue to be a global health issue in the future [26]. Because COVID-19-related mortality and morbidity have increased throughout the pandemic, there is a greater demand for ventilators. Hospitals have been burdened with a healthcare problem as a result of their inability to provide ventilator assistance to the ARDS patients who need it. A straightforward, manually operated bag valve mask (BVM) can also be used to ventilate patients in place of modern ventilators, which are computerized devices. A BVM must deliver between 500 and 800 ml of air to an average male adult patient's lung in order for it to be successful, while 400 ml might still be sufficient if additional oxygen is given [27]. An acceptable breathing rate is determined by squeezing the bag; for adults, this is 10–12 bpm, whereas for children or infants, this is 20–25 bpm [28]. During this pandemic, we have also been unable to locate any fully protective assisted ventilating equipment that may be employed prior to intubation, while patients are waiting for invasive mechanical ventilation, or after intubation [29]. The Indian Institute of Technology (IIT), Hyderabad, has recommended using an Ambu bag or BVM as an alternative to traditional ventilators for COVID-19 patients in life-threatening situations [30].

6.2. Difference in Manual Ventilation

6.2.1. *Self-Inflating Ventilation Bag vs. Free-Flow Inflating Bag*

In the case of manually operated ventilators with a BVM device, a good mask seal against the face is required so that the pressure required to inflate the lungs can be generated. At the same time, knowledge of how to use the ventilation device effectively in order to deliver a breath is also required. Here, the differences between ventilation techniques for self-inflating and free-flow ventilation bags are discussed in detail. Understanding these differences is a prerequisite and important for successful manual ventilation of a patient [31–37].

6.2.2. *Spontaneous vs. Manual Ventilation*

With the help of a manual switch, the breathing mechanics can be changed from spontaneous ventilation to manual ventilation.

6.2.2.1. *Spontaneous ventilation*

The diaphragm and the muscles between the ribs (intercostal muscles) contract during inhaling in spontaneous breathing (Figure 6.1). The chest cavity is made larger by the movement of the ribs in an upward and outward direction caused by the contraction of the intercostal muscles, and it is also made larger by the movement of the diaphragm in a downward direction. Since pressure and volume are inversely related, the inner pressure of a container decreases as its volume rises. The use of a syringe as an example made the entire procedure easier to comprehend. When a syringe's plunger is pulled back, the interior chamber expands, the interior pressure decreases as a result of the lower volume, and fluid is drawn into

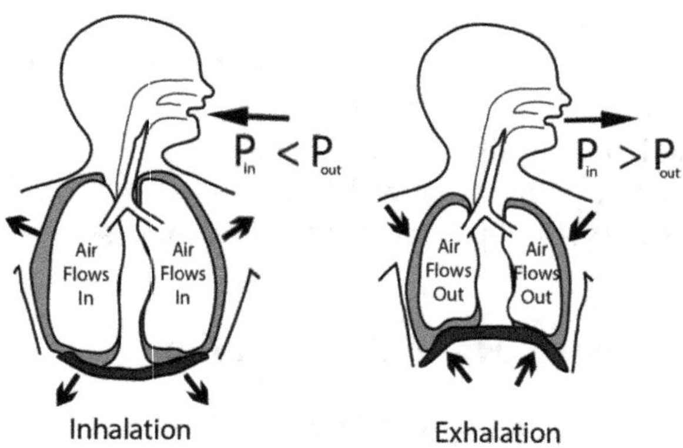

(a) (b)

Figure 6.1. Change in Air Pressure Inside the Thoracic Cavity in Spontaneous Breathing During (a) Inhalation and (b) Exhalation

the chamber. In contrast to this contraction process, chest expansion causes the intrathoracic pressure to fall below atmospheric pressure, which allows air to enter the lungs if the airway is open, until the two pressures are equal. The rib cage is compressed during exhalation by the natural elastic rebound of our chest wall. As the volume lowers and the pressure rises, the diaphragm relaxes and the chest cavity shrinks. Let's use the instance of inserting the syringe plunger inward to better comprehend the same. The remaining air is forced out through an unobstructed airway as the intrathoracic pressure climbs above the atmospheric pressure (minus some oxygen and CO_2 content) [38–43].

6.2.2.2. *Manual ventilation*

Because a patient is mechanically ventilated, gas no longer passively flows into the lungs under the effect of negative intrathoracic pressure, and a positive pressure is necessary to inflate the lungs. As an alternative, we must use a positive pressure to expand the lungs. Manual breathing must lift the chest wall, force the contents of the abdomen and diaphragm downward, and overcome the first alveolar surface tension in order to expand the lungs. In individuals with bronchospasm or pneumonia, it is more difficult to overcome lower lung compliance to produce an adequate tidal volume because their lungs are stiffer. Additionally, a supine patient's higher diaphragm works against them when the bag is squeezed; as a result, more pressure is needed to move the diaphragms — and the abdominal contents beneath them — down and out of the way in a supine patient than in an upright one. Ventilation is also hampered by the patient's overweight or obesity, which adds to the weight of the belly wall and its contents. Manual ventilation must account for the fact that the diaphragm rests higher in the chest, the weight of the abdominal contents and chest wall, and the compliance of the lungs (Figure 6.2).

6.2.2.3. *Bag-valve-mask device*

If the seal is improper, ventilation will not occur. When ventilating a patient, it's crucial to make sure the patient is getting good ventilation. Understanding the variations between the two types of BVM devices in use is the first step in doing this. Figure 6.3 depicts a BVM device, which consists of a flexible bag connected to either an endotracheal tube or ventilation mask by a pressure control valve [44–50].

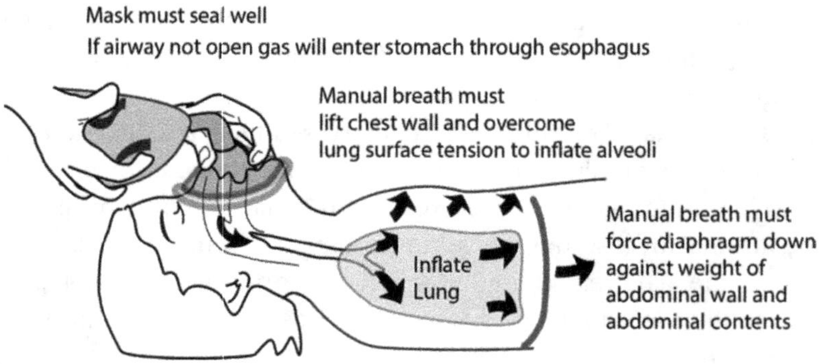

Figure 6.2. Effect of Lung Compliance in Manual Ventilation

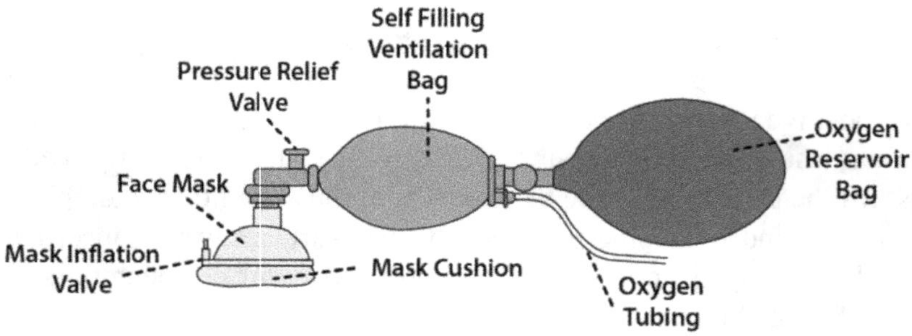

Figure 6.3. Common Parts of BVM Devices

The air is pumped into the lungs through a mask or artificial airway by compressing the bag, which opens valves. On the other hand, by opening the bag, the pressure inside the device is reduced. The patient in this situation exhales passively through a one-way valve. The clinician can usually compensate for the compliance of the patient's lungs (easy inflation) by adjusting the pop-off valve to release at either higher or lower pressures. Additionally, it aids in preventing potential barotrauma by not overpressurizing the lungs.

There are two different kinds of ventilation bags.

6.2.2.4. *Self-inflating ventilation bags*

When we stop squeezing the self-inflating bag, it refills itself. The Ambu bag is the generic name for these self-inflating resuscitation bags. One can expand their lungs by squeezing the bag. If an oxygen source is attached, releasing the pressure allows the bag to refill with both air and oxygen. Without additional oxygen, a self-inflating bag can give an oxygen concentration of 21%, but most sick or injured individuals need more oxygen than that. The bag has the ability to maintain O_2 levels of 40–60% when oxygen is supplied at a rate of 10–12 liters per minute. By adding a reservoir bag and running O_2 at 12–15 liters per minute, the concentration can be increased by up to 100%, provided the reservoir is given sufficient time to fill. A flow inflation bag, on the other hand, is continuously filled with 100% oxygen.

Using a self-inflating bag requires that the bag be squeezed in a way that maximizes oxygen concentration. When the bag is empty and abruptly allowed to reload, it usually does so with ambient air rather than oxygen, whose inflow duration is constrained. It is best to release pressure on your hand gradually over the course of 3–4 s to allow the bag to fill. Breakage of the mask seal should be prevented during the refilling procedure to prevent the bag from filling with ambient air instead of oxygen. Self-inflating bags cannot be used for successful "blow by" unless they are squeezed to push oxygen toward the patient.

6.2.2.5. *Free-flow inflating ventilation bags*

An empty free-flow inflation ventilation bag resembles a deflated balloon, while a self-inflating bag, when not in use, looks like a softly inflated football. A flow inflating bag must have an uninterrupted fresh oxygen flow. It won't automatically refill after the oxygen source is gone or disconnected. In order to prevent bag deflation that prevents the patient from breathing, it is important to maintain a good seal between the ventilation mask and the face. Even flow-inflation or inflow-dependent bags are often employed on anesthetic machines and in other ICU-like settings because they enable finer control of tidal volume, better ventilation monitoring, and higher FiO_2. However, they can be more challenging to utilize. The flow-inflating bag's softness makes it easy to perceive changes in resistance and lung compliance.

While ventilating a neonate with a 500 ml bag, extremely fine tidal volume control is achievable, even if it is less than 50 ml. With each breath, the bag partially deflates, which is easily felt and seen before it reinflates by the flow of gas during spontaneous ventilation. The degree of deflation can be used to estimate the tidal volume. If the seal is broken, it is instantly apparent since the bag collapses. Contrary to self-filling bags, which may give you a false sense of security, as they are usually full even when the filling of the lungs is not proper, free-flow inflating bags are filled all the time.

In contrast to self-filling bags, flow inflation bags cannot deliver positive pressure without a tight seal of the mask against the face. The flow inflation bag deflates like a huge balloon when there is a poor seal. Emergency ventilation in flow inflation bags is typically delivered by self-inflating bags because flow inflation bags depend on an oxygen source and require more training. It doesn't matter what kind of ventilation gear is being utilized; what matters is making sure that the ventilation being done is sufficient. Chest rise, mask fogging, breath sounds, and end-tidal CO_2 should always be periodically checked. One should keep a close eye on things when performing ventilation because changes in mask fit, head position, lung compliance, patient position, and your own hand fatigue might affect ventilation [51–56].

6.3. History of Ambu Bag

Following their early work on a suction pump, the Danish anesthetist Henning Ruben and the German engineer Dr. Holder Hesse created the idea of the BVM in 1953. Their resuscitator, later known as an air-shields manual breathing unit bag, or Ambu bag, was put into production starting in 1956 by their own business, also known as Ambu. As a result, all bag valve masks are now referred to together under the trademarked name Ambu.

6.3.1. *How to Use AMBU Bags*

A child or adult who is not breathing or is having difficulty breathing on his or her own is helped with a particular kind of BVM. A procedure known as "bagging" is followed, in which the bag is held with the hand to continuously give oxygen to a person's lungs. The components of a BVM are a face mask and a non-rebreathing valve. A source of oxygen is connected to the bag's other end (or air). Manual holding keeps the mask in place. Before an intubation can be performed, the bag is squeezed to give the patient breathing through their mouth and nose (Figure 6.4).

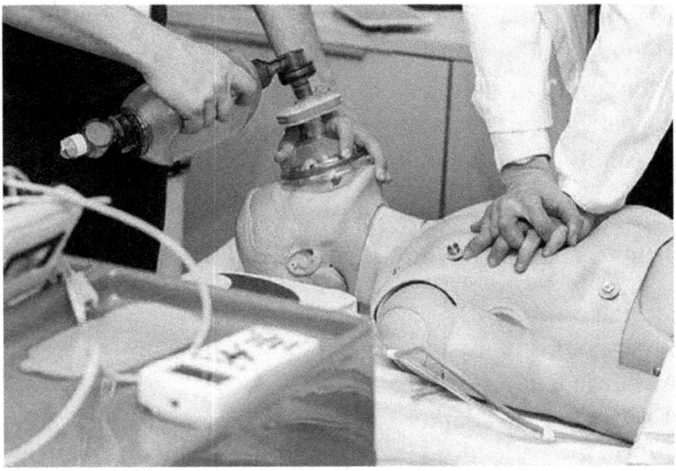

Figure 6.4. Way of Using AMBU Bag

Table 6.1. Required Size of Parts of Ambu Bag for Different Age Groups

	Adult	Child	Infant
Sizes	1,600 ml	500 ml	500 ml
Mask	Size 4	Size 2	Size 1
Reservoir	1,600 ml bag	500 ml bag	500 ml bag
Tubing	Having suitable connectors at both ends for easy and safe connections		

6.3.2. *Components of Ambu Bag*

The BVM consists of a pliable air chamber. The chamber, which has a diameter comparable to an American football, is attached to the face mask through a shutter valve. Air is forced into the patient's lungs by the machine when the face mask is securely fastened and the air chamber, or "bag," is squeezed. The patient's lungs inflate with air through the one-way valve, and the bag self-inflates when it is released, drawing in ambient air or low-pressure oxygen flow from a regulated cylinder. The size requirements for various Ambu bag components for patients in various age groups are listed in Table 6.1.

➢ **Indications:**

- a severe asthma attack, respiratory failure, or drug overdose;
- brain injury or imminent drowning;
- if a patient starts breathing on his own but it is labored or shallow, an Ambu bag may be used to make the breathing more effective.

6.3.3. *How to Use an Ambu Bag Manual Resuscitation BVM*

➢ **Assemble equipment:**

- If required, supply oxygen with proper tubing.
- As needed, adjust the mask's size.
- If necessary, an adaptor for a tracheostomy tube is used.
- Inform the patient of the operation if it is not an emergency.
- If an oxygen order is placed, secure the oxygen tubing to the resuscitation bag.
- Attach a tracheostomy tube to a resuscitation bag.
- If the patient is able to breathe on his or her own, match the patient's breaths with the manual breaths. Squeeze the resuscitation bag to provide air as the patient starts to breathe.
- Just enough bag pressure should cause the chest to elevate.
- The resuscitation bag should be squeezed at a regular interval to administer the recommended number of breaths per minute if the patient is unable to breathe on his or her own.
- Disconnect the tracheostomy tube from the resuscitation bag.
- Thoroughly wash your hands.

➢ **Instructions:**

- If the patient isn't breathing, one should first try to find a technique to get him to do so. If he continues to be non-responsive, use the Ambu bag.
- Before utilizing the device, one should make sure the patient's airways are clear of mucus.
- The patient must be lying flat on his back with his nose pointed upward. To assist in keeping this position, one can place a roll of a small towel beneath his shoulders.
- The face mask attached to the bag adaptor, the oxygen tubing linked to the regulator on the tank, and the oxygen tubing linked

to the bag must be verified. Hold the mask toward the patient's face while squeezing the bag firmly with your thumb to apply pressure.

- The mask should be worn tightly over the patient's mouth and nose. Make sure the chest of the patient rises when the Ambu bag is pressed, as it would if he were breathing normally.
- Adult patients should squeeze the Ambu bag once every 5–6 s, compared to pediatric patients, who should squeeze it every 2–3 s. If the patient is still having trouble breathing, keep bagging him up until either aid arrives or he starts breathing properly.
- If the chest still doesn't rise, try moving the patient's head or the mask before adding a little more pressure to the Ambu bag. The patient should be revived if the condition continues and the patient's chest still does not rise. If this is the case, there must be an obstruction in the patient's airway.

➢ **Risks:**

- Hyperventilation can be brought on by squeezing the bag too quickly, and it can also result in respiratory alkalosis. By compressing the bag at the proper rate, this situation could be mitigated or avoided.
- Cardiac danger can be reduced, and the trapped air in the stomach can be evacuated by inserting a nasal gastric tube.

6.4. Background of the Proposed System

COVID-19 continues to spread worldwide. SARS-CoV-2 causes COVID-19. Wuhan, China, discovered 2019-nCoV in December 2019. The WHO declared a pandemic on March 11, 2020 [57–62]. Symptoms of COVID-19 may appear 2 to 14 days after exposure. The incubation period is the period of time following exposure but prior to the onset of symptoms. Even before exhibiting symptoms, COVID-19 can still spread (pre-symptomatic transmission). Respiratory issues, fever, coughing, and shortness of breath are a few common signs and symptoms. In severe situations, infection might result in pneumonia, severe acute respiratory syndrome, and sometimes even death. As a result of their direct involvement in the diagnosis, treatment, and care of COVID-19 patients, healthcare workers become the subject of primary concern when treating such infectious diseases. As a result, they are constantly susceptible to infection and have a very high risk of contracting the SARS-CoV2 virus. Many of the difficulties that hospitals and their workers face can be solved without the use of technology [63–70].

The new strain of the virus in the second wave was found to be more powerful and infectious, which left younger and healthier people in critical conditions as compared to the first wave of COVID-19, leaving people gasping for breath. There was shortage of enough doctors or healthcare workers to handle the surge of COVID-19 cases; this also had repercussions on the patients who were suffering from other critical illnesses as they faced undue delays in getting treatment.

The second COVID-19 wave in India exposed how ill-prepared the country's healthcare systems were to deal with the sudden rise in daily cases. The unexpected increase in cases collapsed the country's healthcare system. There was a shortage of hospital beds, oxygen, and medications, particularly in India. The shortage of oxygen supplies was also definitely a drawback. During the first wave, there was a shortage of PPE kits, and in the second wave, more deaths were recorded due to the shortage of oxygen supplies and the hunt for medications by hospitals and patients' relatives. Cases were recorded in mass every day, and in search of oxygen, people rushed to the hospitals, where they faced the problem of the unavailability of ICU beds. People were struggling for breath. Because of limited resources and facilities to meet the emergency, a lot of people died. The major reason was that due to the unavailability of hospital beds, people were bound to get treated at home, and when the situation worsened, they had to rush to the hospitals. However, in most of the cases, they could not make it because of the scarcity of oxygen cylinders [71–76].

6.4.1. *Features of the Proposed System*

In the current work, we have developed an IOT-based wireless healthcare remote sensing device/system which could provide real-time online information about the patients under observation either in home isolation or in the hospital. Doctors at the hospital and attendants at home can monitor their COVID-19 patients remotely. The system is smart enough to continuously sense the patient's oxygen level, body temperature, and pulse rate. All the readings taken can be displayed on an LCD and monitored remotely using IoT. The major advantage of the system is that monitoring of all parameters can be done from anywhere in the world because of its IoT technology. The system not only measures the oxygen level of the patient but also provides support to the patient in home isolation. The system is equipped with an Ambu bag so that the patient struggling with a lower oxygen level can buy enough time to reach the hospital safely. This Ambu bag is operated automatically by a DC motor. The frequency of inhalation can also be adjusted in the system [77–84].

Numerous features built into the system increase its usefulness. Since COVID-19 is a highly infectious disease, the patient is being monitored remotely so that hospital staff and home caregivers have at least some interactions with the patient and have a lower risk of contracting the illness. Through the use of a NodeMCU Wi-Fi module, this system is connected to the internet and is capable of displaying real-time data that can be viewed and accessed via the internet using IOT technology from any location around the globe. The following is a list of the system's features:

- As the disease is highly infectious, remote monitoring and management become a necessity. If the patient is not critical, they can also monitor parameters such as oxygen level, pulse value, and body temperature on their own.
- With the help of IoT technology, the doctors or attendants at hospitals and caregivers at homes can also monitor the above-mentioned parameters remotely since frequent monitoring of oxygen level and body temperature is required in COVID-19 patients.
- If the patient is in home isolation and suffocates suddenly due to a drop in oxygen level, then there is provision for the Ambu bag that can be used to save the patient's life as they are rushed to the hospital. The Ambu bag is operated automatically, not manually; hence, it is user friendly.
- The frequency of inhalation in the Ambu bag is also adjustable according to the requirements of the patient's age group.

A block diagram of the proposed system is shown in Figure 6.5.

The main components which are used in this system are listed as follows:

- Ambu bag assembly
- NodeMCU
- Pulse oximeter Sensor MAX30100

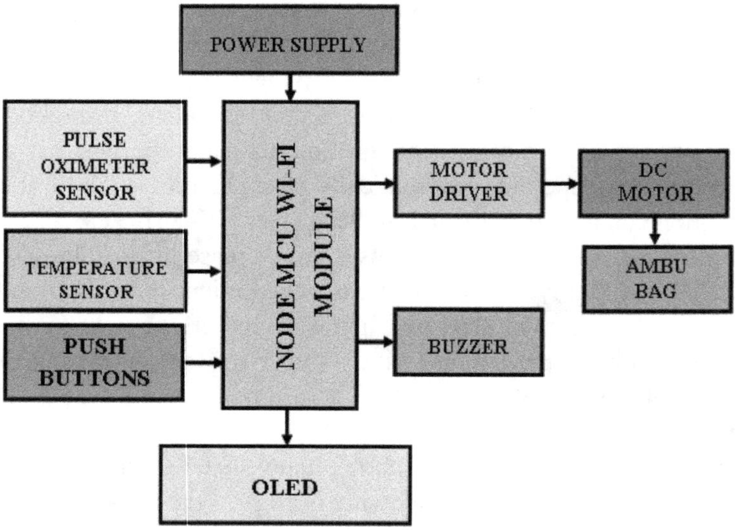

Figure 6.5. Block Diagram of the Proposed System

- Temperature sensor DS18B20
- OLED
- DC motor 12V/1A
- Buzzer

The descriptions of some of the major components of the system are as follows.

6.4.1.1. *NodeMCU ESP8266 pin out*

Although NodeMCU ESP8266-based boards originally shipped with the LUA scripting language for programming, since the Arduino IDE is the most popular tool used by electronics hobbyists to program development boards, the ESP8266 package must be loaded into the Arduino IDE for ESP-based boards. We needed GPIO pin mapping (Figure 6.6) for the NodeMCU pin out, which is marked on the board as D0–D8 but is used in code since the NodeMCU ESP8266 boards were initially built for a different architecture but later integrated with the Arduino IDE.

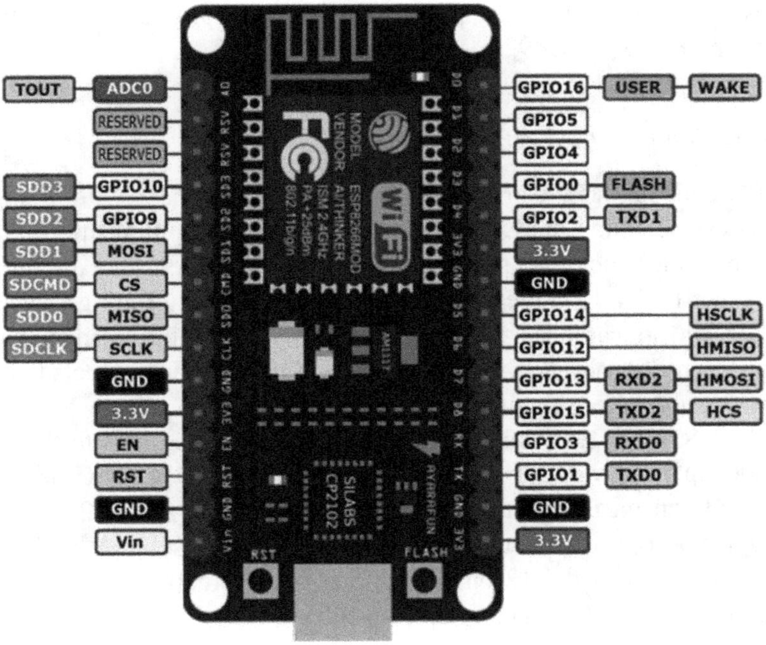

Figure 6.6. Pin Mapping of NodeMCU

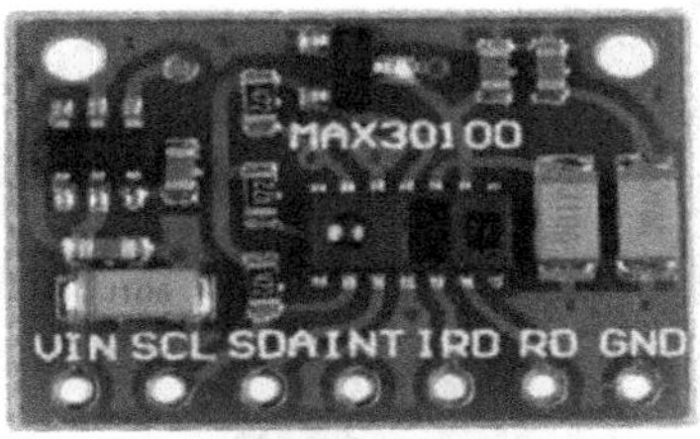

Figure 6.7. MAX30100 Pulse Oximeter Sensor

6.4.1.2. *MAX30100 pulse oximeter sensor*

Figure 6.7 shows the MAX30100 sensor module. It is a simple module that interacts with the microcontroller via an I2C interface. It gives the linked microcontroller information about the SpO_2 and pulse. Basically, this sensor is employed to measure oxygen saturation. As a result, this module can be used to non-invasively monitor the blood's oxygen saturation level and pulse rate. It makes use of optical components and photodetectors, with IR LEDs modulating the LED pulses. The LED current ranges from 0 to 50 mA. The power supply range for this pulse oximeter sensor is 1.8–5.5 V [85–92].

6.4.1.3. *Temperature sensor DS18B20*

Dallas Semiconductor Corp. makes a temperature sensor with a single-wire interface called the DS18B20. For two-way communication with a microcontroller, the novel single-wire interface just needs one digital pin. The sensor typically comes in two forms. A TO-92 packaging transistor has the same appearance as a standard transistor. The other is in the shape of a waterproof probe, which is more suited for measuring objects that are far away, submerged, or underground [93–98].

The DS18B20 temperature sensor shown in Figure 6.8 is fairly accurate and operates without any additional parts. It has a 0.5°C accuracy range and can measure temperatures between −55°C and +125°C. User configuration options for the temperature sensor's resolution include 9, 10, 11, or 12, with the 12-bit resolution (i.e., a precision of 0.0625°C) being the default setting at startup [99–103].

6.4.1.4. *OLED display*

The 0.96-inch SSD1306 mode with a 12,864-pixel OLED display utilized for this project is depicted in Figure 6.9.

Because the OLED display doesn't need a backlight, there is excellent contrast in low light. Additionally, compared to other displays, it

Figure 6.8. MAX30100 Pulse Oximeter Sensor

Figure 6.9. OLED Display

uses less electricity [104–112]. There are four pins on the display that are utilized. The I2C protocol is used for communication. Figure 6.10 displays the proposed system's circuit diagram. The circuit diagram of the system was built with a NodeMCU board. Since NodeMCU cannot provide the required amount of current to operate the DC motor, an L298 motor driver module is used for operating the Ambu bag with the help of a 12 V DC motor. This circuitry has three push buttons: the first is used for any panic/emergency situation, the second is for adult mode, and the third is for child mode. If any panic/emergency occurs, then the patient will press the panic button, which causes the buzzer to beep, and a notification/message will be sent to the concerned doctor or person using the Blynk app.

The second and third push buttons are used to set up the frequency of the Ambu bag in adult and child modes. Through a pulse width modulation (PWM) signal, we can change the voltage level of the DC motor and consequently operate the Ambu bag at different frequencies according to the different age groups. We can monitor the pulse rate and SpO_2 with the help of the MLX30100 pulse sensor module. The temperature of the patient can be measured using the DS18B20 temperature sensor. All the measured values will be displayed on the OLED. The MLX30100 pulse sensor module and OLED are connected with the I2C protocol using the SDA and SCL pins of the NodeMCU's D2 and D1 pins, respectively. All the sensor data will be transmitted over the Blynk server using IoT, which can be accessed from anywhere in the world.

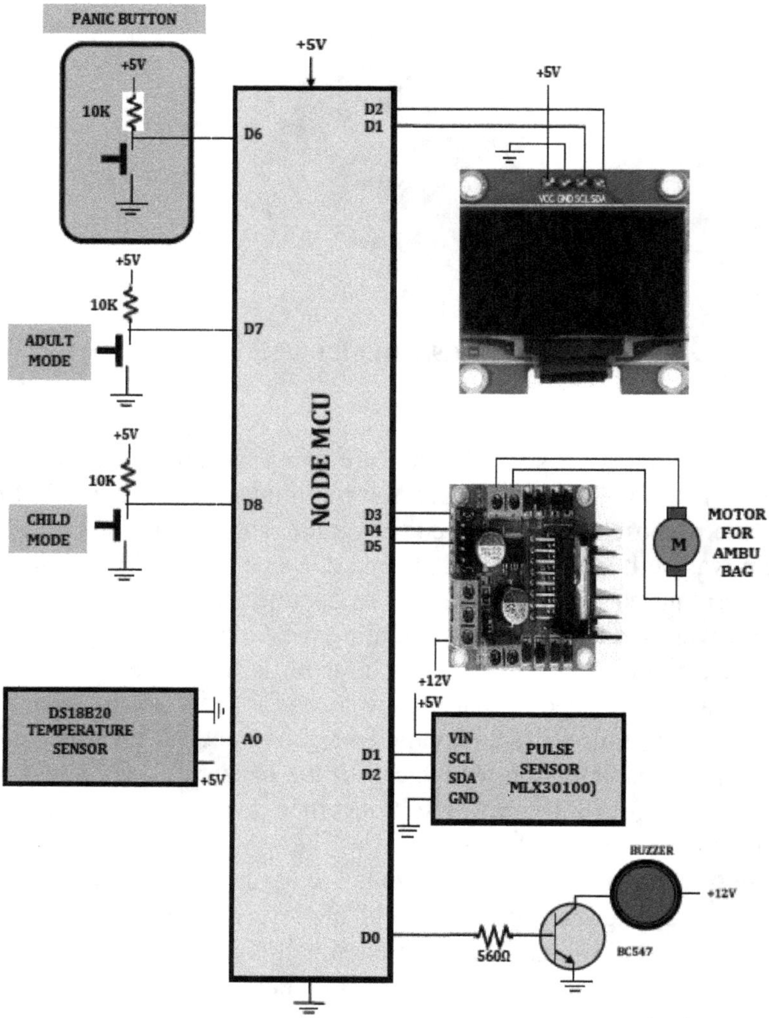

Figure 6.10. Circuit Diagram of the Proposed System

6.4.2. *Setting up Blynk Application*

Download the Blynk application from the Play Store or App Store for Android or iOS, respectively. Enter your email address and a password to register with the Blynk IoT cloud, and then, click on the "New Project" button to give the project a name. For instance, enter "IoT pulse oximeter" as the name, choose the NodeMCU board, and then pick "Wi-Fi" as the connection type. Lastly, press the "Create" button, as displayed in Figure 6.11(a).

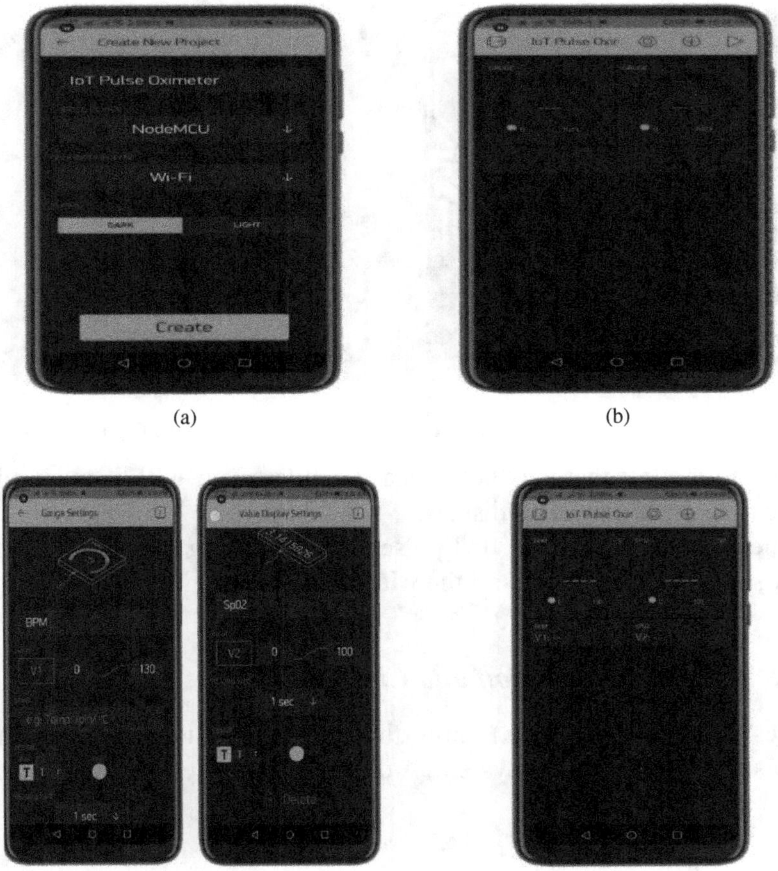

Figure 6.11. (a) Setting up the Blynk app. (b) and (c) Setting up Gauge and Display Widget. (d) Final User Interface of Blynk Server

The registered email address will receive the Blynk authentication token. This token will be necessary later when programming. To add two gauges and two value-display widgets, tap the plus (+) icon on the main screen, as illustrated in Figure 6.11(b).

As indicated in Figure 6.11(c), one gauge will show the BPM values, while the other gauge will show the oxygen level (SpO_2).

Choose the virtual pin V1 for BPM, and enter a value between 0 and 130.

Choose the virtual pin V2 for the oxygen level, and enter a value between 0 and 100.

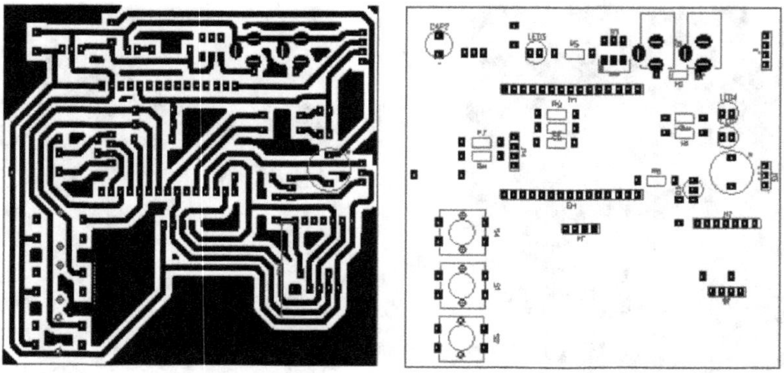

(a) (b)

Figure 6.12. PCB Prepared Using Dip Trace: (a) Bottom View, (b) Top View

Due to the fact that they both behave similarly, we will likewise add identical values to the value display.

The final interface of the "IoT pulse oximeter" using ESP8266 and the Blynk server after setting up all the widgets is shown in Figure 6.11(d).

6.4.3. *PCB Implementation and Circuit*

Figures 6.12(a) and 6.12(b) show the bottom and top views of the PCB used in the proposed system. A dip trace was used to prepare this PCB.

6.5. Results and Discussion

The patient management issue with COVID-19 is addressed in this study in an electronic and environmentally responsible manner. The proposed system was put to the test in real-time following its development, and the outcomes were good and appropriate. As is common knowledge, it is essential to often check the temperature and pulse rate of COVID-19 patients, and the system is able to do so precisely utilizing IoT and the Blynk server. The system has three push buttons: the first for a panic situation, the second for adult mode, and the third for child mode. By pressing the panic button, a patient can alert the doctor or concerned person of any panic/emergency situation using IoT. The adult and child mode push buttons are used to set the frequency with which the Ambu bag should be pressed according to the patient's age group. The Ambu bag is operated

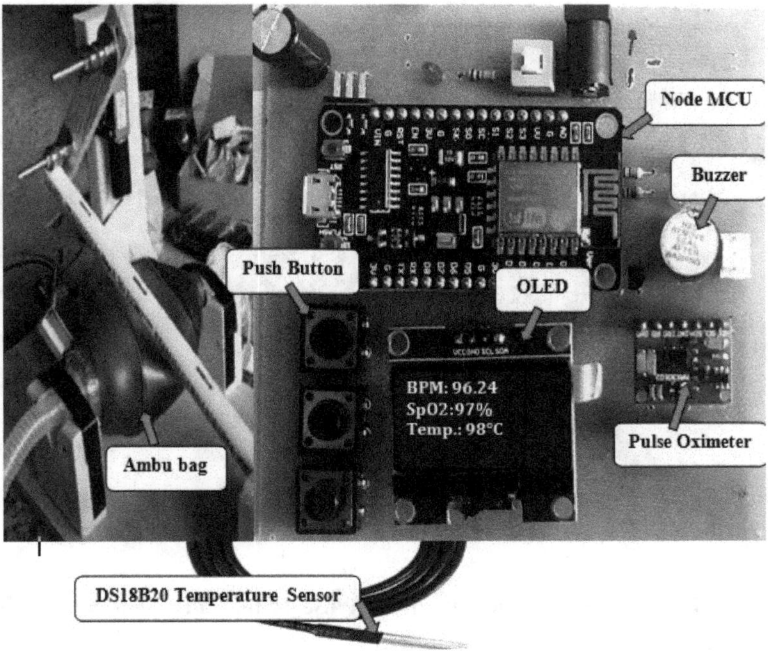

Figure 6.13. Original Picture of Developed Proposed System

by a DC motor. By using a PWM signal, we can adjust the frequency of pressing the Ambu bag for adult/child mode [113–118].

Figure 6.13 shows the original picture of the proposed system. The user would be able to see the heart rate, SpO_2, and temperature of the patient in the displayed results. This system transfers the data through IoT to the Blynk server, which is one of its crucial parts (Figure 6.14). The system was found to work successfully, and all its sensors also worked satisfactorily. Five individuals, ranging in age from 23 to 56, used the gadget during the testing phase. For each feature included in the system, the exact values were obtained. Table 6.2 lists the SpO_2 level, pulse rate, and temperature measurements for five separate users.

The majority of patients had an SpO_2 level of 97, which is near the average values (Table 6.2). The measured values for the various participants showed that they had comparable pulse rates. For many test subjects, different physiological data were measured. Comparing these measured values with those of other commercially available equipment, they were all extremely accurate [119, 120]. If the patient is in home

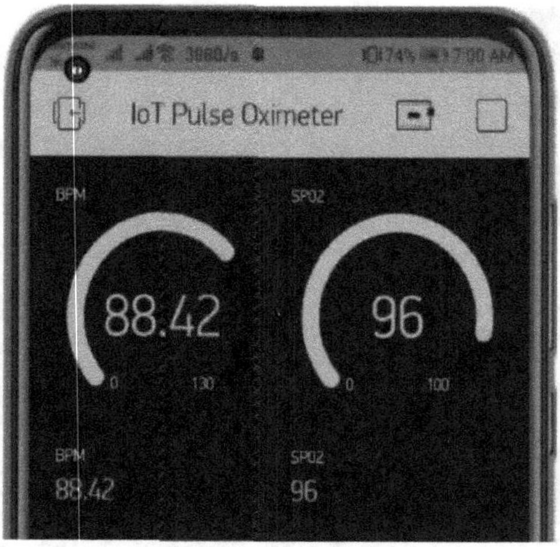

Figure 6.14. Original Picture of Displayed Sensor Data over Blynk Server

Table 6.2. Measured Values of SpO_2, Pulse Rate, and Temperature for Five Different Patients/Users

Patient/User	Age	SpO_2 (%)	Pulse (bpm)	Temperature (°C)
Patient 1	25	97	75	37
Patient 2	32	97	73	36
Patient 3	34	93	70	40
Patient 4	56	97	74	37
Patient 5	23	97	75	40

isolation and suffocates suddenly due to a drop in oxygen level, then there is provision for the Ambu bag that can be used to save the patient's life as he or she is rushed to the hospital. The Ambu bag is operated automatically, not manually; hence, it is user friendly. The frequency of inhalation in the Ambu bag is also adjustable according to the requirements of the patient's age group [121–125].

6.6. Conclusions

The COVID-19 pandemic has resulted in a global health catastrophe, and thousands of people are dying every day. We can minimize the fatality rate by administering proper treatment and making use of technology. During the second wave of COVID-19, the major problem was the inadequate supply of oxygen. India faced severe consequences as the young population bore the brunt of the wave. Hence, it becomes necessary to come up with future control strategies. In this work, we proposed a cost-effective and user-friendly IOT-based life-saving device equipped with an Ambu bag for COVID-19 patients. The proposed system offers an excellent solution to the problem of breathlessness faced by COVID-19 patients, who were left gasping for breath due to the scarcity of oxygen cylinders during the pandemic. The Ambu bag which works as a life-saving device when a patient suffers from a sudden drop in oxygen levels and is being rushed to the hospital if oxygen cylinder is not available. The system is capable of measuring/monitoring pulse rate, SpO_2 level, and temperature, which is very important for COVID-19 patients. The system was tested with five patients, and the measured values of all parameters were found to be in good agreement with the values measured using available devices on the market. Future IoT-based sophisticated features can be added to the system to increase its utility. To sum up, this approach is crucial for prolonging the lives of patients with COVID-19 across the world. Future upgrades could include the capability of tracking additional physiological functions of the human body.

References

1. Wang, C., Horby, P. W., Hayden, F. G., and Gao, G. F. (2020). A novel coronavirus outbreak of global health concern. *The Lancet*, *395*(10223), 470–473.
2. Rothe, C., Schunk, M., Sothmann, P., Bretzel, G., Froeschl, G., Wallrauch, C., Zimmer, T., Thiel, V., Janke, C., Guggemos, W., Seilmaier, M., Drosten, C., Vollmar, P., Zwirglmaier, K., Zange, S., Wölfel, R., and Hoelscher, M. (2020). Transmission of 2019-nCoV infection from an asymptomatic contact in Germany. *New England Journal of Medicine*, *382*(10), 970–971.
3. Zou, L., Ruan, F., Huang, M., *et al.* (2020). SARS-CoV-2 viral load in upper respiratory specimens of infected patients. *New England Journal of Medicine*. Doi: 10.1056/NEJMc2001737.
4. El-Rashid, N., El-Sappagh, S., Islam, S. M. R., El-Bakry, H. M., and Abdelrazek, S. (2020). End-to-end deep learning framework for coronavirus (COVID-19) detection and monitoring. *Electronics*, *9*(1439), 1–25.
5. Hongru, D. and Goyea, T. (2020). *Novel Coronavirus (COVID-19) Cases*. Johns Hopkins University: Baltimore, MD.
6. Hypoxemia: Symptoms, Causes, Treatments. Cleveland Clinic. 2020. https://my.clevelandclinic.org/health/diseases/17727-hypoxemia.

7. Pourhomayoun, M., Alshurafa, N., Dabiri, F., Ardestani, E., Samiee, A., Ghasemzadeh, H., and Sarrafzadeh, M. (2016). Why do we need a remote human-health monitoring system? A study on predictive analytics for heart failure patients. In *11th International Conference on Body Area Networks*, Turin, Italy, pp. 171–172.
8. Serpanos, D. and Wolf, M. (2019). IoT system architectures. In *Internet-of-Things (IoT) Systems Architectures, Algorithms, Methodologies*. Springer International Publishing: Switzerland.
9. Vital signs (body temperature, pulse rate, respiration rate, blood pressure). Johns Hopkins Medicine. https://www.hopkinsmedicine.org/health/conditions-and-diseases/vitalsigns-body-temperature-pulse-rate-respiration-rate-blood pressure.
10. Pulse Oximetry Training Manual. World Health Organization. 2011. https://www.who.int/patientsafety/safesurgery/pulse_oximetry/who_ps_pulse_oxymetry_training_manual_en.pdf?ua=1.
11. Rizzo, A. Silent hypoxia and its role in COVID-19 detection. https://www.news-medical.net/news/202006003/silenthypoxia-anditsrole-in-Covid-19-detection.aspx.
12. Al Harun, M. A., Hossain, M. M., Bari, M. A., Rubel, N. A. S., Karim, M. E., Siddiquee, N., Hossain, M. D., Sultana, F., Taous, A., Islam, A. K. M. M., Haque, A. H. M. A., Haque, M. M. U., Murshedm, K. M. M., Atiqullah, S., Hoque, A. E., and Abdullah, M. (2020). Pulse oximetry is essential in home management of elderly COVID-19 patients. *Bangladesh Journal of Otorhinolaryngology*, *26*(1), 55–67.
13. PulseOximeter Price in Bangladesh. BMA Bazar. https://bmabazar.com/product-tag/pluse-oximeter-price-inbangladesh/?fbclid=IwAR0j6Qt5CLT46gIPo3p_upeRqWFASL80A8Z01uI0FwnIYVHik1q8NRxXo.
14. Deluxe handheld pulse oximeter. Concord Health Supply. https://www.concordhealthsupply.com/Deluxe-Handheld-Pulse-Oximeter-p/cci-300m.htm?fbc.
15. Wrist worn pulse oximeter. Concord Health Supply. https://www.concordhealthsupply.com/Wrist-Oximeter-p/75006.html.
16. Evans, M. (2020). COVID-19: How to Treat Coronavirus at Home. Patient, UK.https://patient.info/news-and-features/covid-19-how-to-treat-coronavirus-at-home.
17. Arulananth, T. S. and Shilpa, B. (2017). "Fingertip based heart beat monitoring system using embedded systems. In *Proceedings International conference of Electronics, Communication and Aerospace Technology (ICECA)*, Coimbatore, India, pp. 227–230.
18. Khan, M. M. (2020). IoT based smart healthcare services for rural unprivileged people in Bangladesh: Current situation and challenges. In *1st International Electronic Conference on Applied Science*, MDPI, Switzerland, pp. 1–6.

19. Islam, M. M., Rahmanand, A., and Islam, M. R. (2020). Development of smart healthcare monitoring system in IoT environment. *SN Computer Science*, *1*(3), 1–11.
20. Srividya, B. and Satyanarayana, V. (2018). Personal lung function monitoring system for asthma patients using internet of things (IOT). *International Journal of Research in Electronics and Computer Engineering*, *6*, 966–971.
21. Srinivasan, P., Ayub Khan, A., Prabu, T., Manoj, M., Ranjan, M., and Karthik, M. (2020). Heart beat sensor using fingertip through Arduino. *Journal of Critical Reviews*, *7*(7), 1058–1060.
22. Pierleoni, P., Pernini, L., Belli, A., and Palma, L. (2014). An android based heart monitoring system for the elderly and for patients with heart disease. *International Journal of Telemedicine and Applications*, *2014*. Article ID 625156.
23. Prasath, J. S. (2013). Wireless monitoring of heart rate using microcontroller. *International Journal of Advanced Research in Computer Science and Electronics Engineering (IJARCSEE)*, *2*(2), 214–219.
24. Tamilselvi, V., Sribalaji, S., Vigneshwaran, P., Vinu, P., and GeethaRamani, J. (2020). IoT based health monitoring system. In *2020 6th International Conference on Advanced Computing and Communication Systems (ICACCS)*, Coimbatore, India.
25. Reza, T., Shoilee, S. B. A., Akhand, S. M., and Khan, M. M. (2017). Development of android based pulse monitoring system. In *2017 Second International Conference on Electrical, Computer and Communication Technologies (ICECCT)*, Coimbatore, India.
26. Sutradhar, I., Das, G. R., Hasan, M., Wazib, A., and Sarker, M. (2019). Prevalence and risk factors of chronic obstructive pulmonary disease in Bangladesh: A systematic review. *Cureus*, *11*(1), 3970–3978.
27. Limmer, D. and O'Keefe, M. F. (2005). *Emergency Care*, E. T. Dickinson (ed.) (10th edn.). Pearson, Prentice Hall: Upper Saddle River, New Jersey.
28. https://www.ahajournals.org/doi/full/10.1161/circ.102.suppl_1.1-22 (accessed 29 March 2022).
29. JumlongkulArnon. (2021). Automated AMBU ventilator with negative pressure headbox and transporting capsule for COVID-19 patient transfer. *Frontiers in Robotics and AI*, *7*.
30. https://mumbaimirror.indiatimes.com/coronavirus/news/bjp-mlc-prasad-lad-demands-the-opening-of-parels-gandhi-hospital/articleshow/74907481.cms (accessed 29 March 2022).
31. Jain, N. and Raj, B. (2018). Parasitic capacitance and resistance model development and optimization of raised source/drain SOI FinFET structure for analog circuit applications. *Journal of Nanoelectronics and Optoelectronins*, *13*, 531–539.

32. Singh, S., Vishvakarma, S. K., and Raj, B. (2018). Analytical modeling of split-gate junction-less transistor for a biosensor application. *Sensing and Bio-sensing*, *18*, 31–36.
33. Gopal, M. and Raj, B. (2013). Low power 8T SRAM cell design for high stability video applications. *ITSI Transaction on Electrical and Electronics Engineering*, *1*(5), 91–97.
34. Raj, B., Mitra, J., Bihani, D. K., Rangharajan, V., Saxena, A. K., and Dasgupta, S. (2010). Analysis of noise margin, power and process variation for 32 nm FinFET based 6T SRAM cell. *Journal of Computer (JCP)*, *5*(6), 1–8.
35. Sharma, D., Mehra, R., and Raj, B. (2021). Comparative analysis of photovoltaic technologies for high efficiency solar cell design. *Superlattices and Microstructures*, *153*, 106861.
36. Kaur, P., Buttar, A. S., and Raj, B. (2021). A comprehensive analysis of nanoscale transistor based biosensor: A review. *Indian Journal of Pure and Applied Physics*, *59*, 304–318.
37. Yadav, D., Raj, B., and Raj, B. (2020). Design and simulation of low power microcontroller for IoT applications. *Journal of Sensor Letters*, *18*, 401–409.
38. Singh, S. and Raj, B. (2020). A 2-D analytical surface potential and drain current modeling of double-gate vertical t-shaped tunnel FET. *Journal of Computational Electronics*, *19*, 1154–1163.
39. Singh, J. and Raj, B. (2019). An accurate and generic window function for non-linear memristor model. *Journal of Computational Electronics*, *18*(2), 640–647.
40. Kaur, M., Gupta, N., Kumar, S., Raj, B., and Singh, A. K. (2021). Comparative RF and crosstalk analysis of carbon based nano interconnects. *IET Circuits, Devices & Systems*, *15*(6), 493–503.
41. Kandasamy, N., Ahmad, F., Ajitha, D., Raj, B., and Telagam, N. (2020). Quantum dot cellular automata based scan flip flop and boundary scan register. *IETE Journal of Research*, *66*, 535–548.
42. Sharma, S. K., Raj, B., and Khosla, M. (2019). Enhanced photosensivity of highly spectrum selective cylindrical gate $In_{1-x}Ga_xAs$ nanowire MOSFET photodetector. *Modern Physics Letter-B*, *33*(12), 1950144.
43. Singh, J. and Raj, B. (2019). Design and investigation of 7T2M NVSARM with enhanced stability and temperature impact on store/restore energy. *IEEE Transactions on Very Large Scale Integration Systems*, *27*(6), 1322–1328.
44. Bhardwaj, A. K., Gupta, S., Raj, B., and Singh, A. (2019). Impact of double gate geometry on the performance of carbon nanotube field effect transistor structures for low power digital design. *Computational and Theoretical Nanoscience*, *16*, 1813–1820.

45. Jain, N. and Raj, B. (2019). Thermal stability analysis and performance exploration of asymmetrical dual-k underlap Spacer (ADKUS) SOI FinFET for security and privacy applications. *Indian Journal of Pure & Applied Physics (IJPAP)*, *57*, 352–360.
46. Singh, A., Khosla, M., and Raj, B. (2019). Design and analysis of dynamically configurable electrostatic doped carbon nanotube tunnel FET. *Microelectronics Journal*, *85*, 17–24.
47. Jain, N. and Raj, B. (2019). Dual-k spacer region variation at the drain side of asymmetric SOI FinFET structure: Performance analysis towards the analog/RF design applications. *Journal of Nanoelectronics and Optoelectronics*, *14*, 349–359.
48. Singh, J., Sharma, S., Raj, B., and Khosla, M. (2018). Analysis of barrier layer thickness on performance of $In_{1-x}Ga_xAs$ based gate stack cylindrical gate nanowire MOSFET. *Journal of Nanoelectronics and Optoelectronics (JNO)*, *13*, 1473–1477.
49. Jain, N. and Raj, B. (2018). Analysis and performance exploration of high-k SOI FinFETs over the conventional low-k SOI FinFET toward analog/RF design. *Journal of Semiconductors (JoS)*, *39*(12), 124002.
50. Goyal, C., Ubhi, J. S., and Raj, B. (2018). A reliable leakage reduction technique for approximate full adder with reduced ground bounce noise. *Journal of Mathematical Problems in Engineering*, *2018*, 16. Article ID 3501041.
51. Anuradha, J. S., Raj, B., and Khosla, M. (2018). Design and performance analysis of nano-scale memristor-based nonvolatile SRAM. *Journal of Sensor Letter*, *16*, 798–805.
52. Wadhwa, G. and Raj, B. (2018). Parametric variation analysis of charge-plasma-based dielectric modulated JLTFET for biosensor application. *IEEE Sensor Journal*, *18*(15), 6070–6077.
53. Singh, J. and Raj, B. (2018). Comparative analysis of memristor models for memories design. *Journal of Semiconductors (JoS)*, *39*(7), 074006.
54. Yadav, D., Chouhan, S. S., Vishvakarma, S. K., and Raj, B. (2018). Application specific microcontroller design for IoT based WSN. *Sensor Letter*, *16*, 374–385.
55. Singh, G., Sarin, R. K., and Raj, B. (2018). Fault-tolerant design and analysis of quantum-dot cellular automata based circuits. *IEEE/IET Circuits, Devices & Systems*, *12*, 638–64.
56. Singh, J. and Raj, B. (2018). Modeling of mean barrier height levying various image forces of metal insulator metal structure to enhance the performance of conductive filament based memristor model. *IEEE Nanotechnology*, *17*(2), 268–267.
57. Jain, A., Sharma, S., and Raj, B. (2018). Analysis of triple metal surrounding gate (TM-SG) III–V nanowire MOSFET for photosensing application. *Opto-electronics Journal*, *26*(2), 141–148.

58. Jain, A., Sharma, S., and Raj, B. (2016). Design and analysis of high sensitivity photosensor using cylindrical surrounding gate MOSFET for low power sensor applications. *Engineering Science and Technology*, *19*(4), 1864–1870.
59. Singh, A., Khosla, M., and Raj, B. (2017). Analysis of electrostatic doped Schottky barrier carbon nanotube FET for low power applications. *Journal of Materials Science: Materials in Electronics*, *28*, 1762–1768.
60. Kumar, G. S., Singh, A., and Raj, B. (2018). Design and analysis of gate all around CNTFET based SRAM cell design. *Journal of Computational Electronics*, *17*(1), 138–145.
61. Singh, G. P., Sohi, B. S., and Raj, B. (2017). Material properties analysis of graphene base transistor (GBT) for VLSI analog circuits. *Indian Journal of Pure & Applied Physics (IJPAP)*, *55*, 896–902.
62. Singh, A., Khosla, M., and Raj, B. (2016). Comparative analysis of carbon nanotube field effect transistor and nanowire transistor for low power circuit design. *Journal of Nanoelectronics and Optoelectronics*, *11*, 388–393.
63. Kumar, S. and Raj, B. (2017). Estimation of stability and performance metric for inward access transistor based 6T SRAM cell design using n-type/p-type DMDG-GDOV TFET. *IEEE VLSI Circuits and Systems Letter*, *3*(2), 25–39.
64. Sharma, S., Kumar, A., Pattanaik, M., and Raj, B. (2013). Forward body biased multimode multi-threshold CMOS technique for ground bounce noise reduction in static CMOS adders. *International Journal of Information and Electronics Engineering*, *3*(3), 567–572.
65. Singh, H., Kumar, P., and Raj, B. (2013). Performance analysis of majority gate SET based 1-bit full adder. *International Journal of Computer and Communication Engineering* (IJCCE), *2*(4), 2933–2940.
66. Bhardwaj, A. K., Gupta, S., and Raj, B. (2020). Investigation of parameters for schottky barrier (SB) height for schottky barrier based carbon nanotube field effect transistor device. *Journal of Nanoelectronics and Optoelectronics*, *15*, 783–791.
67. Bansal, P. and Raj, B. (2017). Memristor: A versatile nonlinear model for dopant drift and boundary issues. *Journal of Computational and Theoretical Nanoscience (JCTN)*, *14*(5), 2319–2325.
68. Jain, N. and Raj, B. (2016). An analog and digital design perspective comprehensive approach on Fin-FET (Fin-field effect transistor) technology — A review. *Reviews in Advanced Sciences and Engineering (RASE)*, *5*, 1–14.
69. Sharma, S., Raj, B., and Khosla, M. (2017). Subthreshold performance of $In_{1-x}Ga_xAs$ based dual metal with gate stack cylindrical/surrounding gate nanowire MOSFET for low power analog applications. *Journal of Nanoelectronics and Optoelectronics*, *12*, 171–176.

70. Raj, B., Saxena, A. K., and Dasgupta, S. (2009). Analytical modeling for the estimation of leakage current and subthreshold swing factor of nanoscale double gate FinFET device. *Microelectronics International*, *26*, 53–63.
71. Soniya, S. S., Wadhwa, G., and Raj, B. (2019). An analytical modeling for dual source vertical tunnel field effect transistor. *International Journal of Recent Technology and Engineering (IJRTE)*, *8*(2), 603–608.
72. Singh, S. and Raj, B. (2019). Design and analysis of hetrojunction vertical T-shaped tunnel field effect transistor. *Journal of Electronics Material*, *48*(10), 6253–6260.
73. Goyal, C., Ubhi, J. S., and Raj, B. (2019). A low leakage CNTFET based inexact full adder for low power image processing applications. *International Journal of Circuit Theory and Applications*, *47*(9), 1446–1458.
74. Raj, B., Saxena, A. K., and Dasgupta, S. (2008). A compact drain current and threshold voltage quantum mechanical analytical modeling for FinFETs. *Journal of Nanoelectronics and Optoelectronics (JNO)*, *3*(2), 163–170.
75. Wadhwa, G. and Raj, B. (2020). An analytical modeling of charge plasma based tunnel field effect transistor with impacts of gate underlap region. *Superlattices and Microstructures*, *142*, 106512.
76. Singh, S. and Raj, B. (2020). Modeling and Simulation analysis of SiGe hetrojunction double GateVertical t-shaped tunnel FET. *Superlattices and Microstructures*, *142*, 106496.
77. Singh, A., Saini, D. K., Agarwal, D., Aggarwal, S., Khosla, M., and Raj, B. (2016). Modeling and simulation of carbon nanotube field effect transistor and its circuit application. *Journal of Semiconductors (JoS)*, *37*, 074001.
78. Jain, N. and Raj, B. (2016). Device and circuit co-design perspective comprehensive approach on FinFET technology — A review. *Journal of Electron Devices*, *23*(1), 1890–1901.
79. Kumar, S. and Raj, B. (2016). Analysis of ION and ambipolar current for dual-material gate-drain overlapped DG-TFET. *Journal of Nanoelectronics and Optoelectronics*, *11*, 323–333.
80. Anjum, N., Bali, T., and Raj, B. (2013). Design and simulation of handwritten multiscript character recognition. *International Journal of Advanced Research in Computer and Communication Engineering*, *2*(7), 2544–2549.
81. Sharma, S., Raj, B., and Khosla, M. (2016). A gaussian approach for analytical subthreshold current model of cylindrical nanowire FET with quantum mechanical effects. *Microelectronics Journal*, *53*, 65–72.
82. Singh, K. and Raj, B. (2015). Performance and analysis of temperature dependent multi-walled carbon nanotubes as global interconnects at different technology nodes. *Journal of Computational Electronics*, *14*(2), 469–476.

83. Kumar, S. and Raj, B. (2015). Compact channel potential analytical Modeling of DG-TFET based on evanescent–mode approach. *Journal of Computational Electronics*, *14*(2), 820–827.
84. Singh, K. and Raj, B. (2015). Temperature dependent modeling and performance evaluation of multi-walled CNT and single-walled CNT as global interconnects. *Journal of Electronic Materials*, *44*(12), 4825–4835.
85. Sharma, V. K., Pattanaik, M., and Raj, B. (2014). INDEP approach for leakage reduction in nanoscale CMOS circuits. *International Journal of Electronics*, *102*(2), 200–215.
86. Singh, K. and Raj, B. (2015). Influence of temperature on MWCNT bundle, SWCNT bundle and copper interconnects for nanoscaled technology nodes. *Journal of Materials Science: Materials in Electronics*, *26*(8), 6134–6142.
87. Anjum, N., Bali, T., and Raj, B. (2013). Design and simulation of handwritten Gurumukhi and Devanagri numerical recognition. *International Journal of Computer Applications*, *73*(12), 16–21.
88. Khandelwal, S., Gupta, V., Raj, B., and Gupta, R. D. (2015). Process variability aware low leakage reliable nano scale DG-FinFET SRAM cell design technique. *Journal of Nanoelectronics and Optoelectronics*, *10*(6), 810–817.
89. Sharma, V. K., Pattanaik, M., and Raj, B. (2014). ONOFIC approach: Low power high speed nanoscale VLSI circuits design. *International Journal of Electronics*, *101*(1), 61–73.
90. Khandelwal, S., Raj, B., and Gupta, R. D. (2015). FinFET based 6T SRAM cell design: Analysis of performance metric, process variation and temperature effect. *Journal of Computational and Theoretical Nanoscience*, *12*, 2500–2506.
91. Singh, S., Yadav, S., Rahul, J., Srivastava, A., and Raj, B. (2015). Impact of HfO_2 in graded channel dual insulator double gate MOSFET. *Journal of Computational and Theoretical Nanoscience*, *12*(6), 950–953.
92. Sharma, V. K., Pattanaik, M., and Raj, B. (2014). PVT variations aware low leakage INDEP approach for nanoscale CMOS circuits. *Microelectronics Reliability*, *54*, 90–99.
93. Raj, B., Saxena, A. K., and Dasgupta, S. (2013). Quantum mechanical analytical modeling of nanoscale DG FinFET: Evaluation of potential, threshold voltage and source/drain resistance. *Journal of Material Science in Semiconductor Processing*, *16*(4), 1131–1137.
94. Gopal, M., Prasad, S. S. D., and Raj, B. (2013). 8T SRAM cell design for dynamic and leakage power reduction. *International Journal of Computer Applications*, *71*(9), 43–48.
95. Pattanaik, M., Raj, B., Sharma, S., and Kumar, A. (2012). Diode based trimode multi-threshold CMOS technique for ground bounce noise reduction in static CMOS adders. *Advanced Materials Research*, *548*, 885–889.

96. Raj, B., Saxena, A. K., and Dasgupta, S. (2011). Nanoscale FinFET based SRAM cell design: Analysis of performance metric, process variation, underlapped FinFET and temperature effect. *IEEE Circuits and System Magazine*, *11*(2), 38–50.
97. Sharma, V. K., Pattanaik, M., and Raj, B. (2013). Leakage current ONOFIC approach for deep submicron VLSI circuit design. *International Journal of Electrical, Computer, Electronics and Communication Engineering*, 7(4), 239–244.
98. Chawla, T., Khosla, M., and Raj, B. (2021). Design and simulation of triple metal double-gate germanium on insulator vertical tunnel field effect transistor. *Microelectronics Journal*, *114*, 105125.
99. Kaur, P., Gill, S. S., and Raj, B. (2022). Comparative analysis of OFETs materials and devices for sensor applications. *Journal of Silicon*, *14*, 4463–4471.
100. Sharma, S. K., Kumar, P., Raj, B., and Raj, B. (2022). $In_{1-x}Ga_xAs$ double metal gate-stacking cylindrical nanowire MOSFET for highly sensitive photo detector. *Journal of Silicon*, *14*, 3535–3541.
101. Raj, B., Saxena, A. K., and Dasgupta, S. (2009). Analytical modeling of quasi planar nanoscale double gate FinFET with source/drain resistance and field dependent carrier mobility: A quantum mechanical study. *Journal of Computer (JCP)*, *4*(9), 1–8.
102. Bhushan, S., Khandelwal, S., and Raj, B. (2012). Analyzing different mode FinFET based memory cell at different power supply for leakage reduction. In *7th International Conference on Bio-Inspired Computing: Theories and Application, (BIC-TA 2012) Advances in Intelligent Systems and Computing*, Springer, India, Vol. 2, pp. 89–100.
103. Singh, J. and Raj, B. (2018). Temperature dependent analytical modeling and simulations of nanoscale memristor. *Engineering Science and Technology*, *21*, 862–868.
104. Singh, S., Bala, S., Raj, B., and Raj, B. (2020). Improved sensitivity of dielectric modulated junctionless transistor for nanoscale biosensor design. *Sensor Letter*, *18*, 328–333.
105. Kumar, V., Vishvakarma, S. K., and Raj, B. (2020). Design and performance analysis of ASIC for IoT applications. *Sensor Letter*, *18*, 31–38.
106. Jaiswal, A., Sarin, R. K., Raj, B., and Sukhija, S. (2018). A novel circular slotted microstrip-fed patch antenna with three triangle shape defected ground structure for multiband applications. *Advanced Electromagnetic (AEM)*, *7*(3), 56–63.
107. Wadhwa, G. and Raj, B. (2018). Label free detection of biomolecules using charge-plasma-based gate underlap dielectric modulated junctionless TFET. *Journal of Electronic Materials (JEMS)*, *47*(8), 4683–4693.

108. Singh, G., Sarin, R. K., and Raj, B. (2017). Design and performance analysis of a new efficient coplanar quantum-dot cellular automata adder. *Indian Journal of Pure & Applied Physics (IJPAP)*, *55*, 97–103.
109. Singh, A., Khosla, M., and Raj, B. (2017). Design and analysis of electrostatic doped Schottky barrier CNTFET based low power SRAM. *International Journal of Electronics and Communications*, *80*, 67–72.
110. Kaur, P., Pandey, V., and Raj, B. (2020). Comparative study of efficient design, control and monitoring of solar power using IoT. *Sensor Letter*, *18*, 419–426.
111. Bhardwaj, A. K., Gupta, S., and Raj, B. (2020). Development & analysis of compact model for double gate Schottky barrier CNTFET. *Journal of Nanoelectronics and Optoelectronics*, *15*, 1199–1208.
112. Wadhwa, G., Kamboj, P., Singh, J., and Raj, B. (2020). Design and investigation of junctionless DGTFET for biological molecule recognition. *Transactions on Electrical and Electronic Materials*, *22*, 282–289.
113. Chawla, T., Khosla, M., and Raj, B. (2020). Optimization of double-gate dual material GeOI-vertical TFET for VLSI circuit design. *IEEE VLSI Circuits and Systems Letter*, *6*(2), 13–25.
114. Verma, S. K., Singh, S., Wadhwa, G., and Raj, B. (2020). Detection of biomolecules using charge-plasma based gate underlap dielectric modulated dopingless TFET. *Transactions on Electrical and Electronic Materials (TEEM)*, *21*, 528–535.
115. Jain, N. and Raj, B. (2017). Impact of underlap spacer region variation on Electrostatic and analog/RF performance of symmetrical high-k SOI FinFET at 20 nm channel length. *Journal of Semiconductors (JoS)*, *38*(12), 122002.
116. Singh, S. and Raj, B. (2020). Analytical modeling and simulation analysis of T-shaped III–V heterojunction vertical T-FET. *Superlattices and Microstructures*, *147*, 106717.
117. Singh, G., Sarin, R. K., and Raj, B. (2017). Design and analysis of area efficient QCA based reversible logic gates. *Journal of Microprocessors and Microsystems*, *52*, 59–68.
118. Singh, A., Khosla, M., and Raj, B. (2016). Compact model for ballistic single wall CNTFET under quantum capacitance limit. *Journal of Semiconductors (JoS)*, *37*, 104001.
119. Singh, Sonal., Khosla, M., Wadhwa, G., and Raj, B. (2021). Design and analysis of double-gate junctionless vertical TFET for gas sensing applications. *Applied Physics A*, *127*(16), 2.
120. Singh, I., Raj, B., Khosla, M., and Kaushik, B. K. (2020). Potential MRAM technologies for low power SoCs. *SPIN*, *10*(4), 2050027.
121. Masud, M., Gaba, G. S., Alqahtani, S., Muhammad, G., Gupta, B. B., Kumar, P., and Ghoneim, A. (2020). A lightweight and robust secure key

establishment protocol for internet of medical things in COVID-19 patients care. *IEEE Internet of Things Journal*, *8*(21), 15694–15703.

122. Chopra, M., Singh, S. K., Gupta, A., Aggarwal, K., Gupta, B. B., and Colace, F. (2022). Analysis & prognosis of sustainable development goals using big data-based approach during COVID-19 pandemic. *Sustainable Technology and Entrepreneurship*, *1*(2), 100012.
123. Rahman, M. A., Hossain, M. S., Alrajeh, N. A., and Gupta, B. B. (2021). A multimodal, multimedia point-of-care deep learning framework for COVID-19 diagnosis. *ACM Transactions on Multimedia Computing Communications and Applications*, *17*(1s), 1–24.
124. Jha, S., Goyal, M. K., Gupta, B., and Gupta, A. K. (2021). A novel analysis of COVID 19 risk in India incorporating climatic and socioeconomic Factors. *Technological forecasting and social change*, *167*, 120679.
125. Zhou, Z., Gaurav, A., Gupta, B. B., Hamdi, H., and Nedjah, N. (2021). A statistical approach to secure health care services from DDoS attacks during COVID-19 pandemic. *Neural Computing and Applications*, *12*, 1–14.

https://doi.org/10.1142/9789811279539_0007

Chapter 7

VANETs: Introduction, Communication Protocols, and Challenges

Shivani Sharma*,†, Sachin Kumar‡, and Sateesh Kumar Awasthi§

Department of Electronics and Communication Engineering, Dr. B.R. Ambedkar National Institute of Technology, Jalandhar 144011, India

†*shivanis.ec.19@nitj.ac.in*

‡*sachin.kumar915@gmail.com*

§*awasthisk@nitj.ac.in*

Abstract

The management of increased demand for the utilization of available road infrastructure thanks to urbanization that has attracted the interest of researchers in the domain of Intelligent Transportation System (ITS). This helps the commuter to travel in congestion-free, accident-free, and efficient road scenarios. Vehicular *ad hoc* networks (VANETs) form a vital element of the ITS. VANET is defined as a subcategory of mobile *ad hoc* networks (MANETs) with some predefined distinct properties. Unlike MANET, VANET has no constraint on the node's battery life, but there is a bandwidth constraint in this scenario. VANET supports two types of communication: either among vehicles or among vehicles

*Corresponding author.

and infrastructure. The key objectives of VANETs are to modify public safety and transportation efficiency in the highly dynamic and mobile scenario. This chapter provides detailed information regarding the VANETs and their characteristics, network layer-wise communication protocols, shortcomings, features, and applications, and finally the challenges and future scope.

Keywords: VANETs, ITS, RSS, communication overhead, data dissemination, latency

7.1. Introduction

Urbanization has an adverse effect on the utilization of road infrastructure during peak hours, thus leading to traffic congestion, accidents, pollution, and wastage of fuel. Congestion can be resolved in two ways: (1) by building new roads and (2) by optimizing the use of existing infrastructure. The former is not a feasible solution because of limited resources, so the latter approach acts as a major contributor in this domain. This can be achieved with the deposition of an intelligent transportation system (ITS) which is defined as the integration of information revolution, communication, embedded servers, and control systems to enhance the utilization, safety, and productivity within the traffic scenario. Broadly, all the ITS applications fall under six categories [1]: advanced public transportation system (APTS), advanced traffic management system (ATMS), automated highway system (AHS), advanced traveler information system (ATIS), commercial vehicle control (CVC), and incident management system (IMS). The vehicular *ad hoc* network (VANET) is the main component of ITS. VANET may be defined as a self-organized, distributed, and highly mobile network for vehicular communication through wireless channels. VANETs are the subcategory of mobile *ad hoc* networks (MANETs) with some specific characteristics [1, 2]. Figure 7.1 shows the VANET scenario. In vehicular networks, mobility is restricted according to the terrain profile, type of road and direction, road traffic rules and regulations, and commuters' behavior. Due to these restrictions, the existing communication protocols for MANETs need enhancement when used in a VANET scenario. The vehicle's direction can be predicted up to some extent by various means, and the power-related issue can be resolved by the batteries in the vehicles [3].

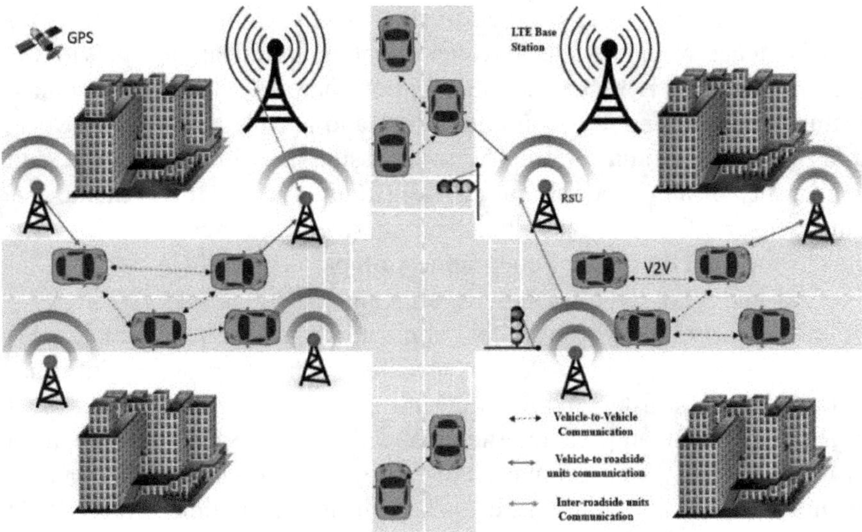

Figure 7.1. VANET Scenario [1]

Academia and industry scholars have done a lot of research on VANET as it is considered as the major element of ITS. The final aim of ITS is to improve commuter safety and provide reliable and efficient traffic management and services to commuters [4]. ITS makes use of either the conventional approach, i.e., cellular communication, or VANET communication. In the former approach, the vehicle can transmit and receive the information via cellular access points. Here, the common issues are as follows: all the vehicles will receive the same information in broadcast mode; even if the commuters are not interested in that information, they will get the information; the issue of cell breathing is also very common in the highly-dense scenario. To resolve this issue, VANET communication, which enables vehicles to obtain information on demand, was introduced. With the advancement in technology, the integration of cellular as well as VANET communication protocols was achieved for the data-gathering process [5]. Static road side sensors (RSS) are responsible for the establishment of the link for wireless communication between mobile nodes (vehicles) and static infrastructure, i.e., vehicle-to-infrastructure (V2I) communication. At present, there is a lot of deployment needed for the successful implementation of ITS. As this deployment is very expensive and futuristic, vehicle-to-vehicle (V2V) communication comes into existence when vehicles have no direct communication link with the RSS. Then there is hybrid communication that makes use of combined features of V2V and V2I communication and is called vehicle-to-everything (V2X) communication. The basic assumption includes the default availability of an on-board unit (OBU) (transceiver + GPS + memory + other sensors) for the human–machine interface within every node (vehicle) for the functionality of communication protocols.

V2I and V2V have various benefits and functionalities. While V2V offers support to emergency safety applications, such as collision avoidance, broadcasting accident occurrences or other interruptions, information on business tasks, and avoiding infrastructure deployment fees and user charges, V2I supports internet applications, gathering and estimation of traffic globally via aggregation of information at various data centers, technical simplicity, guaranteed QoS, professional maintenance, global coverage, and reliability.

Our major goal is to describe the set of rules for guaranteed on-time transmission of information (such as road congestion, location-based services, weather, and road conditions) to commuters in a cost-efficient way. For instance, V2V communication helps the commuter to reroute himself after getting the information of congestion occurred ahead. These types of protocols act as a key element in several other applications like searching for hotel/parking lots/petrol pumps, traffic flow prediction, speed limit warning, emergency vehicle assistance, cooperative adaptive cruise control, fleet management, mobile commerce, map and multimedia update, and many more. In one-hop V2V communication, every vehicle broadcasts the beacon message periodically and this process is known as diffusion. The beacon message contains information regarding the vehicle's geographical location, spot speed, direction, time stamp, and so on and is transmitted via technology named dedicated short-range communication (DSRC) [6]. Figure 7.2. shows the DSRC/WAVE architecture in VANET. In any application, data are collected from the nearby vehicles and then stored in the neighboring tables in the form of aggregated data which further forwards the table to the nearby nodes after updating their respective tables.

VANET-related applications are broadly categorized into two subparts: safety and non-safety [7, 8]. The safety messages require real-time message delivery to commuters with minimum end-to-end delay.

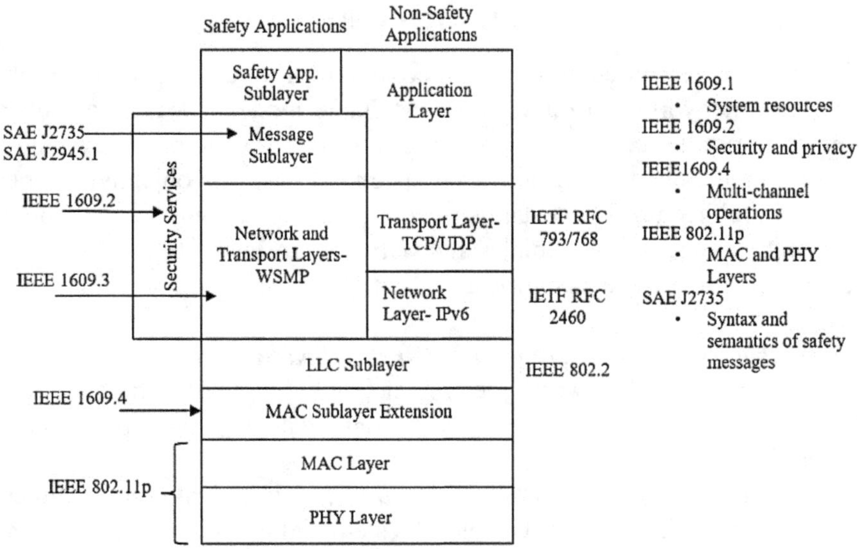

Figure 7.2. DSRC/WAVE Architecture

They primarily depend on the monitoring of target vehicles and nearby vehicles as well as on road conditions. In case of any emergency, all the nodes within the network exchange act as beacons and help each other by cooperation. The safety applications include pre/post-crash intimation, cooperative collision avoidance, and rollover prior intimation [9, 10]. After the detection of the unusual condition, the event-based message is generated and forwarded via the relevant part of the vehicular network at utmost priority. The objective of non-safety applications is to provide entertainment and traffic efficiency information to commuters. These are also known as comfort services as well as generalized information services. Traffic efficiency application includes the selection of the best alternate route, network optimal trip for commuters, and minimum trip time route [8]. The objective of non-safety messages is not only to provide information to the huge count of commuters over a wide area but also to provide services that target the smaller region to enable cooperative driving [11]. This chapter provides a brief introduction to the VANET as a key element of ITS, layer-wise communication protocols in the VANET scenario, their features and shortcomings, and the challenges faced by VANETs.

7.2. VANET and Its Characteristics

The enhancement in cellular mobile communication and latest evolution in *ad hoc* networks permit various installable models for the vehicular domain in rural, urban, and freeway highway scenarios to provide a number of applications with several QoS specifications. The objective of the VANET architecture is to establish communication among the neighboring vehicular nodes, and vehicles and stationary infrastructure, which can be classified into three categories:

(1) *Vehicle-to-infrastructure (V2I)/centralized ad hoc network*: It permits communication between the vehicle and the fixed infrastructure. This category majorly focuses on data gathering and processing applications.
(2) *Vehicle-to-vehicle (V2V)/distributed ad hoc network*: It permits direct communication among vehicles located within the communication range without any infrastructure support. This category majorly targets security, dissemination, and safety applications.
(3) *Vehicle-to-everything (V2X)/hybrid ad hoc network*: It allows a direct line-of-sight communication among the vehicles as well as information exchange between vehicles and fixed infrastructure either in a one-hop or a multi-hop manner.

VANET as a subpart of MANET employs some similar characteristics, such as lower bandwidth, omnidirectional broadcast, and short-term communication range. Along with this, VANETs have some specific features as follows:

- *Dynamic topology*: The spot speed of the vehicles and the radio transmission characteristics are the two major factors that make the vehicular network highly dynamic. Vehicles have huge relative speeds of the order of 50 km/h and 100 km/h in the urban and highway scenarios, respectively. This implies that vehicles can preferably connect or leave the network within a very less time span, consequently leading to fast and frequent variations in topology.
- *Frequent disconnections*: The highly dynamic topology of vehicular networks consequently leads to frequent variations in the connectivity, leading to a discontinuity in the availability of links among vehicles during information exchange [12].

- *Restricted mobility and prediction*: Vehicle mobility pattern is restricted by the roads, highways and streets, traffic conditions, speed limit, commuter behavior, weather, and traffic lights. With prior knowledge of mobility patterns, the feasibility of the prediction of the vehicle's future location increases.
- *Propagation models*: Broadly, there are three operating scenarios of VANETs: highway, urban, and rural. In the highway scenario, the free-space propagation model is generally used; here, the signal faces interference due to the reflection induced by the walls/panels along the road network. In the city, the surroundings introduce communication complexity because of variations in vehicle count (density) and the terrain profile acting as a hindrance in the propagation of signal. Further hurdles/obstacles introduce the multi-path, fading, and shadowing effects. The free-space propagation model is generally not suitable for such environments. In the rural scenario, because of the complex topographic structure (hills, forests, valleys, fields, etc.), the factors such as reflection and attenuation act as a vital part of signal propagation. Thus, the free-space propagation model doesn't work well here. The propagation model in the vehicular network considers the impact of interference on communication from the vehicles as well as the deployed infrastructure.

7.3. Layer-Wise Protocols for VANETS

Protocols for vehicular scenarios may be defined as a set of rules that defines the communication among vehicles and between the fixed infrastructure and vehicles in consideration of the above-mentioned VANET characteristics. This section provides a brief introduction to the protocols designed for the VANET as per standard network layer architecture.

7.3.1. *Physical Layer-Based Protocols*

The physical layer has to mandatorily make the consideration of the Doppler frequency shift and multi-path fading as a result of the faster mobility of vehicles along the road. Test scenarios for V2V communication make use of infrared and radio waves [13]. Radio waves include millimeter waves, micro, and very high-frequency waves for V2V communication. VHF and microwave are used for information broadcast, whereas millimeter waves and infrared are preferable for line-of-sight communication. Specifically, in the case of VANETs, the dedicated short-range communication (DSRC) architecture is designed for short-range as well as medium-range data transmission and reception. DSRC is dedicated to public safety as well as private applications. In the USA, the authorized body named Federal Communications Commission (FCC) assigned a bandwidth of up to 75 MHz for DSRC in the range of 5.850–5.925 GHz, whereas another body named European Telecommunications Standards Institute (ETSI) assigned a bandwidth of 70 MHz for DSRC in the range of 5.855–5.925 GHz [14]. DSRC allows vehicular mobility of up to 200 km/h and data rate and coverage range of 6–27 Mbps and 300–1,000 m, respectively.

Wireless access in vehicular environment (WAVE) is another name for DSRC, which is the enhancement of IEEE 802.11, nowadays known as IEEE802.11p, responsible for defining the interface techniques and functions controllable by the MAC layer [15].

The operational bandwidth is further partitioned into 06 Service Channels (SCH) and 01 Control Channel (CCH). Each channel has a width of 10 MHz and is dedicated to a specific set of ITS applications [16]. The entire spectra of DSRC are sub-divided into segments of 50 ms. The message transmission is divided into two categories: low-priority messages or data dissemination messages via SCH and high-priority messages or safety messages via CCH. Vehicles present in the network monitor these data packets, and if the CCH is in active mode, then all the vehicles are restricted to stop their general data transmission and reception for the CCH time stamp for the reception/transmission of safety messages via CCH. Initially, DSRC was designed for the data exchange between vehicles and static infrastructure. Further, within IEEE technical committee, the WAVE protocol suggests improvements in the physical as well as MAC layer of the IEEE 802.11 standards to ensure sustenance of the ITS services. This encompasses the information exchange among fast-moving vehicular nodes, vehicles and fixed infrastructure at the 5.9 GHz band.

7.3.2. *MAC Layer-Based Protocols*

The MAC layer is responsible for efficient, reliable, and unbiased access to the channel. The transmission via the MAC protocol is application-specific. For example, the transmission and reception of messages in safety applications must have low latency with a high success rate.

It requires efficient channel sharing, which is difficult to manage in the vehicular network because of dynamic topology and fast mobility.

The MAC layer protocols suffer from a hidden terminal problem in the VANET scenario, which is very common in the environment where nodes are in long platoons, further resulting in a reduction in the data transfer rate [17]. This point is of great importance as the availability of multimedia-related applications to commuters in the vehicular network requires a higher packet transmission rate. Further, in the case of a vehicular network, the bandwidth needs to be shared by all the vehicles interacting within the coverage range. In this section, we further discuss the MAC layer protocols for VANETs present in the literature.

Owing to the use of OFDM technology and carrier sense techniques to control the channel accessibility and collision avoidance, respectively, the IEEE 802.11p protocol was proposed to meet the specifications of V2V and V2I/I2V communications that include lower latency and higher reliability as the key specifications. The objective is to enable an efficient communication setup with minimum overhead and simplified basic service set (BSS) operations via IEEE 802.11p in a purely *ad hoc* manner. For instance, consider the case of the United States where vehicle infrastructure integration (VII) action proposed that the data dissemination regarding the incident occurrence must be communicated via VANET to all the enabled vehicles in the range of 500 m within 500 ms [18].

Similar to the IEEE 802.11 standard, in Ref. [19], the authors proposed a directional MAC (D-MAC) protocol with two distinct schemes. In this protocol, the acknowledgment message has been sent immediately after the data transmission; if the node has the knowledge of ongoing data transmission within two terminals, then that node itself will not perform the data transmission. D-MAC scheme-1 and D-MAC scheme-2 make use of directional RTS frames and directional + omnidirectional RTS frames, respectively. The ground ideology behind the D-MAC protocol is that the defective directional antenna at any terminal can be replaced with the other directional antenna to allow data transmission. The main objective of this protocol is the reduction in packet collision and enhancement in channel reuse.

However, some of the MAC protocols make use of the ALOHA concept for the definition of data transmission scheduling. ADHOC-MAC in Ref. [20] makes use of the reliable reservation ALOHA (RR-ALOHA) technique, which is a decentralized channel reservation and assignment protocol for the generation of reliable one-hop broadcast channels and the basic channel (BCH). The BCH makes use of the signaling information to resolve the issue of hidden as well as exposed terminal and allow the deployment of broadcast services. The ground principle is to contain the information of the slots within the previous interval preserved by the periodic transmission of frame information (FI) by each terminal. The ADHOC-MAC protocol operates independent of the physical layer, but its major drawback is that the channel utilization is very inefficient. The count of vehicles that can be communicated within a range cannot exceed the count of time intervals present in the frame stamp. In this category, the protocol named vehicular cooperative media access control (VC-MAC) that makes use of cooperative communication has been proposed [21]. To maximize the network throughput, the dissemination via fixed access terminal is performed on the predefined generalized condition during information download that all the commuters are concerned with the same data. During transmission, the count of vehicles may not be able to receive accurate information due to inefficient wireless channels. In this case, vehicles that have the reception of accurate information act as the relay terminal to forward the information to the nearby vehicles. Furthermore, to decrease the count of interference and collisions, protocol considers a few vehicles as a club of good relay terminals. In consideration of various types of control channels, the MP-MAC protocol has been proposed in Ref. [22]. It is based on the technique that describes the criteria for packet transmission. The highest priority is given to the safety packets, and the least priority is given to the control packets. It makes use of a multipriority Markov approach for optimal channel use on the basis of network traffic. In addition to this, the protocol incorporates the p-persistent MAC approach for the reduction of the probability of packet collision during data transmission. Table 7.1 shows the comparison of the above-discussed MAC-layer protocols based on characteristics, channel access mechanisms, and their advantages and disadvantages.

7.3.3. *Network Layer-Based Protocols*

At the network layer, the key responsibility of routing protocols is to deploy the techniques to allow uninterrupted and reliable communication. VANETS supports several communication paradigms, including the following: (1) unicast communication, (2) multicast/geocast communication, and (3) broadcast communication.

Table 7.1. Comparison of Various MAC Layer Protocols

Protocol	Characteristics	Channel Access Technique	Pros	Cons
IEEE 802.11p	Amendment in standard IEEE 802.11 protocol	CSMA/CA	Reliable and latency-efficient protocol	Lag in quality of service is not efficient for real traffic scenarios
DMAC	Use of directional antenna	CSMA-Oriented	Significant improvement in performance and collision count reduction	The assumption is that every terminal has knowledge of the geographical location
ADHOC MAC	Guaranteed efficient QoS	RR-ALOHA	Overcome the issue of the hidden terminal and make a significant reduction in the packet collision count	The number of time slots is fixed
VC-MAC	Utilizes the spatial reusability concept under the broadcast scenario	C-ALOHA	Significant improvement in the system throughput and reduction in packet collision	Specifically for broadcast scenario
MP-MAC	Prioritizes the data packets as a higher priority for safety data packets and after that control packets	p-persistent MAC scheduling	Reliable/efficient-safety message transmission and reduction in the channel collision	Suitable only for single-hop communication

The key objective of *unicast communication* is to transmit information from the source to the destination node in a vehicular scenario through a chain of multiple nodes, i.e., multi-hop communication. However, the destination can be either at a fixed position or an approximated position within the coverage range. Unicast communication is considered as a useful mode in the vehicular scenario, but multicast communications are way more suitable than the former when it comes to disseminating the message to multiple nodes within the network. *Multicast/geocast communication* can be defined as the data exchange between one source node and multiple destination nodes. Geocast is considered as a dedicated type of multicast addressing, where the information is forwarded to the number of targets located in a specified geographical location, generally with effect to the location of the source node, whereas *broadcast communication* allows the source vehicle to send data to all the nearby vehicles. The received message is forwarded to the destination nodes through another broadcast. The discovery mode in various routing protocols uses the advantage of the broadcast paradigm to explore the feasible route between source and destination [23, 24].

Broadly, the multi-hop routing strategies fall under two categories: topology-based and location/position-based [25–29]. In topology-based protocols, packet transmission relies on available information about the communication route. Here, each node discovers and maintains its routing table. Topology-based routing protocols have two subcategories: table-driven/proactive, and on-demand/reactive routing. Position-based routing protocols follow the intuition of prior knowledge of the geographical position of the source, neighboring nodes, and target. This category is further subdivided into three types: (1) delay-resistant, (2) non-delay-resistant, and (3) hybrid.

In contrast to non-delay-resistant protocols, delay-resistant protocols require continuous network connectivity and are best suited for highly populated vehicular scenarios. Protocols under the category of the hybrid approach are considerable in partial network connectivity.

Initially, the same routing protocols designed for MANET (like AODV, DSR, and DSDV) were incorporated for the VANET scenario [30, 31]. But these protocols did not perform well in the VANET scenario because of dynamic topology and fast movement of vehicles. Due to the continuous mobility profile of vehicles, the position-based routing protocols are best suited in the VANET environment because of the fact of increased availability of inbuilt tracking systems in vehicles along with the improved location accuracy up to several feet [32]. In addition to this, the position-based routing protocols neither maintain nor exchange the channel state information like proactive and reactive routing and are proven to be a promising and robust approach for highly dynamic VANET scenarios.

The authors suggested a position-based rerouting technique for the MANET scenario named GPSR that relies on the concept of greedy forwarding [33]. The proposed protocol has one route discovery phase that results in increased delay in the network. Furthermore, the dynamic mobility of vehicles introduces the routing loops during the perimeter phase of GPSR. The advancement in GPSR was achieved with the introduction of GPCR and GPSRJ+ in Refs. [34, 35], respectively. Since both the designed protocols are position-based, they were restricted to the local profile of the network. The D-Greedy/D-MinCost [36] and VADD [37] protocols also fall under the same category. These are designed for error detection during route discovery in GPSR. These protocols are responsible to make the decision of forwarding or storing message packets until the suitable node is found. The designed protocols reduce the transmission delay and calculate the estimated end-to-end delay on the basis of the vehicle's speed and the total number of intersections. CAR [38] and A-STAR [39] routing protocols make the utilization of road traffic information for efficient message transmission. These routing protocols address the issue of network connectivity instead of end-to-end delay. The author proposed a cross-multi-layer position-based communication protocol named PROMPT [40]. The key objective is to improve the end-to-end latency with the help of route information collected by vehicles via periodic beacon message transmission.

In the category of geocast routing, various routing protocols are present in the literature. Two distinct approaches were proposed in Ref. [44] and [45] for the dissemination of data. Here, one message is forwarded to a group of vehicles as per requirement. These routing protocols rely on the geographical position of vehicles, i.e., the position of the vehicles determines the necessity of the information for the particular vehicle. For example, commuters must have prior knowledge of the occurrence of any unusual incident within the region they are approaching. The former approach transmits a beacon message to the vehicles present within the area of interest. This area defines the relevance of the message to the vehicles. One more approach in this category is IVG [44] which identifies the area of interest based on speed, geographical position, and driving direction of the vehicular node.

Caching is another vast concept that falls in the category of geocast routing [46] whose objective is to improve the packet delivery rate. The idea is to introduce a cache segment to hold the message until a new node is discovered. Further, the simulation results validated the significant improvement in throughput.

In the category of broadcast communication, numerous routing protocols were proposed in the literature. BROADCOMM uses the hierarchical approach for routing in an emergency vehicle, in which broadcast occurs by the vehicles present at two levels [47]. The objective is to enhance the QoS parameters in broadcast communication. Another proposed algorithm under this category named UMB was proposed in Ref. [48]. UMB protocol addresses the issue of the hidden terminal, broadcast storm, and reliability in the case of multi-hop broadcast communication in city scenarios. With the help of this protocol, efficient channel utilization along with a high delivery rate of the message can be achieved. V-TRADE and HV-TRADE are the two strategies proposed in Ref. [49]; while the former uses the knowledge of GPS and vector distance to broadcast the information, the latter makes use of location history for guaranteed reliable message broadcast. HyDia broadcasting technique was proposed in Ref. [50] for highway scenarios. The protocols have two modules: one is the broadcast suppression technique and another one is the store-and-carry forward technique for guaranteed message transmission. The application of this protocol is limited to the highway environment. Furthermore, the extension of this protocol named VoV is proposed in Ref. [51] for the urban area with dynamic traffic conditions. The additional module of the rate-controlling system was introduced for the adaptability of the radio channel.

Table 7.2 shows the comparison of these protocols on various features, such as network architecture, communication mode, forwarding mechanism, scenario type, and applications.

7.3.4. *Transport and Application Layer Protocols*

As mentioned earlier, VANETs have the characteristics of continuous connectivity and dynamic topology variations. Similar to MANETs, VANETs have a more precise and predictable mobility profile.

The fast-moving vehicle interaction with fixed infrastructure to download the data in the VANET scenario requires a higher data rate that consequently leads to huge losses and results in performance degradation of

Table 7.2. Comparison of Various Network Layer Protocols

MAC Routing Protocol	Network Architecture	Communication Paradigm involved	Forwarding Mechanism	Road Type	Applications	Corns
GPSR	Distributed		Greedy forwarding		CBR data traffic	Because of the greedy forwarding mechanism, the applicability is restricted to the urban scenario, as it does not support direct communication
GPCR			Message forwarding	Urban city	—	Not suitable for lower density environment
GPRSJ+		Unicast	Greedy forwarding		CBR data traffic	Requires more simulation efforts in realistic scenarios
D-Greedy/D-MinCost	Hybrid		Multi-hop data muling and forwarding		—	Makes use of global data only while making a routing decision
VADD			Predicted packet forwarding			Selection of an upstream link is difficult
A-STAR	Distributed		Message forwarding along with traffic information		CBR data traffic	Limited to the city area

(Continued)

Table 7.2. (*Continued*)

MAC Routing Protocol	Network Architecture	Communication Paradigm involved	Forwarding Mechanism	Road Type	Applications	Corns
CAR			Message forwarding			The approach is dependent on historic data (traffic density and relative mobility)
PROPT	Centralized		Position-based message forwarding	Grid scenario	Variable data traffic	It is difficult to simulate a realistic scenario
MURU [41]			Expected discontinuous connections (heuristic)	Grid scenario	CBR data traffic	Difficult to update the overhead in terms of EDD metric
GyTAR [42]			Street awareness-oriented message forwarding	Urban city		Restricted to urban scenario due to greedy forwarding mechanism
Direct Message [43]	Distributed		Message forwarding	Urban city along with accident information	—	Packet forwarding is decided on the basis of the degree of centrality for a node

IVG		Geocast	Message forwarding	Urban as well as rural scenario	—	Generalized solution on the basis of GPS device
Caching Geocast			Store–carry forwarding	For all scenarios	—	Working is affected by network topology
BROADCOM		Broadcast	Virtual cell-oriented message forwarding	Fixed grid scenario	Information broadcast	Performance evaluation on the basis of historic data
UMB	Hybrid		Message forwarding	Urban city	CBR data traffic	Suited for highly dense scenario
V-TRADE/ HV-TRADE	Distributed		Virtual cell-oriented message forwarding	For all scenarios		Overhead issue because of selection of forwarding node at each level
HyDi	Distributed		Distance-based forwarding	Highway scenario	—	Suitable for highway environment only
VoV	Distributed		Location and distance-based	For all scenarios	Video transmission	Focuses solely on video dissemination

both TCP as well as UDP protocols [52]. In vehicular networks, numerous applications demand almost similar services to those offered by TCP, which includes efficient and on-time information delivery.

TCP has its performance limitations in a wireless network having the characteristics of higher mobility and topology change. To resolve this issue, authors proposed a transport protocol named vehicular transport protocol (VTP) for unicast communication-specific applications in the VANET scenario. It connects the network and utilizes the statistical data for improvement in connection during connection disruption. Its blueprint relies on the route features that are useful for transport protocol in VANET. Another protocol based on the principle of TCP named MCTP has been proposed in Ref. [53]. The key objective is to provide node-to-node reliable quality service between the vehicle and the fixed access point. The above-discussed protocols are proposed for unicast routing-based applications [54]. There are several VANET applications that require a multicast mode of communication, which demands new techniques and do not rely on conventional transport layer protocols. But it is difficult to design a reliable multicasting communication-based transport protocol as it is independent of a specific state.

The protocol designed for the application layer should reduce the processing delay, which acts as a key consideration during the dissemination of critical information to an emergency vehicle and in the case of delay-sensitive applications. The former relies on the geographical position of the source event and the velocity and location of the interested recipients. The application layer protocols need to work in conjunction with the fixed real-time stamp for guaranteed transmission and reception in due time. However, in the latter scenario, the VANET should consider small latency in the case of infotainment applications that includes the availability of real-time multimedia data to the commuters. These are also proposed for the development of marketing tools in the business sector. Consider the case of hotels, parking lots, gas-filling stations, and many more that can broadcast the data in the VANET and then the willing commuters can send a request for more detailed information. These protocols have the application in the business transaction domain; these protocols must be reliable and delay-efficient. In this area, a protocol named vehicular information transfer protocol (VITP) has been designed [55]. The objective is to create a distributed *ad hoc* infrastructure to provide the services in the VANET scenario. The key idea of the functionality is similar to HTTP/location-knowledge stateless transport protocol.

7.4. Features and Applications of VANETs

In this section, we focus on the details of the features of VANETs and the applications supported by VANET scenarios.

7.4.1. *Features of VANETs*

VANETs may be defined as the combinations of vehicles (nodes), which behave asymmetrically unlike wireless nodes in MANET. Thus, various parameters affect the layout implementation of the VANET environment. The traffic on a highway is straight and easily manageable, whereas in urban scenarios, the situation gets more complicated. Roads are generally hard to manage as in urban areas, these are usually shared by homes, trees, workplaces and other hindrances, which do not make it possible to achieve a direct way of communication in a particular direction. So, it is important to consider the urban city and arterial roads for the development of communication protocols in order to achieve a better packet drop ratio (PDR) compared to conventional techniques [56]. The considerable feature of the vehicle in the VANET environment is the availability of a lot of energy and computational capabilities that are usually lagging in MANET and other *ad hoc* networks known for their inefficient and repository manner. Thus, the computation in VANET entirely depends upon the speed and direction [57–59]. In estimable mobility, the VANET behaves differently compared to other *ad hoc* networks, as the vehicles are moving at high speed in a predefined road profile. Vehicle mobility is predictable as its movement depends on street planning, terrain, and traffic light signals. So, it becomes a necessity to have inbuilt navigation systems in every vehicle [60]. In topological communication, the VANET constitutes communication protocols that require the consideration of geographical location, as real-time data is the basic demand of safe-driving applications. Security applications form another category supported by VANET with minimum latency and less PDR [61, 62]. In a large network, the density within the network becomes a significant aspect of VANET as it considers the features, such as streets, arterial roads, link roads, highways, urban cities, forests, buildings, and many more. Numerous protocols based on the fixed street and the intersections are proposed by the researchers. Variable traffic resulting from traffic jams has a direct impact on network density, subsequently leading to a frequent discontinuity in connection within the network. In consideration of this, the changes in the vehicle density must be featured by VANET communication for the achievement of on-time and error-free routing of packets. In the case of a highly dynamic scenario, the topology of VANET varies very fast with speed. Thus, with the deployment of VANET, the large trajectory of network topology needs to be considered carefully while routing the data packets.

7.4.2. *Applications*

VANETs support various applications that are very diverse. Broadly, these applications are grouped into three categories: (1) safety-oriented, (2) efficiency-based, and (3) infotainment-oriented applications [1]. These categories are elaborately explained in the following:

(1) *Safety-oriented applications*: Safety is a major concern while a driver is driving a vehicle. To protect commuters from any incident or crash, it is mandatory to have the placement of safety-based applications in the scenario [63]. A collision avoidance system (CAS) is one application that helps the driver by providing information about the neighboring vehicles by giving a beep sound to avoid any crash via V2V communication. The pre-crash system is another name for CAS, which makes use of RADAR and other laser-based sensors and cameras to monitor the crash condition, further producing warning signals to apply the brakes in due time [64]. A fuzzy-control algorithm provides safety to vehicles and helps maintain headway to avoid crashes. Lane-changing assistant, blind-spot detections, collision avoidance, and adaptive cruise control are some safety-oriented VANET applications. All these applications are restricted by the time-bound delivery of information, otherwise resulting in the collision of vehicles [65].
(2) *Efficiency-based applications*: All the applications related to vehicle convenience fall under this category. These applications are further categorized into three domains: (1) intersection control, (2) parking monitoring and (3) route navigation [1, 64]. The objective of intersection control services is to monitor and control the traffic at intersection points via an efficient utilization of traffic signals with less waiting duration. Traffic can be controlled either by means of traffic lights or non-traffic light systems. The major challenge in the traffic light system is that the complete knowledge of signal strength schedule planning is a prerequisite. Traffic detectors detect the traffic and

process it on the basis of traffic volume. The current scenario is that V2V or V2I communication is capable of providing the information related to the vehicle's ID, spot speed, and location gathered and collectively offering an efficient and accurate scheduling to the commuter. Non-traffic light-based technique makes use of the controller to control the commuter's behavior; it monitors the trajectory of every vehicle and then assists the commuter to drive in a collision-free scenario [66]. It is difficult to find available parking lots in urban cities, especially near hotels, malls, and complexes. To resolve these issues, parking monitoring is the facility provided to commuters in the VANET scenario. VANET helps commuters with assistance in finding parking space locally [66]. In route navigation, the monitoring on the basis of vehicular networks is to surpass the traces of GPS-based tracking. A system for the information of crowd-sensing road traffic on the basis of an *ad hoc* scenario has been proposed [67, 68]. This approach contributes to the controlling of environmental pollution via rerouting, and prior estimation of the occurrence of traffic jams can help in the reduction of fuel wastage and CO_2 emissions [69, 70].

(3) *Infotainment-oriented applications*: Majorly, these applications include internet services and file sharing among vehicles. These applications send alerts and notifications on the infotainment devices so that the commuter can make use of all available services hands-free. This domain not only just considers entertainment services but also provides the alert that helps the VANET to take the new shape. With the availability of internet connection within the vehicles, numerous inbuilt integrated services for entertainment can be used by commuters [1, 70]. The entertainment and the information feature can be directly incorporated into the vehicle by just connecting the cable with the smartphone on the infotainment panel. VANET communication helps in video sharing among vehicles.

7.5. Challenges and Future Scope

With the knowledge of the challenges and features of the VANET scenario, some future constraints must also be noted to design efficient communication protocols:

(1) **Heterogeneous vehicular network:** Numerous non-interoperable networking technologies are integrated with the fast deployment and

availability of the mobile computing architecture. Consequently, this leads to seamless interconnectivity in the influence of complex dynamic network topology in the form of Quality of Service (QoS), security, node addressing, and billing. Therefore, it is quite possible to expect that the future generation of ITS reflects an enhanced heuristic approach to the networking solution. This requires support for the various existing interconnected wireless networks to allow universal and ubiquitous access to internet services [71].

(2) **Localized system:** Safety applications in the VANET scenario emphasize the requirement for highly reliable and accurate location systems. The basic solution to the location technique for the VANET scenario is to integrate a tracking device in every vehicle. There are some issues with satellite-based positioning systems (like GPS, Galileo, and NAVIC), such as the unavailability of data all the time due to reflection and poor signal strength. In addition to this, these systems are prone to spoofing and blocking types of attacks. Furthermore, it has a precision error of 10–30 m, which does not fulfil the need for critical safety applications and therefore demands the development of a new positioning system. Numerous location techniques were designed for the calculation of the position of dynamic nodes, such as dead reckoning, image/video processing, map matching, cellular localization, differential GPS technique, and localization services [71]. All the above-mentioned techniques have their pros and cons, and there is no such technique that can fulfil all the specifications of critical safety applications at the instant, such as the availability of data anywhere at any time along with higher accuracy and efficient location estimation. The efficient and ubiquitous location system used by vehicles in the VANET scenario must be achieved by the integration of various approaches and data fusion. Specific characteristics of VANETs like dynamic mobility constraints restricted by road profile, commuter's behavior, and spot speed of vehicle affect the network topology, which further leads to the broadcast of historic location information. However, some approaches to enhance the beacon frequency create redundant overhead and disrupt the transmission of information. So, the analytical study of approaches for the prediction of the location of vehicles during the time duration becomes a better alternative in the location systems.

(3) **Management and storage of data:** As mentioned above, we can think of a huge heterogeneous vehicular network with billions of

vehicles, which will generate a large amount of distributed information that needs to be stored in a generalized way and distributed in the VANETs. The features mentioned in Refs. [72, 73], large scale in terms of the size of the network and count of generated data, as well as the inbuilt properties of VANETs demand a new and upgraded challenge in the management of data in this scenario.

(4) **Interrupt-free communication:** Current issues, like higher latency and less reliable message delivery, are usually there in sparse scenarios. To enhance the reliability of message delivery, various solutions utilized the approach of carry-and-forward, which consequently leads to an increase in latency. These problems can be solved by new advanced data communication techniques in heterogeneous scenarios. The commuter's behavior can be considered as another alternative to enhance the carry-forward method for the reduction of packet delivery duration.

(5) **Monitoring a target:** Communication is considered a basic aspect of the network, and in the case of VANETs, it depends on the geographical location of vehicles. Thus, the track of target movement acts as a fundamental operation in the VANETs for communication approaches as well as for the services and applications that can be benefited from such kind of data [74]. Monitoring demands the creation of an algorithm for the identification of routes followed by vehicles in the vehicular network and can estimate the next location when required.

(6) **Dynamic network density:** In urban cities, there are a number of vehicles in a very small area. Such a scenario demands the development of a channel-access control protocol for packet collision avoidance and packet transmission errors. In the case of the highway scenarios, the network topology is sparse and the connectivity is intermittent. Such scenarios demand the development of protocols having the knowledge of such disconnections. At the same time, vehicles moving in both scenarios must be adaptable to the variations introduced in network density to achieve a better data transfer rate.

(7) **Network segmentation:** Network segmentation is a challenging task while designing a network as it causes some nodes out of reach from the network. It may present in less traffic or in rural areas. It is assumed that the initial stage of deployment of VANET access points, where a small number of vehicles will be incorporated with transceivers, further leads to the more frequent segmentation of the

network [75]. Naïve protocols relying on the topological information within the node are not suited to the VANET scenario.

(8) **Security and privacy:** Various network security problems are similar to those of naïve wireless networks although the security issues in VANETs are unique and intrinsic because of network size, dynamic topology change, relatively higher mobility, and various types of services and applications in conjunction with the conflict in requirements that will be given to such networks. Beyond these challenges, there is always a trade-off between authentication/non-repudiation and privacy [76].

The next main issue is to isolate the system and information from attackers to protect the integrity of information. Some of the features of VANET cause obstruction to meeting the security specifications that further demand the design of novel protocols with the characteristics of low overhead, least-hop count communication, pre-stored aggregated information, and optimal data dissemination solutions [76, 77]. Along with the existing protocols, there is also a requirement for designing new security protocols for heterogeneous VANET scenarios.

(9) **Geographical addressing:** The geographical location of the vehicle is the primary parameter required in a number of applications to achieve data communication, which requires the geographical address of the node. Three geographical addressing categories were highlighted in Refs. [77, 78]: application layer, multicast, and unicast IP to deal with the GPS unique address. Furthermore, providing vehicle mobility trajectory and commuter behavior, monitoring and processing location address to estimate the future location of the vehicle, is a very challenging task in the VANET environment.

(10) **Protocols standardization:** In VANETs, there are different types of vehicles, such as cars, buses, trucks, motorcycles, and taxis. In this case, it is required to establish communication among all and this can only be achievable when all the vehicles are configurable with the same protocol. This requires standardization and can be achieved with the help of industry, academia, and government [79–81].

(11) **Interconnection with other networks:** We can consider the interaction of drivers/commuters in a VANET along with the people, services, and applications within some other networks [82–85]. This type of interconnection helps to provide better services in terms of weather conditions, global routes, and traffic conditions.

All the data can be gathered from the sensors, the internet, and various other services.

7.6. Conclusion

VANET is defined as a basic building block for ITS, smart fixed infrastructure, and smart vehicles. The advent of VANET is characterized by equipped vehicles that have the capability to enable communication and are self-organized and adaptive into the mesh network, creating a number of applications that make driving safer and more efficient on road by decreasing crashes, travel time, and traffic congestion. However, VANETs are more likely to be the most realistic interpretation of any *ad hoc* network. Certain features of VANETs tend to produce particular types of networking issues, which demand fully distributed approaches. There are some additional challenges faced during the design of any protocol in VANETs in conjunction with the challenges already there in MANETs. Specifically, vehicular mobility tends to create dynamic scenarios along with the topology change, consequently leading to a very small span of multi-hop communication. In this particular case, the approaches require the knowledge of the system's state even though the local information is inefficient because of sudden and frequent variations in network topology. There are numerous existing research challenges that need to be addressed in several domains that are incorporated with real-world deployment as the techniques demand the acceptance of innovative techniques. In the period of last decade, there were noticeable research advancements in the domain of VANETs and integrated technologies that attract the research community's interest in transportation, networking, and wireless communication. Various automobile companies, research and government organizations now contribute to the evaluation, design, creation, and future of VANET architecture, which comes along with the opportunities of interconnected vehicular nodes and fixed infrastructure. A very basic and fundamental point in all the points related to vehicular networks is the unique kind of algorithm design for the VANET scenario. This leads to the aspect of the discussion on the main features of vehicular networking, architectural detail, networking layer constraints, issues, and future perspectives. We hope the brief discussion of the VANET scenario will make a significant contribution to protocol and application designing for improvement in services provided by the network having distinct features and assist the commuter in deciding the secure traveling path.

References

1. Sharma, S. and Awasthi, S. K. (2022). Introduction to intelligent transportation system: Overview, classification based on physical architecture, and challenges. *International Journal of Sensor Networks*, *38*(4), 215–240.
2. Ros, F. J., Ruiz, P. M., Sanchez, J. A., and Stojmenovic, I. (2009). Mobile ad hoc routing in the context of vehicular networks. In S. Olariu and M. Weigle (Eds.), *Handbook of Vehicular Networks*, pp. 1–48. Chapman & Hall/CRC/T&F: London.
3. Broustis, I. and Faloutsos, M. (2008). Routing in vehicular networks: Feasibility, modeling, and security. *International Journal of Vehicular Technology*, *2008*, 1–8.
4. Zheng, K., Zheng, Q., Chatzimisios, P., Xiang, W., and Zhou, Y. (2015). Heterogeneous vehicular networking: A survey on architecture, challenges, and solutions. *IEEE Communications Surveys & Tutorials*, *17*(4), 2377–2396.
5. Odeling, M. and Eneration, G. (2008). Inter-vehicle communication systems: Survey. *IEEE Communications Surveys & Tutorials*, *10*(2), 48–69.
6. Li, Y. (March 2010). An overview of the DSRC/WAVE technology. In *Proceedings of International Conference on Heterogeneous Networking for Quality, Reliability, Security and Robustness*, November 17–19, 2010, Vol. 15, pp. 9–21. Springer: Berlin, Heidelberg.
7. Bernsen, J. and Manivannan, D. (2009). Unicast routing protocols for vehicular ad hoc networks: A critical comparison and classification. *Pervasive and Mobile Computing*, *5*, 1–18.
8. Casteigts, A., Nayak, A., Casteigts, A., Nayak, A., and Stojmenovic, I. (2011). Communication protocols for vehicular ad hoc networks. *Wireless Communications and Mobile Computing*, *11*(5), 567–582.
9. Biswas, S., Tatchikou, R., and Dion, F. (2006).Vehicle-to-vehicle wireless communication protocols for enhancing highway traffic safety. *IEEE Communications Magazine*, *44*(1), 74–82.
10. Ding, Y., Wang, C., and Xiao, L. (2007). *A Static-node Assisted Adaptive Routing Protocol in Vehicular Networks*. ACM VANET: New York, pp. 59–68.
11. Willke, T., Tientrakool, P., and Maxemchuk, N. F. (2009). A survey of inter-vehicle communication protocols and their applications. *IEEE Communications Surveys & Tutorials*, *11*(2), 3–20.
12. Villas, L. A., Boukerche, A., Araujo, R. B., Loureiro, A. A. F., and Ueyama, J. (2013). Network partition-aware geographical data dissemination. In *IEEE International Conference on Communications (ICC)*, June 9, 2013, pp. 1439–1443.
13. Papadimitratos, P., De La Fortelle, A., Evenssen, K., Brignolo, R., and Cosenza, S. (2010). Vehicular communication systems: Enabling

technologies, applications, and future outlook on intelligent transportation. *Communications Magazine*, *47*(11), 84–95.

14. Kenney, J. (2011). Dedicated short-range communications (DSRC) standards in the United States. *Proceedings of IEEE*, *99*(7), 1162–1182.
15. Jiang, D. and Delgrossi, L. (2008). IEEE 802.11p: Towards an international standard for wireless access in vehicular environments. In *Proceedings of Vehicular Technology Conference, VTC Spring*, May 11, 2008, pp. 2036–2040.
16. European Telecommunications Standards Institute, Intelligent Transport Systems (ITS) (2013). Radiocommunications equipment operating in the 5.855 MHz to 5.925 MHz frequency band, Technical Report, ETSI, France.
17. Almalag, M. S., Weigle, M. C., and Olariu, S. (2013). *MAC Protocols for VANET. Mobile Ad hoc Networking: Cutting Edge Directions*, pp. 599–618. John Wiley & Sons, Inc.: New York, NY.
18. Menouar, H., Filali, F., and Lenardi, M. (2006). A survey and qualitative analysis of mac protocols for vehicular ad hoc networks. *IEEE Wireless Communications*, *13*(5), 30–35.
19. Ko, Y.-B., Shankarkumar, V., and Vaidya, N. H. (2000). Medium access control protocols using directional antennas in ad hoc networks. In *Proceedings of the 19th Annual Joint Conference of the IEEE Computer and Communications Societies (INFOCOM'00)*, March 26, 2000, pp. 13–21.
20. Borgonovo, F., Capone, A., Cesana, M., and Fratta, L. (2004) Ad hoc mac: New mac architecture for ad hoc networks providing efficient and reliable point-to-point and broadcast services. *Wireless Network*, *10*(4), 359–366.
21. Zhang, J., Zhang, Q., and Jia, W. (2009). VC-MAC: A cooperative mac protocol in vehicular networks. *IEEE Transactions on Vehicular Technology*, *58*(3), 1561–1571.
22. Shao, C., Leng, S., Zhang, Y., and Fu, H. (2014). A multi-priority supported medium access control in vehicular ad hoc networks. *Computer Communications*, *39*, 11–21.
23. Villas, L. A., Boukerche, A., Maia, G., Pazzi, R. W., and Loureiro, A. A. F. (2014). DRIVE: An efficient and robust data dissemination protocol for highway and urban vehicular ad hoc networks. *Computer Networks*, *75*, 381–394.
24. Villas, L. A., Boukerche, A., Ramos, H. S., de Oliveira, H. A. B. F., de Araujo, R. B., and Loureiro, A. A. F. (2013). DRINA: A lightweight and reliable routing approach for in-network aggregation in wireless sensor networks. *IEEE Transactions on Computers*, *62*(4), 676–689.
25. Fonseca, A. and Vazo, T. (2013). Applicability of position-based routing for VANET in highways and urban environment. *Journal of Network and Computer Applications*, *36*(3), 961–973.

26. Jerbi, M., Senouci, S.-M., Rasheed, T., and Ghamri-Doudane, Y. (2008) Towards efficient geographic routing in urban vehicular networks. *IEEE Transactions on Vehicular Technology*, *58*(9), 5048–5059.
27. Lee, K. C., Lee, U., and Gerla, M. (2010). Survey of routing protocols in vehicular ad hoc networks. In M. Watfa (Ed.), *Advances in Vehicular Ad-hoc Networks: Developments and Challenges*, pp. 149–170. IGI Global.
28. Nithya Darisini, P. and Kumari, N. (2013). A survey of routing protocols for VANET in urban scenarios. In *Proceedings of International Conference on Pattern Recognition, Informatics and Mobile Engineering (PRIME)*, February 21, 2013, pp. 464–467.
29. Slavik, M. and Mahgoub, I. (2014). Spatial distribution and channel quality adaptive protocol for multihop wireless broadcast routing in VANET. *IEEE Transactions on Mobile Computing*, *12*(4), 722–734.
30. Perkins, C. and Royer, E. (1999). Ad-hoc on-demand distance vector routing. In *Proceedings of the Second IEEE Workshop on Mobile Computing Systems and Applications (WMCSA'99)*, February 1999, pp. 90–100.
31. Johnson, D., Maltz, D., and Broch, J. (2001). DSR: Dynamic source routing protocol for multi-hop wireless ad hoc networks. *Ad Hoc Network*, *1*, 139–172.
32. Li, F. and Wang, Y. (2007). Routing in vehicular ad hoc networks: A survey. *IEEE Vehicular Technology Magazine*, *2*(2), 12–22.
33. Karp, B. and Kung, H. (2000). GPSR: Greedy perimeter stateless routing for wireless networks. In *Proceedings of the 6th Annual International Conference on Mobile Computing and Networking*, August 1, 2000, pp. 243–254.
34. Lochert, C., Mauve, M., Fler, H. and Hartenstein, H. (2005). Geographic routing in city scenarios. *ACM SIGMOBILE Mobile Computing and Communications*, *9*(1), 69–72.
35. Lee, K., Haerri, J., Lee, U., and Gerla, M. (2007). Enhanced perimeter routing for geographic forwarding protocols in urban vehicular scenarios. In *Proceedings of IEEE Globecom Workshops*, November 26, 2007, pp. 1–10.
36. Skordylis, A., and Trigoni, N. (2008). Delay-bounded routing in vehicular ad-hoc networks. In *Proceedings of the 9th ACM International Symposium on Mobile Ad hoc Networking and Computing (MobiHoc'08)*, May 26, 2008, pp. 341–350.
37. Zhao, J. and Cao, G. (2013). VADD: Vehicle-assisted data delivery in vehicular ad hoc networks. *IEEE Transactions on Vehicular Technology*, *57*(3), 1910–1922.
38. Naumov, V. and Gross, T. (2007). Connectivity-aware routing (car) in vehicular ad-hoc networks. In *Proceedings of the 26th IEEE International Conference on Computer Communications*, May 6, 2007, pp. 1919–1927.
39. Seet, B.-C., Liu, G., Lee, B.-S., Foh, C.-H., Wong, K.-J., and Lee, K.-K. (2004). A-star: A mobile ad hoc routing strategy for metropolis vehicular communications. In *Proceedings of the 3rd International IFIP-TC6 Networking Conference (NETWORKING'04), Volume 3042 of Lecture Notes*

in Computer Science, May 9–14, 2004, pp. 989–999. Springer: Berlin, Heidelberg.

40. Jarupan, B. and Ekici, E. (2010). Prompt: A cross-layer position-based communication protocol for delay-aware vehicular access networks. *Ad Hoc Network*, *8*, 489–505.
41. Mo, Z., Zhu, H., Makki, K., and Pissinou, N. (2006). Muru: A multi-hop routing protocol for urban vehicular ad hoc networks. In *Proceedings of the 3rd Annual International Conference on Mobile and Ubiquitous Systems*, July 17, 2006, pp. 1–8.
42. Jerbi, M., Senouci, S.-M., Meraihi, R., and Ghamri-Doudane, Y. (2007). An improved vehicular ad hoc routing protocol for city environments. In *Proceedings of the IEEE International Conference on Communications*, June 24, 2007, pp. 3972–3979.
43. Naumov, V., Baumann, R., and Gross, T. (2006). An evaluation of inter-vehicle ad hoc networks based on realistic vehicular traces. In *Proceedings of the 7th ACM International Symposium on Mobile Ad Hoc Networking and Computing (MobiHoc'06)*, May 22, 2006, pp. 108–119.
44. Bachir, A. and Benslimane, A. (April 2003). A multicast protocol in ad hoc networks inter-vehicle geocast. In *Proceedings of the 57th IEEE Semiannual Vehicular Technology Conference, VTC 2003-Spring*, April 22, 2003, Vol. 4, pp. 2456–2460.
45. Briesemeister, L., Schafers, L., and Hommel, G. (2000). Disseminating messages among highly mobile hosts based on inter-vehicle communication. In *Proceedings of the IEEE Intelligent Vehicles Symposium*, Vol. 4, October 5, 2000, pp. 522–527.
46. Maihofer, C. and Eberhardt, R. (2004). Geocast in vehicular environments: Caching and transmission range control for improved efficiency. In *Proceedings of IEEE Symposium Intelligent Vehicles*, June 14, 2004, pp. 951–956.
47. Durresi, M., Durresi, A., and Barolli, L. (2005). Emergency broadcast protocol for inter-vehicle communications. In *Proceedings of the 11th International Conference on Parallel and Distributed Systems*, Vol. 2, July 20, 2005, pp. 402–406.
48. Korkmaz, G., Ekici, E., Özgüner, F., and Özgüner, U. (2004). Urban multihop broadcast protocol for inter-vehicle communication systems. In *Proceedings of the 1st ACM International Workshop on Vehicular Ad hoc Networks*, October 1, 2004, pp. 76–85.
49. Sun, M.-T., Feng, W.-C., Lai, T.-H., Yamada, K., Okada, H., and Fujimura, K. (2000). Gps-based message broadcasting for inter-vehicle communication. In *Processing of the International Conference on Parallel Processing*, August 21, 2000, pp. 279–286.
50. Maia, G., Villas, L. A., Boukerche, A., Viana, A. C., Aquino, A. L. L., and Loureiro, A. A. F. (2013). Data dissemination in Urban Vehicular Ad hoc

Networks with diverse traffic conditions. In *IEEE Symposium on Computers and Communications (ISCC)*, July 7, 2013, pp. 459–464.

51. Maia, G., Rezende, C., Villas, L. A., Boukerche, A., Viana, A. C., Aquino, A. L., and Loureiro, A. A. (2013). Traffic aware video dissemination over vehicular ad hoc networks. In *Proceedings of the ACM International Conference on Modeling, Analysis and Simulation of Wireless and Mobile Systems (MSWiM '13)*, November 3, 2013, pp. 419–426.
52. Ott, J. and Kutscher, D. (2004). Drive-thru internet: IEEE 802.11b for "automobile" users. In *Proceedings of the 23rd Annual Joint Conference of the IEEE Computer and Communications Societies*, March 7, 2004, Vol. 1.
53. Bechler, M., Jaap, S., and Wolf, L. (2005). An optimized TCP for internet access of vehicular ad hoc networks. In *Proceedings of the 4th International IFIP-TC6 Networking Conference (NETWORKING'05), Volume 3462 of Lecture Notes in Computer Science*, May 2–6, 2005, pp. 869–880. Springer: Berlin, Heidelberg.
54. Liu, J. and Singh, S. (2001). ATCP: TCP for mobile ad hoc networks. *IEEE Journal on Selected Areas in Communications*, *19*(7), 1300–1315.
55. Iqbal, M. D. D. S. and Iftode, T. N. L. (2005). VITP: An information transfer protocol for vehicular computing. In *Proceedings of the 2nd ACM International Workshop on Vehicular Ad Hoc Networks (VANET'05)*, September 2, 2005, pp. 30–39.
56. Luo, Y., Zhang, W., and Hu, Y. (2010). A new cluster-based routing protocol for VANET. In *2010 Second International Conference on Networks Security, Wireless Communications and trusted Computing*, April 24, 2010, Vol. 1, pp. 176–180.
57. Mershad, K., Artail, H., and Gerla, M. (2012). ROAMER: Roadside units as message routers in VANETs. *Ad Hoc Networks*, *10*(3), 479–496.
58. Tsiachris, S., Koltsidas, G., and Pavlidou, F. N. (2013). Junction-based geographic routing algorithm for vehicular ad hoc networks. *Wireless Personal Communications*, *71*(2), 955–973.
59. Hamid, U. Z. A., Zamzuri, H., and Limbu, D. K. (2019). Internet of vehicle (IoV) applications in expediting the implementation of smart highway of autonomous vehicle: A survey. *Performability in Internet of Things*, *2018*, 137–157.
60. Li, N., Martinez-Ortega, J. F., Diaz, V. H., and Fernandez, J. A. S. (2018). Probability prediction-based reliable and efficient opportunistic routing algorithm for VANETs. *IEEE/ACM Transactions on Networking*, *26*(4), 1933–1947.
61. Kaiwartya, O., Abdullah, A. H., Cao, Y., Altameem, A., and Mukesh. (2016). Internet of vehicles: Motivation, layered architecture, network model, challenges, and future aspects. *IEEE Access*, *4*, 5356–5373.

62. Boussoufa-Lahlah, S., Semchedine, F., and BouallouchéMedjkoune, L. (2018). Geographic routing protocols for Vehicular Ad hoc NETworks (VANETs): A survey. *Vehicular Communications*, *11*, 20–31.
63. Wu, W., Yang, Z., and Li, K. (2016). Internet of Vehicles and applications. In *Internet of Things: Principles and Paradigms*, pp. 299–317. Morgan Kaufmann: Burlington, MA, USA; Elsevier: Amsterdam, Netherlands.
64. Zeadally, S., Hunt, R., Chen, Y. S., Irwin, A., and Hassan, A. (2012). Vehicular ad hoc networks (VANETS): Status, results, and challenges. *Telecommunication Systems*, *50*(4), 217–241.
65. Seth, I. (2020). A survey on driver behavior detection techniques. *International Journal of Scientific & Technology Research*, *7*(3), 401–404.
66. Hamid, U. Z. A., Zamzuri, H., and Limbu, D. K. (2019). Internet of vehicle (IoV) applications in expediting the implementation of smart highway of autonomous vehicle: A survey. *Performality in Internet of Things*, *2018*, 137–157.
67. Verroios, V., Efstathiou, V., and Delis, A. (2011). Reaching available public parking spaces in urban environments using ad hoc networking. In *IEEE 12th International Conference on Mobile Data Management*, Lulea, Sweden, Vol. 1, pp. 141–151.
68. Lu, R., Lin, X., Zhu, H., and Shen, X. (2009). SPARK: A new VANET-based smart parking scheme for large parking lots. In *IEEE INFOCOM*, Rio de Janeiro, Brazil, pp. 1413–1421.
69. Chim, T. W., Yiu, S. M., Hui, L. C. K., and Li, V. O. K. (2012). VSPN: VANET-based secure and privacy-preserving navigation. *IEEE Transactions on Computers*, *63*(2), 510–524.
70. Yang, F., Wang, S., Li, J., Liu, Z. and Sun, Q. (2014). An overview of internet of vehicles. *China Communications*, *11*(10), 1–15.
71. National Science Foundation (2012). http://1.usa.gov/16kJQ3T.
72. Lindner, W. and Madden, S. (2004). Data management issues in disconnected sensor networks. Informatik 2004, Informatik verbindet, Band 2, Beiträge der 34. Jahrestagung der Gesellschaft für Informatik eV (GI'04), pp. 349–354.
73. Wolfson, O., Xu, B., and Cho, H. J. (2009). Multimedia traffic information in vehicular networks. In *Proceedings of the 17th ACM SIGSPATIAL International Conference on Advances in Geographic Information Systems, (GIS'09)*, pp. 480–483. ACM: New York, USA.
74. Thomaidis, G., Vassilis, K., Lytrivis, P., Tsogas, M., Karaseitanidis, G. and Amditis, A. (2011). Target tracking and fusion in vehicular networks. In *Intelligent Vehicles Symposium (IV), IEEE*, June 5, 2011, pp. 1080–1085.
75. Guo, J. and Balon, N. (2006). *Vehicular Ad Hoc Networks and Dedicated Short-range Communication*. Technical Report, University of Michigan, Dearborn, pp. 1–56.

76. Stampoulis, A. and Chai, Z. (2007). A survey of security in vehicular networks. http://zoo.cs.yale.edu/~ams257/projects/wireless-survey.pdf (accessed 28 November, 2011).
77. Raya, M., Papadimitratos, P., and Hubaux, J.-P. (2006). Securing vehicular communications. *IEEE Wireless Communication Special Issue on Inter-Vehicular Communications*, *13*(5), 8–15.
78. Imieliński, T. and Navas, J. C. (1999). GPS-based geographic addressing, routing, and resource discovery. *Communication of the ACM*, *42*, pp. 86–92.
79. Navas, J. C. and Imielinski, T. (1997). Geocast — Geographic addressing and routing. In *Proceedings of the 3rd Annual ACM/IEEE International Conference on Mobile Computing and Networking, (MobiCom'97)*, pp. 66–76. ACM: New York, NY, USA.
80. Karagiannis, G., Altintas, O., Ekici, E., Heijenk, G., Jarupan, B., Lin, K. and Weil, T. (2011). Vehicular networking: A survey and tutorial on requirements, architectures, challenges, standards and solutions. *IEEE Communications Surveys & Tutorials*, *13*(4), 584–616.
81. Kumar, S., Verma, A., Verma, P. K., and Verma, R. (2022). A comparative study of energy efficient collision avoidance protocols for WBAN. In *IEEE 6th Conference on Information and Communication Technology*, November 2022, pp. 1–5.
82. Gaurav, A., Gupta, B. B., Peñalvo, F. J. G., Nedjah, N., and Psannis, K. (2022). DDoS attack detection in vehicular ad-hoc network (VANET) for 5G networks. In *Security and Privacy Preserving for IoT and 5G Networks: Techniques, Challenges, and New Directions*, pp. 263–278. Springer: Berlin/Heidelberg, Germany.
83. Gaurav, A., Gupta, B. B., Castiglione, A., Psannis, K., and Choi, C. (2020). A novel approach for fake news detection in vehicular ad-hoc network (VANET). In *Computational Data and Social Networks: 9th International Conference (CSoNet 2020),* Dallas, TX, USA, December 11–13, 2020, pp. 386–397. Springer International Publishing.
84. Liang, Y., Liu, Y., and Gupta, B. B. (2022). PPRP: Preserving-privacy route planning scheme in VANETs. *ACM Transactions on Internet Technology*, *22*(4), 1–18.
85. Chui, K. T., Kochhar, T. S., Chhabra, A., Singh, S. K., Singh, D., Peraković, D., Almomani, A., and Arya, V. (2022). Traffic accident prevention in low visibility conditions using VANETs cloud environment. *International Journal of Cloud Applications and Computing (IJCAC)*, *12*(1), 1–21.

https://doi.org/10.1142/9789811279539_0008

Chapter 8

Performance Analysis of Clustering Algorithms in Wireless Camera Sensor Networks for Energy-Efficient Cooperative Communication

Nishant Tripathi[†,‡,§,¶], **Charanjeet Singh**[*,†,||], and **Kamal Kumar Sharma**[†,§,**]

[†]*Department of Electronics and Communication Engineering, Lovely Professional University, Phagwara, Punjab, India*
[‡]*PSIT, Kanpur, Uttar Pradesh, India*
[§]*Ambala College of Engineering and Applied Research, Ambala, Haryana, India*
[¶]*phd.nishant17@gmail.com*
[||]*rcharanjeet@gmail.com*
[**]*kamalsharma111@gmail.com*

Abstract

Generally, the current application areas in communication are completely dependent on the establishment and operation of (camera-based/non-camera-based) wireless sensor networks for cooperative communication (CC). CC, being an ideal sub-part of the upcoming 6G telecom

[*]Corresponding author.

infrastructure, ensures that optimum operations can be analyzed and evaluated based on the lifeline of the model of a network. It should not be very power-consuming (i.e., it must be energy efficient with better throughput), and its reliability for the transmission of data packets should be optimal. The majority of researchers working in sensor-based fields have shifted their attention to the application domains of 6G (preferably in the CC domain), with the aim of optimizing the energy, power, and life span of sensor-based networks while also taking reliability into consideration (camera-based or non-camera-based). In light of this, this chapter gives a broader analytical review of CC based on recent research, followed by a discussion on research gaps and performance issues in current wireless networks leading up to CC. Finally, the implementation of several clustering schemas for data transmission in MATLAB is described, and their performances are compared in terms of energy efficiency.

Keywords: Cooperative communication, clustering, energy efficiency, 6G, wireless camera sensor networks

8.1. Introduction

Wireless camera sensor networks (WCSNs) [1, 3] are networks of unattended, ultrasmall, minimum-power member sensor nodes (MSNs) that communicate with one another. These new-age wireless networks can collaborate with upcoming G and 6G communication modules and beyond. When sensors are deployed in a random fashion across an investigation area, their partial dispensation, wireless-based communication, and energy store competences send sensed data to a sink or base station (BS) for analysis. The improvement of technologies and devices has shaped many opportunities for the efficient use of resources in critical environments. In this context, wireless sensor networks (WSNs) have brought about a sea change. With the introduction of this technology, it became possible to collect and deliver useful information to the intended recipient. Applications such as battlefield surveillance, smart offices, and traffic monitoring can be effectively monitored with the help of such systems. For the purpose of gathering useful information, WSNs are composed of several unmanned and extremely small, compact, and low-power-consumption MSNs that are scattered throughout a region of interest (Figure 8.1), such as an inaccessible area or a disaster zone. Numerous research opportunities have arisen with the growth of small MSNs capable of sensing, processing

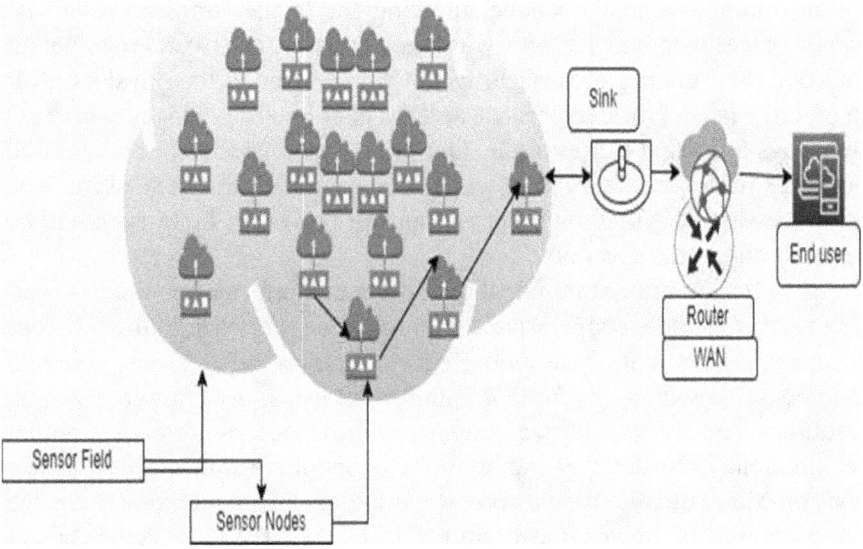

Figure 8.1. Wireless Network Setup (Camera Based/Non-Camera-Based)

expressive information, and interactivity with their destination. These battery-powered sensor nodes have tremendous partial processing and storage capabilities. Because WSNs are exposed to dynamic environments, connectivity loss between nodes may occur as a result of such a configuration, resulting in a decrease in network performance. To improve performance, energy-efficient protocols must be designed, extending the network's life. When a critical node fails to operate properly, the performance of centralized algorithms degrades significantly, resulting in a severe protocol failure. In comparison to traditional methods, distributed protocols are more efficient at handling failures and may be a more appropriate solution. For network energy efficiency, energy-efficient clustering structured routing protocols capable of data aggregation are being developed. Localized algorithms can operate within a cluster without the need to wait for control messages to arrive, thereby reducing the overall delay. Furthermore, when compared to centralized algorithms, the scalability of these localized algorithms is significantly improved.

Transmit diversity necessitates several transmitter antenna systems in most cases. Many wireless devices, however, can only have one antenna due to their small size or complexity. Cooperative communication (CC) is a relatively new form of communication that allows single-antenna mobiles to share their antennas in a multi-user environment and create a virtual multiple-antenna sender, allowing for greater transmit diversity. Users of cellular and *ad hoc* wireless networks can work together to improve their quality of service, which is measured at the physical layer in bit error rates, block error rates, and the likelihood of outages. There are two ways to look at power usage. One way is that more power is required because, in cooperative mode, each user transmits for both users. The other standpoint is that the baseline transmit power for both users will be reduced due to the diversity.

It's also important to think about how much information the base station needs to handle cooperative communication. For each of the previous schemes, the amount of additional information provided varies. The goal is to come up with a system that is fair to all users, uses minimal system resources, and works with the system's multiple access protocol. Another requirement is to develop methods for cooperative transmission power control. Much of what we've done so far assumes that everyone sends the same amount of power. Furthermore, in CDMA-based systems, power control is essential for managing the near–far effect and minimizing interference. As a result, power control schemes that work well in a cooperative communication context are essential.

To create heterogeneous networks, small cells are stacked on top of macro base station (MBS) tiers. This contributes to meeting the growing demand for faster, higher-quality communication. Small cells can help reduce MBS loads, but they are not always successful. Because of the high potential of small cell density, collaboration is becoming increasingly important. Software-defined networking (SDN), for example, is a wired network technology that can also be used in cellular networks. The understanding of multi-tier networks is aided by SDN.

Wireless networks face two major challenges: energy efficiency and the recovery of energy from the environment. Energy efficiency refers to the amount of data that can be sent per unit of energy and is measured in bits per Joule.

A number of factors influence the stability and performance of a wireless network. Mobile users' location, distance, and connectivity are just a few of the variables that can change frequently in adaptive topology networks. Due to the Doppler shift, different distances, and different obstacles, the link quality of users of centralized wireless networks may change over time as they move into a cell. The MAC layer parameters, as well as how users collaborate, select relays, and assign power, may all be affected by this mobility feature. Relay selection is one of the most serious problems with mobile networks.

In this chapter, we compare the MATLAB outputs of clustering algorithms for WSNs/WCSNs based on the following metrics: time slot of no node to die out for stability calculation, module under supervision lifespan, and effective efficiency. Cluster heads (CHs) receive information from sensor nodes, and it is their responsibility to transmit this information to a sink located further along the path in a field using several jumps. Clustering algorithms for sensor networks, such as low-energy adaptive clustering hierarchy (LEACH), distributed energy-efficient clustering (DEEC), TEEN, stable election protocol (SEP), and EESAA [1, 2, 4, 6], have achieved reasonable goals in terms of improved performance of networks under a variety of criteria and circumstances [1, 2, 4, 6–9]. This chapter discusses the details of the different clustering schemes, namely LEACH [1], DEEC [1], SEP [1, 2], and their modified versions. The MATLAB outputs of each scheme are presented in separate sections to evaluate their performance. The final outcomes of the study are written in such a way that they will provide researchers with enough material to understand all of the multilevel distributed clustering schemes with varying degrees of heterogeneity. The chapter begins with the basics of WCSNs and their deployment issues. This will be followed by an introduction to clusterings and their advantages. Following the two, we discuss different algorithms and then the simulations of most commonly used clustering algorithms. The final part is an analytical conclusion based on the application areas of clustering algorithms and the best performing algorithm.

8.2. Literature Review of Cooperative Communication

Wang and Zhang [3] proposed the alternating quadrature differential binary phase shift keying/direct-sequence code division multiple access (AQDBPSK/DS-CDMA) scheme that aims at greater energy efficiency and interference mitigation for three-dimensional (3D) clustered wireless communications sensor networks, with the minimum target-object coverage rate constraint applying the NEW LEACH architecture for energy effectiveness and interference alleviation following the Nakagami-m model.

Nosratinia *et al.* [6] proposed a concept of CC that uses single-antenna devices in a multi-user environment to share their antennas and generate a virtual multiple-antenna transmitter, allowing them to achieve transmit diversity.

Parmar and Pirishothm [7] discussed WSNs in environmental monitoring applications. The sensor data in these applications may include light intensity, temperature, pressure, humidity, and their variations, which may be managed by vital clustering methods for the networks, such as LEACH, TEEN, and APTEEN.

Kiwan and Morgan [8] worked on the concept of hierarchical network for large networks and thoroughly discussed the methodology with the protocols, providing an adequate explanation for the advantages and disadvantages of each protocol. In this paper, hierarchical network routing is explained and presented as a solution for large network problems using an example. Particular hierarchical network routing algorithms are concisely compared. The paper also gave a short review of a number of clustering algorithms.

Rathi *et al.* [9] proposed to extend the lifetime of the sensor network through the approaches of gradient-based routing and hierarchical techniques, which show superior scalability and proficient communication. According to the authors' description, hierarchical routing upholds the energy intake of sensor nodes and performs data aggregation, which helps in lessening the number of transmitted messages to the base station. The problem of single-path routing approach has been covered by considering multipath routing.

Jain and Chaudhary [10] compared LEACH and PEGASIS [1] in terms of energy efficiency and life span. Their comparison criterion was based on energy efficiency in LEACH and power optimization in PEGASIS, and their study showed a trade-off between drastic improvements in the life span of individual SNs and those with the overall network.

Hooggar *et al.* [13] proposed a new 3D coverage algorithm for three-dimensional WSNs. The presented algorithm was based on our 3D clustering algorithm. Full coverage and overlap avoidance were the two proposed coverage methods. In the normal network implementation, 59–89% energy savings were achieved, while the proposed coverage algorithm showed different effects: 14–75% energy savings after changing the cluster size.

Abakumov and Koucheryavy [14] proposed an MCA algorithm that supports USNs. The authors' approach and the main goal were to maximize coverage during extended time attainment for such USNs. The basic criteria for assessing the LEACH protocol they adopted were coverage area and the number of connected nodes in the current round. The proposed algorithm was found to be better than LEACH for 100% and 90% coverages.

Hai *et al.* [15] proposed a method of sensor-energy optimization in 3D WSNs by determining an optimal topology of sensors that prolonged the network lifetime and reduced their energy expenditure. The mathematical model that was proposed for clustering in 3D WSNs used the Lagrange multiplier method, and solutions to the model of cluster centers resulted in a new algorithm, named FCM-3 WSN. The authors also provided experimental confirmation that FCM-3 WSN provided better results than LEACH, centralized LEACH, single-hop clustering, energy-efficient protocol (SCEEP), hybrid LEACH (H-LEACH), as well as *K*-means and fuzzy *C*-Means (FCM).

Ghosal *et al.* [18] developed a method of lifetime maximization in WSNs by devising a novel dynamic clustering. The authors provided an optimal clustering technique based on an analysis in which the optimal clustering radius of each level was computed using ADMM followed by the development of OPTIC, an on-demand, distributed optimal clustering algorithm. Experimental results also demonstrated that the work scheme increased the network lifetime without compromising the network performance metrics. Further, the authors measure the performance of the designed clustering algorithm.

Xie and Zhang [2] proposed a 3D coverage scheme using WCSNs, aiming to maximize the lifespan by minimizing energy consumption. The authors built up a model for the field of view (FOV) of our 3D WCSN, deriving the analytical expressions that describe the minimum number of arbitrarily organized camera sensor nodes to ensure the 3D coverage rate. The authors worked on the new LEACH architecture by applying heterogeneous sensors.

Amis *et al.* [20] presented the recent research results on data routing in sensor networks and classified the approaches into three main categories, namely data-centric, hierarchical, and location-based. The authors also gave hybrid protocols that fit into more than one category.

Brust and Baker [21, 22] worked on the optimization of the cluster head selection technique in the LEACH protocol, in which a proper cluster head prediction model was developed with optimum output for the concerned network. The simulated results also provided validation for the same in optimum energy utilization by the sensor nodes and for the overall sensor network.

There are several more articles on this topic, some of which are summarized in Table 8.1.

Table 8.1. Important Publications/Researches in the Field of Clustering in WSN

S. No.	Key Areas of Research	Author(s)	Publisher or Journal	Year of Research
1	MIMO-OFDM-based approach for 3D clustering, prevention by exposure path [16]	Wang *et al.*	IEEE Transactions	2020
2	Energy-aware clustering techniques, Lion Pride Optimizer (LPO) algorithm, and fuzzy logic used for network design [2]	Tabatabaei *et al.*	Springer	2019
3	AQ-DBPSK modulation or DS-CDMA multiple access technique is used, least coverage rate constraint [4]	Wang and Zhang	IEEE Transactions	2014
4	Cross-layer-sensing clustering based on fog computing [2]	Sun *et al.*	Special Section on Green Internet of Things, IEEE Access	2019
5	MIMO-OFDM technique in CC [29]	Wang and Zhang	*Proc. IEEE Wireless Communication*	2015
6	Genetic algorithm using dynamic clustering [35]	Yuan *et al.*	*Journal of Network and Systems Management Volume*, Springer	2019
7	Review of LEACH, TEEN, and APTEEN [10]	Parmar and Pirishothm	*International Journal of Science and Research*	2015
8	Expanded version of coverage in strict surroundings [14]	Elhoseny *et al.*	Springer	2018
9	Application area survey of WSNs [36]	Rathi *et al.*	*International Journal of Distributed and Parallel Systems*	2012
10	Comparison between LEACH and PEGASIS [20]	Jain and Chaudhary	*International Journal of Computer Science and Mobile Computing*	2017

Table 8.1. (*Continued*)

S. No.	Key Areas of Research	Author(s)	Publisher or Journal	Year of Research
11	Flying ubiquitous sensor networks as a queuing system [1]	Kirichek *et al.*	*International Conference on Advanced Communication Technology*, ICACT 2015, Phoenix Park, Korea	2015
12	New 3D method to make clusters and improve efficiency of network [1]	Attarzadeh *et al.*	*Global Journal of Computer Science and Technology*	2011
13	Energy-efficient 3D coverage for clustering [12]	Hooggar *et al.*	*Journal of Academic and Applied Studies*	2013
14	The cluster head selection algorithm in the 3D USN [15]	Abakumov *et al.*	*International Conference on Advanced Communication Technology*, ICACT 2014, Phoenix Park, Korea.	2014
15	Novel fuzzy clustering is used [10]	Hai *et al.*	*Applied Soft Computing Journal*	2017
16	Descriptive survey of WSNs [3]	Akyildiz *et al.*	Elsevier	2002
17	A distributed clustering algorithm guided by the base station to extend the lifetime of WSNs [9]	Yuste-Delgado *et al.*	Elsevier	2016
18	Distributed on-demand clustering algorithm for lifetime optimization in wireless sensor networks [1, 2]	Ghosal *et al.*	Wireless Sensor Networks, *Journal of Parallel and Distributed Computing*	2020
19	Minimum coverage rate constraint is used for a better clustering methodology [22, 14, 30]	Xie and Zhang	Globecom 2013 — *Ad Hoc* and Sensor Networking Symposium	2013

(*Continued*)

Table 8.1. (*Continued*)

S. No.	Key Areas of Research	Author(s)	Publisher or Journal	Year of Research
20	Routing protocols for WSNs [1, 7, 13]	Shabbir *et al.*	IntechOpen	2017
21	Detailed aspects of cooperative distributed MIMO channels [1, 31]	Spagnolini	IEEE	2007
22	Theoretical model for CC [1, 2]	Nosratinia *et al.*	IEEE	2004
23	Detailed survey of hierarchical routing [1, 13]	Kiwan and Morgan	IEEE Canadian Conference of Electrical And Computer Engineering (CCECE)	2013
24	Clustering algorithm for 3D wireless mobile sensor networks [1, 2]	Abakumov *et al.*	Springer International Publishing, Switzerland	2015
25	Detailed explanations of current and future of WSNs	Pathan *et al.*	CRC Press	2012
26	Overlapping clustering method	Niu *et al.*	IEEE Explore	2016
27	Cluster-based energy-efficient transmission using a new hybrid compressed sensing in WSN	Liu *et al.*	IEEE Explore	2016

8.3. Issues in Clustering for Developing Better WSN/WCSN for CC/CWC

The primary objective of clustering is to minimize the power needed for communication to save the battery life of a sensor network, beyond which it also offers the advantages listed in Figure 8.2 [37–51]. There are several issues for clustering which are very important to consider in designing a clustering protocol for any WSN/WCSN for CC:

(1) longer lifetime of SNs and the overall network,
(2) balanced load for proper power distribution across different SNs,

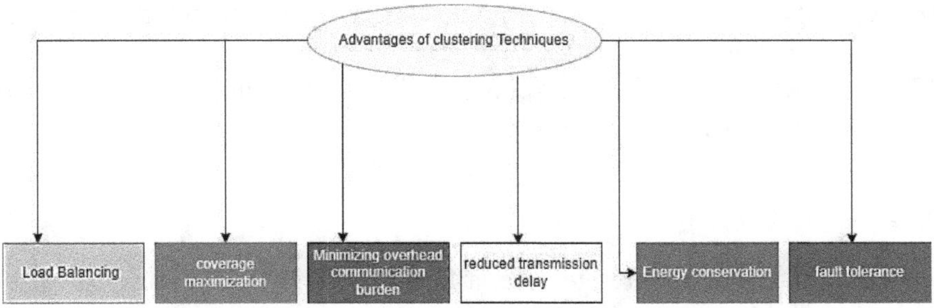

Figure 8.2. Various Advantages of Clustering

(3) scalability,
(4) cost-effective on different trade-off logics,
(5) security,
(6) compatibility of heterogeneous networking environments with homogeneous ones.

8.3.1. *Research Gap*

(1) Most of the research work covered in the Table 8.1 covers for WSN and not for WCSN (clearly points a research gap). However some work using Heterogenous network has been done (table covered them) but w.r.t WSN only and not with the Wireless camera sensor networks (WCSN). Therefore, a weighted SN can be considered to determine the overall network performance of heterogeneous networks, particularly for WCSNs [52–57].
(2) The distributed clustering approach has been discussed only with a single kind of network scheme (grouping the SNs in equal numbers), so a multilevel-based network from small to large networks, transitioning from 2D to 3D, can be taken into consideration.
(3) Movement-based data aggregation has been addressed, but with limited capacity in homogeneous environments.
(4) Cooperative diversity has been verified with random variables only, not with Gaussian variables or AWGN.
(5) Camera sensor networks have not been tested with hierarchical clustering in the distributed approach for multilevel-based networking, and their optimum power trade-off with respect to the overall WCSN life span has to be tested.

(6) Security features are still a huge concern because of large power requirements.
(7) Load balancing and energy consumption at optimum levels have still not been configured by and large.

8.4. Clustering

Based on the network architecture, a layout, as shown in Figure 8.3, has to be adopted before implementing the model.

8.4.1. *Single or Direct Link Model Network*

- Each MSN and sink establish a direct link for data transmission [58–63].
- Advantage: easier implementation.
- Disadvantage: not scalable.

8.4.2. *Hop-Based or Peer-Based Link Network*

- The MSNs will have route-based transmission capability.
- The model is not scalable as MSNs nearby the BS are used for data routing to other MSNs and the BS.

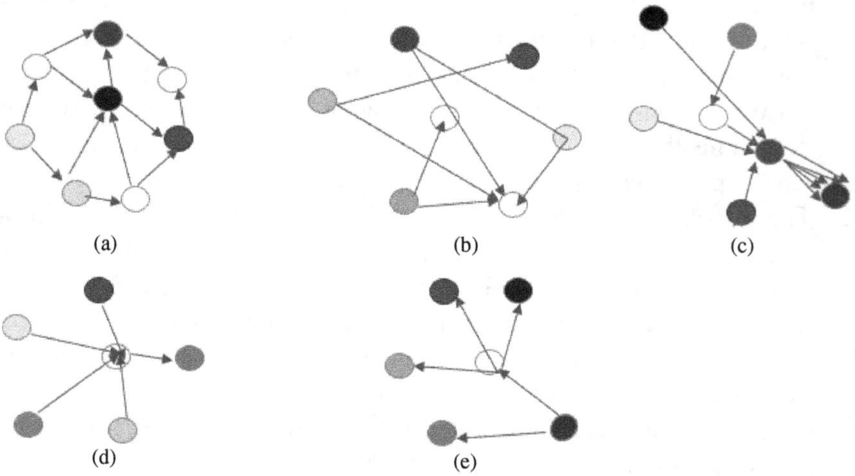

Figure 8.3. Routing Network Architecture. (a) local communication between SNs, (b) link to link communication, (c) convergence, (d) aggregation mode of data transfer, and (e) divergence mode of data transfer

- Advantage: Large network and scalable since traffic in such MSNs may increase.
- Disadvantage: Energy wastage causes low network lifespan.

8.4.3. *Multi-Hoping-Based Network Depending upon Clustering*

- The MSNs form a structure by combining in groups, and then, a group leader (CH) for any cluster (group) will be chosen to transmit data to the BS or sink directly (Figure 8.4). The MSNs transmit their data to the CH, which further assesses the data and sends the processed data collected from all MSNs to the sink/BS [64–68].
- Disadvantage: The energy of the CH will drop faster than other MSNs.

8.4.4. *Multi-Hoping Clustering Based on Dynamic CH Network Model*

- Dynamically changes the role of the CH among other MSNs [69–75].

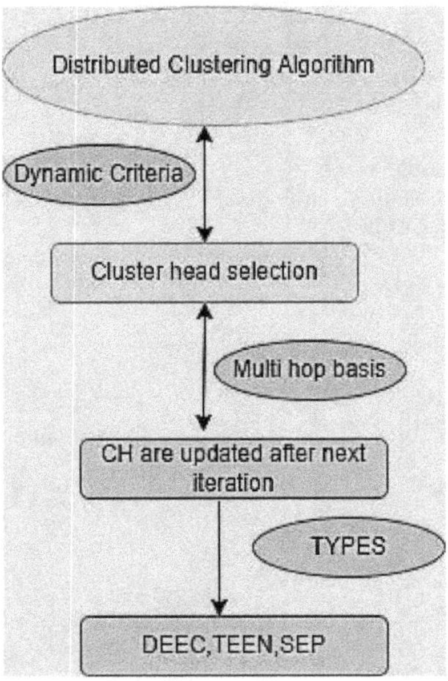

Figure 8.4. Distributed Clustering Algorithm: Basic Approach and Types

8.5. Different Clustering Techniques

The different types of clustering divisions are listed in Figure 8.5, and their key characteristics are summarized as follows.

8.5.1. *Homogeneous Clustering (Single-Hoping-Based): Key Characteristics*

(i) ***LEACH-C*** *(LEACH centralized) [1, 2]*:

- Algorithm is centrally initiated.
- Balanced cluster formation.
- Not energy efficient.
- Not robust for global data of the network [76–81].

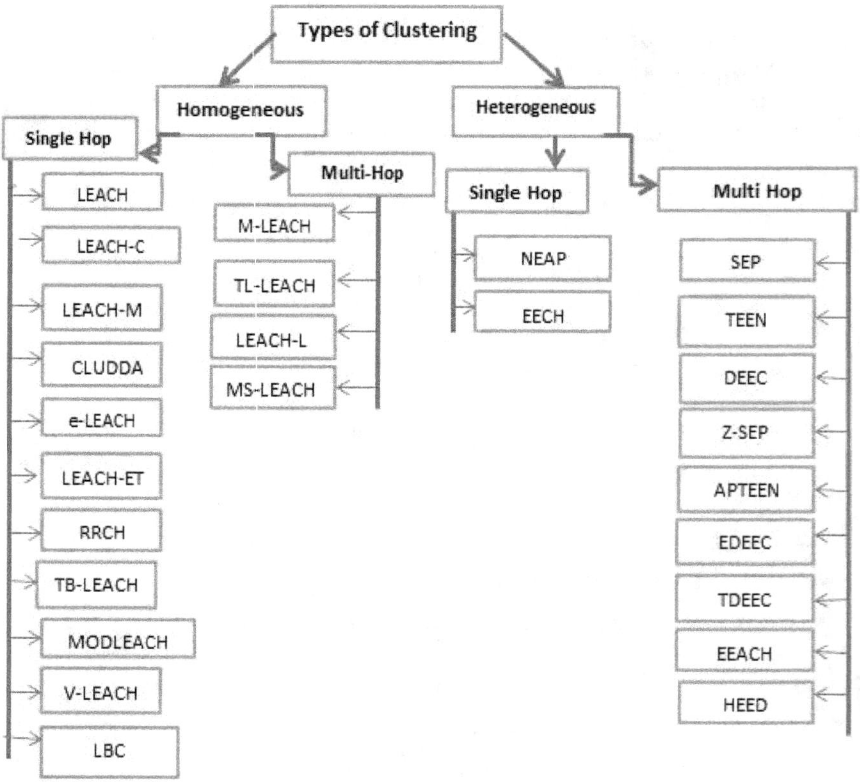

Figure 8.5. Different Types of Clustering Divisions

(ii) ***LEACH-F*** *(LEACH fixed) [1–5]*:
- Fixed cluster formation.
- Energy consumption among MSN is balanced.
- Bypasses additional complexity and difficulty in the setup phase.
- Because of fixed round energy, the information and energy of MSNs are wasted before CH is in the die state.

(iii) ***CLUDDA*** *(clustered diffusion with dynamic data aggregation) [3, 4]*:
- Avoid flooding issue by eliminating unended data transmission.
- Dynamic data aggregation.
- Large time delay.
- Requires larger memory storage [82–89].

(iv) ***s-LEACH*** *(solar-aware LEACH) [4–7]*:
- Solar power is used.
- Improved network life.
- Utilized in centralized as well as distributed algorithms.

(v) ***LEACH-ET*** *(LEACH energy threshold)*:
- Computes rotation policy in LEACH.
- Energy efficiency is slightly better.
- Control message causes a large amount of energy dissipation [90–95].

(vi) ***E-LEACH*** *(energy LEACH)*:
- Two-phase process, with phase one divided into rounds.
- A part of energy is wasted in round phase transmission, which is fixed.

(vii) ***RRCH*** *(round-robin cluster head) [3, 4]*:
- Achieves better efficiency using a single setup procedure.
- Reduces energy consumption.
- Poor load balancing [96–103].
- Additional overhead burden.

(viii) ***TB-LEACH*** *(time-based cluster head selection algorithm for LEACH) [5, 6, 15]*:
- Procedure of selecting CHs is varied.
- Lifetime of networks is enhanced.
- Does not support large networks.
- Load balancing is poor.
- Overhead burden is high.

(ix) ***MLEACH-L** (more energy-efficient LEACH) [3, 6, 19]*:

- Used for dense and vast WSNs.
- Enhances the setup stage and additional overhead.
- Better energy profile.
- Load balancing issues are present [24–34].

(x) ***V-LEACH** (new version LEACH) [1, 9, 10]*:

- Improved results compared to original LEACH [1] because a redundant CH acts as a CH when the former is not working.
- Increased time of setup phase.
- Duplication is possible.

(xi) ***P-LEACH** (partition-based LEACH)*:

- Balanced wasted energy profile.
- Setup enhancement issue is present.

(xii) ***WST-LEACH** (weighted spanning tree for LEACH) [3, 8]*:

- Selecting CHs relies on three conditions, which optimizes transmission path.
- Reduced power dissipation.
- Increased network life.

(xiii) ***EBC** (energy-balanced clustering) [2, 3, 8]*:

- Saves the unnecessary energy consumed throughout reclustering.
- Additional overhead issue.
- Load balancing is still an issue.
- Additional time is required for different stages of transmission.

(xiv) ***LEACH-SC** (LEACH selective cluster) [1–3, 8]*:

- Enhanced scalability.
- Additional overhead issues.
- Load balancing problem.

8.5.2. *Homogenous Multi-Hoping-Based: Key Characteristics*

(xv) ***M-LEACH** (multi-hop LEACH) [2, 3]*:

- Appropriate for huge module network with low density.
- Limited scalability issue.
- For small networks, it does not give optimum solutions.

(xvi) ***TL-LEACH** (two-level LEACH) [1, 3, 11]*:

- Energy consumption is reduced.
- Allocation of energy load among MSNs in dense networks.
- Not too good for vast dense model networks.

(xvii) ***LEACH-L** [1, 11, 21]*:

- Stabilizes the model network load.
- Reduced energy level usage per MSN.
- Additional overhead issues.

(xviii) ***MS-LEACH** (combines multi-hop and single hop) [21]*:

- Reduced consumption of energy due to combination of single- and multi-hopping profiles.
- Bounded or limited scalability.
- Additional overhead issues.

8.5.3. *Heterogeneous Single Hop: Key Characteristics*

(xix) ***EECHE** (energy-efficient cluster head election protocol) [1]*:

- Throughput and lifetime are enhanced as compared to LEACH and SEP.
- Limited scalability compared to SEP.
- Latency is reduced.

(xx) ***NEAP** (novel energy adaptive protocol) [1, 8]*:

- Improved consistency.
- Inadequate scalability.
- Not good for dense networks.

8.5.4. *Heterogeneous Multi-Hop Protocols: Key Characteristics*

(xxi) ***SEP** (stable election protocol) [1]*:

- No requirement of collecting data nearby MSN energy/round.
- For multilevel networks, its performance is poor.

(xxii) ***HEED** (hybrid energy-efficient distributed clustering) [1, 13, 27]*:

- Stabilizes power among MSNs.
- Overhead issues are managed and controlled.
- Latency is not good.
- Limited scalability for the network.

(xxiii) ***EEUC*** *(novel energy-efficient clustering approach) [3, 8]*:

- Minimizes problem using unequal cluster sizes.
- Imbalanced cluster.
- Poor load balancing.
- Better energy efficiency.

(xxiv) ***LEACH-HPR*** *[1, 11]*:

- CH elects the best MSN to stabilize energy consumption profile.
- Additional overhead issue.

(xxv) ***DEUC*** *[1, 2]*:

- Hotspot issue minimized.
- Overhead issue minimized.
- Poor load balancing.
- Better energy consumption profile.

8.6. MATLAB Simulation Outputs of LEACH, M-LEACH, SEP, DEEC, TEEN, and EEACH

This section presents the MATLAB results of different clustering schemes: (i) LEACH (Figure 8.6) and its modified version (Mod-LEACH; Figure 8.7); (ii) DEEC (Figure 8.8) and its modified versions, DDEEC (Figure 8.9), TDEEC (Figure 8.10), and EDEEC (Figure 8.11), with Figure 8.12 and Table 8.2 comparing the different DEEC versions based on a number of parameters; (iii) SEP and its modified version, Zonal SEP, compared with LEACH (Figures 8.13–8.15 and Table 8.3).

8.6.1. *LEACH*

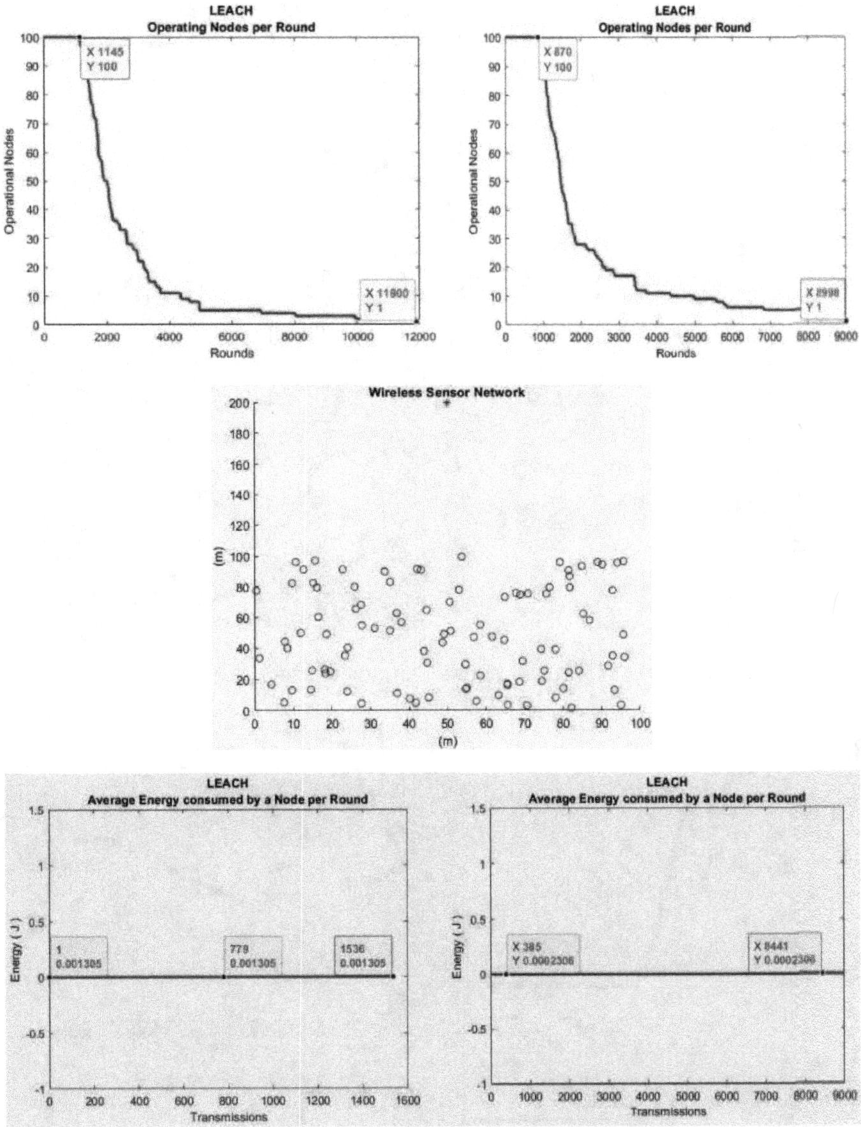

Figure 8.6. MATLAB Simulation Outputs of LEACH Protocols

8.6.2. *Modified LEACH (Mod-LEACH)*

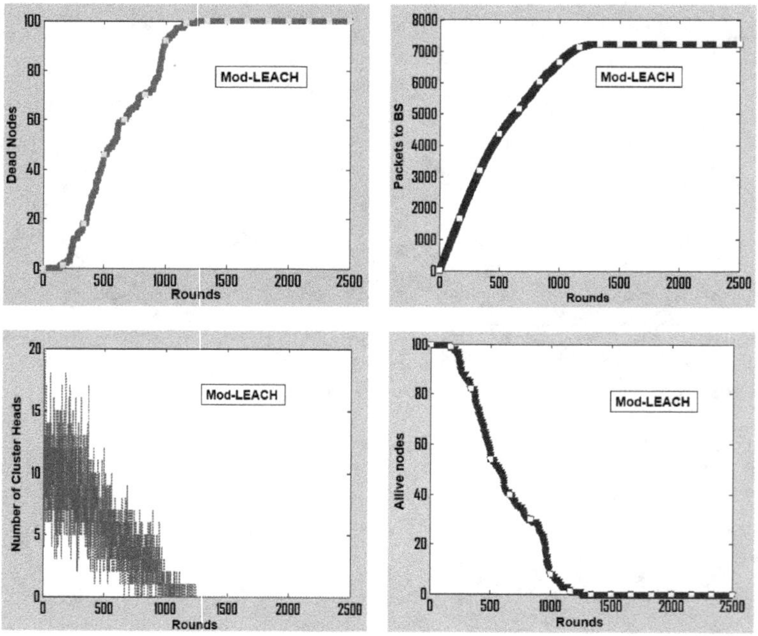

Figure 8.7. MATLAB Simulation Outputs of Mod-LEACH Protocol

8.6.3. *DEEC*

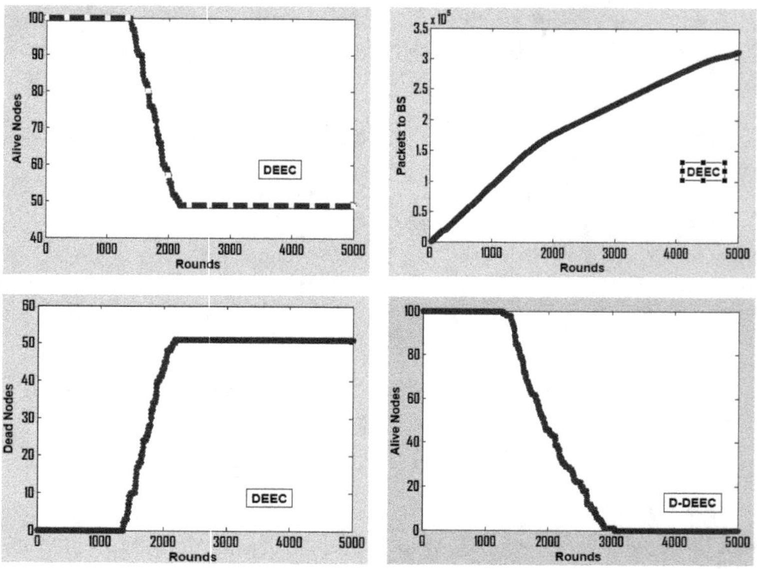

Figure 8.8. MATLAB Output of DEEC

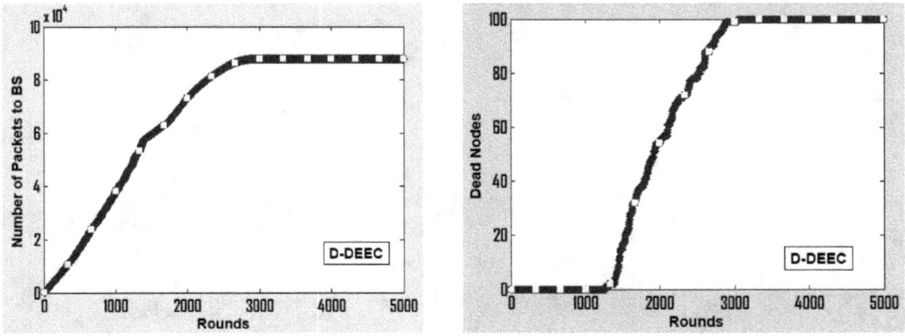

Figure 8.9. MATLAB Output of DDEEC

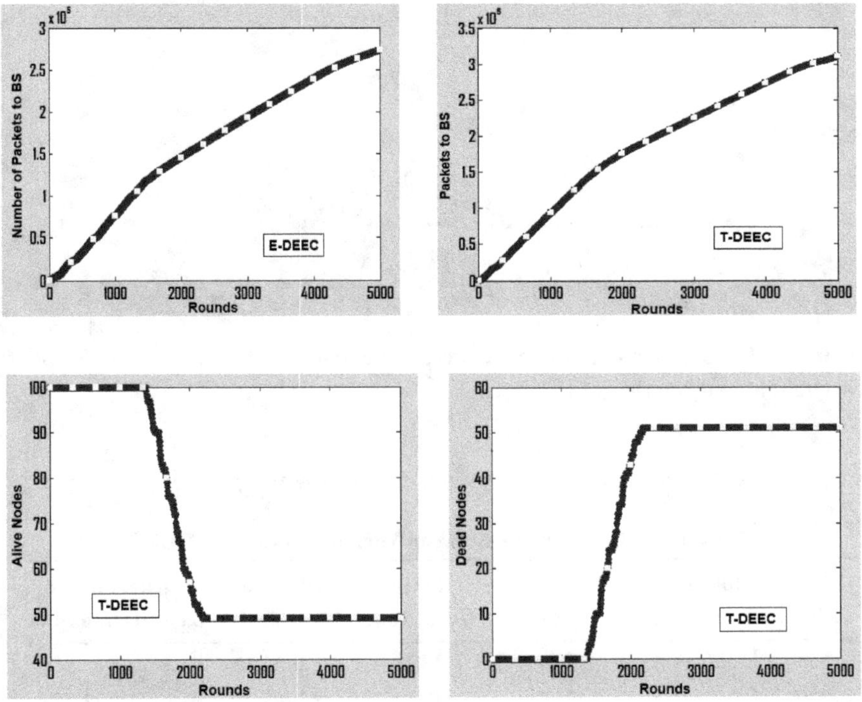

Figure 8.10. MATLAB Output of TDEEC

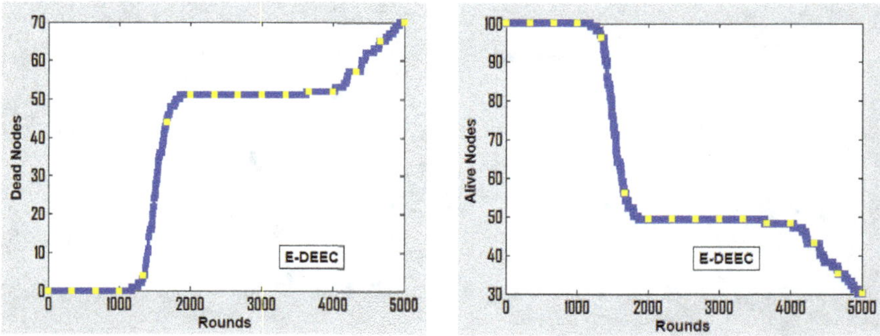

Figure 8.11. MATLAB Output of EDEEC

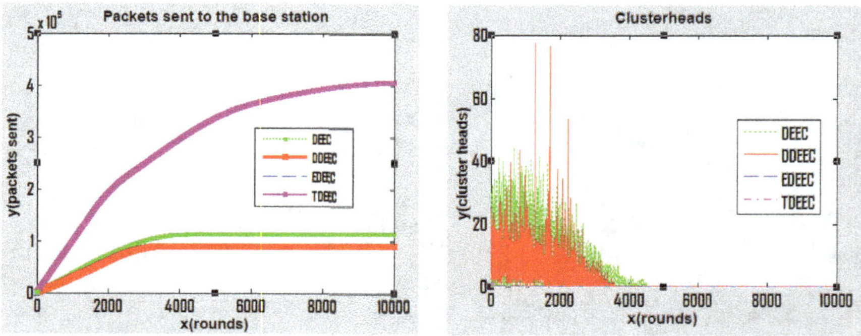

Figure 8.12. Comparison Among Different Versions of DEEC on the Basis of Throughput and Number of CH

8.6.3.1. *Comparison between various versions of DEEC*

Table 8.2. Comparison Between Various Versions of DEEC

S. No.	Clustering Technique	First Dead Node	Tenth Dead Node	All Dead Node	Packets to the Base Station/Sink
1	DEEC	1210	1394	2964	67616
2	D-DEEC	1397	1520	3267	91325
3	T-DEEC	1321	1577	—	314519
4	E-DDEC	1395	1500	—	311039

8.6.4. *Comparison of SEP, Zonal SEP, and LEACH*

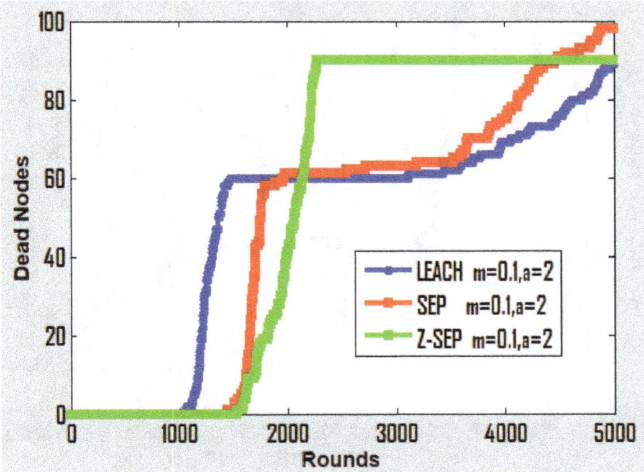

Figure 8.13. Dead Nodes vs. Rounds at m = 0.1 and a = 2

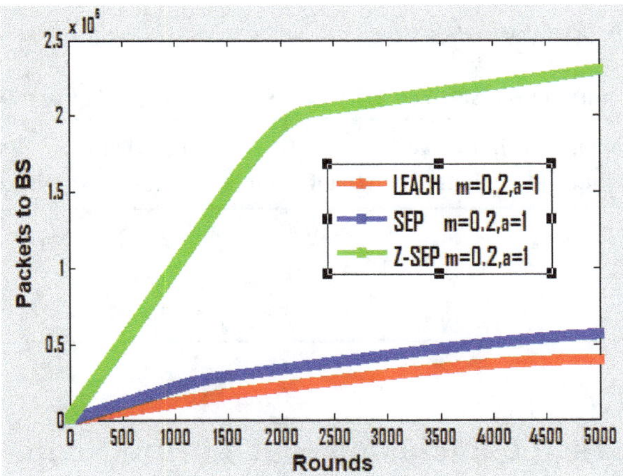

Figure 8.14. Packets to BS vs. Rounds at m = 0.1 and a = 2

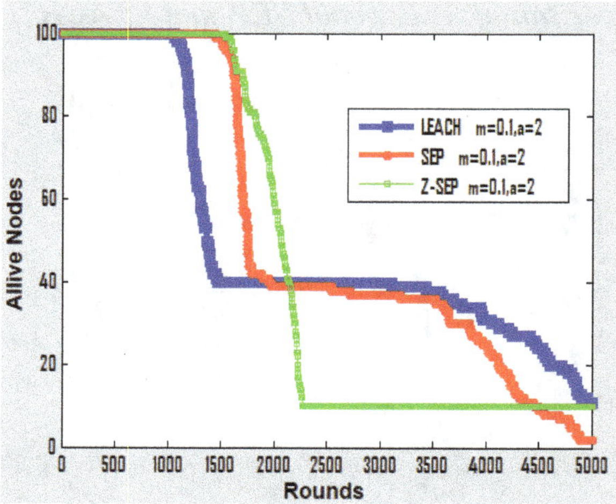

Figure 8.15. Alive Nodes vs. Rounds at $m = 0.1$ and $a = 2$

8.6.4.1. *Comparison of data*

Table 8.3. Comparison Between LEACH, SEP, and Z-SEP at $m = 0.1$ and $a = 2$

S. No.	Clustering Technique	First Dead Node	Tenth Dead Node	Half Dead Node	Packets to the Base Station/Sink
1	LEACH	983	1090	1246	20930
7	SEP	1135	1218	1347	35537
8	Z-SEP	1548	1663	2098	228328

8.7. Analytical Conclusion and Future Scope for More Applications

SEP, LEACH, and Z-SEP [1–3] were analyzed and compared, with Z-SEP demonstrating superior performance against LEACH and SEP. While designing any wireless network module protocol/algorithm, MSN energy, network energy, and overall module network lifespan are the main factors of concern, and all of them should be high. Developing an overall advanced and improvised version of a more energy-efficient algorithm/protocol that distributes network load is a difficult task. Based on this study (Table 8.4), Z-SEP is a good approach to dense networks;

Table 8.4. Analytical Comparison Based on Density, Size, and Level of Heterogeneity for Different Clustering Algorithms

Clustering Technique	Increased Density	Increased Area Size	Increased Heterogeneity
LEACH	Low efficiency	Low efficiency	Low efficiency
MLEACH	Moderate efficiency	Moderate efficiency	Low efficiency
SEP	Better efficiency than all homogeneous algorithms	Better efficiency than all other homogeneous algorithms	Better efficiency than all other homogeneous algorithms
Z-SEP	Better efficiency than SEP	Better efficiency than SEP	Throughput fluctuates, dead nodes are stable, and efficiency is almost similar to that of SEP
EZ-SEP	Better efficiency than Z-SEP	Better efficiency than Z-SEP and TEEN	Efficiency is improved, throughput is not Load balancing problem is present

nevertheless, it does have certain problematic issues. A further improved version of Z-SEP [2, 4, 10] involves an extension of the zonal stable election protocol (EZ-SEP) [1, 3, 23], which is a dual-level protocol incorporating a modified CH selection-based approach, along with node communication with the BS. As a result of this, data are transferred among multiple BS in a mixed zigzag manner, delivering straight to the end point/sink, and the E-advanced MSN utilizes clustering algorithm/technique. This improves the lifespan of the module under supervision by accounting for energy dissipation. Compared to the original Z-SEP, the extended version is better with respect to the overall network lifespan as an enhanced routing algorithm is chosen by electing the CH among MSNs based on residual energy [104–108]. Additionally, the time when the module or network remains stable until the first node dies is improved with the total number of active nodes in the network. Based on the simulation results (refer to Table 8.5 for the input parameters and Table 8.6 for more details), network performance factors, such as the lifespan of modules, packets/MSN transmission to end point/sink, and energy deterioration or consumption, have improved in all the different modified versions of the

Table 8.5. Input Parameters Used in MATLAB

S. No.	LEACH, Mod-LEACH	SEP, ZSEP, EZ-SEP	DEEC, EDEEC	SEP	Mod-LEACH
Field dimensions: x and y maximum (in m)	x_m = 100; y_m = 100;	x_m = 100 y_m = 100	x_m = 100; y_m = 100;	x_m = 100 y_m = 100	x_m = 400 y_m = 400
Initial energy	Eo = 2	Eo = 0.5;	Eo = 0.5	Eo = 0.5;	Eo = 0.5
Energy required to run circuits J/bit	Eelec = 50*10^(−9) ETx = 50*10^(−9) E_{Rx} = 50*10^(−9)	Eelec = 50*10^(−9) ETx = 50*10^(−9) E_{Rx} = 50*10^(−9)	Eelec = 50*10^(−9) ETx = 50*10^(−9) E_{Rx} = 50*10^(−9)	Eelec = 50*10^(−9) ETx = 50*10^(−9) E_{Rx} = 50*10^(−9)	Eelec = 50*10^(−9) ETx = 50*10^(−9) E_{Rx} = 50*10^(−9)
Nodes	n = 100	n = 100	n = 100	n = 100	n = 100
Data aggregation energy J/bit	EDA = 5*10^(−9)	EDA = 5*10^(−9)	EDA = 5*10^(−9)	EDA = 5*10^(−9)	EDA = 5*10^(−9)
Suggested percentage of cluster head	$p = 0.05$	$p = 0.1$	$p = 0.1$	$p = 0.1$	$p = 0.1$
Rmax	5,000	5,000	5,000	5,000	5,000
Emp		0.0013*0.000000000001	0.0013*0.000000000001	0.0013*0.000000000001	0.0013*0.000000000001
E_{fs}		10*0.000000000001	10*0.000000000001	10*0.000000000001	10*0.000000000001
E_{FS1}					$E_{fs}1$ = Efs/10

Table 8.6. Clustering Algorithm Application Areas, Complexity Level in Implementation, and Stability Factors

S. No.	Clustering Technique	Environment	Application Domain (based on Number of Users or Density/Cluster)	Complexity Levels in Implementing	Stability of the Network
1	LEACH	Homogenous	Moderate density passive wireless networks. Best suited for non-urgent transmission. Suitable for a low-density module	Lowest	Moderate
2	PEGASIS	Homogenous	Suitable for low-density networks. Best suited for non-urgent transmission.	Higher than LEACH	Moderate
3	HEED	Homogenous	For medium-density passive and active wireless networks. Well-suited for general data transmission. Suitable for moderate quantity of member sensors in a wireless network.	Moderate	Moderate
4	LEACH-C	Heterogeneous	For medium-density passive and active wireless networks (camera-based also). Best suited for general and urgent data transmission. Suitable for moderate quantity of member sensors in a wireless network.	Moderate	Low
5	Mod-LEACH	Heterogeneous	For medium-density passive and active wireless networks (camera-based also). Better suited than all homogeneous clusterings with general and urgent data transmission. Suitable for moderate quantity of member sensors in a wireless network.	Moderate	Moderate

(*Continued*)

Table 8.6. (*Continued*)

S. No.	Clustering Technique	Environment	Application Domain (based on Number of Users or Density/Cluster)	Complexity Levels in Implementing	Stability of the Network
6	EMod-LEACH	Heterogeneous	For medium-density passive and active wireless networks (camera-based also) Better suited than all homogeneous clustering with general and urgent data transmission. Improved energy efficiency than previous versions of LEACH, moves it ahead in all kinds of application for non-urgent data.	Moderate to high (depending on value of *m*)	moderate
7	DEEC	Heterogeneous	For medium-density passive and active wireless networks (camera-based also) Better suited than all homogeneous clustering with general and urgent data transmission. Improved energy efficiency than previous versions of LEACH, moves it ahead in all kinds of application for non-urgent data.	Higher than EMod-LEACH	moderate
8	EDEEC	Heterogeneous	Better than DEEC, makes it more prominent to use for moderate-density network with passive approach.	Higher than DEEC	low
9	TDEEC	Heterogeneous	Good for medium to large scale of users in a cluster with optimum output	Higher than EMod-LEACH	Moderate
10	TEEN	Heterogeneous	Temperature sensitivity makes it vulnerable for very large number of users in a cluster, so it is good for scattered and low-density networks.	Temperature sensitivity makes its implementation complex	high

11	APTEEN	Heterogeneous	Temperature sensitivity makes it vulnerable for very large number of users in a cluster, so it is good for scattered and low-density networks.	Temperature sensitivity makes its implementation complex	high
12	SEP	Heterogeneous	Stability is good, which makes it suitable for medium to large wireless networks with medium density of users.	Stability for getting a greater number of rounds for first dead node condition, makes its implementation complex	moderate
13	Z-SEP	Heterogeneous	Suitable for passive networks of low density to medium density of users	Stability for getting a greater number of rounds for first dead node condition, makes its implementation complex	moderate
14	EEAHP	Heterogeneous	Provides a moderately good communication setup for moderate to high density of user for a wider area also.	Moderately complex	low to moderate
15	EZ-SEP	Heterogeneous	Effective for high-density wireless network applications.	Stability for getting a greater number of rounds for first dead node condition, makes its implementation complex	Moderate

basic algorithms, namely LEACH transition to MLEACH, or different versions of DDEC, such as EDEEC, TDEEC, DDEEC, or SEP, with their advancement to ZSEP and EZ-SEP. However, if we increase the level of heterogeneity, the density of clusters, or the area of the network, different algorithms behave differently.

References

1. Tripathi, N. and Sharma, K. K. (2022). Distributed and hierarchical clustering techniques comparison in wireless camera sensor networks. In M. S. Kaiser, A. Bandyopadhyay, K. Ray, R. Singh, and V. Nagar (eds.), *Proceedings of Trends in Electronics and Health Informatics. Lecture Notes in Networks and Systems*, Vol. 376. Springer: Singapore.
2. Xie, L. and Zhang, X. (2013). 3D clustering-based camera wireless sensor networks for maximizing lifespan with Minimum coverage rate constraint. In *Proceedings of the IEEE GLOBECOM*, Atlanta, GA, USA, December 9–13, 2013, pp. 298–303.
3. Wang, J. and Zhang, X. (2014). AQ-DBPSK/DS-CDMA based energy-efficient and interference-mitigation scheme for 3D clustered WCSNs with minimum coverage rate constraint. In *Proceedings of the IEEE MILCOM*, Baltimore, MD, USA, October 6–8, 2014, pp. 305–310.
4. Del Coso, A., Spagnolini, U., and Ibars, C. (2007). Cooperative distributed MIMO channels in wireless sensor networks. *IEEE Journal on Selected Areas in Communications*, *25*(2), 402–414.
5. Wang, J. and Zhang, X. (2015). Cooperative MIMO-OFDM based multi-hop 3D clustered wireless camera sensor networks. In *Proceedings of the IEEE Wireless Communication Network Conference (WCNC)*, New Orleans, LA, USA, March 2015, pp. 1350–1355.
6. Nosratinia, A., Hunter, T. E. and Hedayat, A. (2004). Cooperative communication in wireless networks, in *IEEE Communications Magazine*, *42*(10), 74–80.
7. Parmar, J. and Pirishothm, A. (2015). Study of wireless sensor networks using leach-teen and Apteen routing protocols. *International Journal of Science and Research (IJSR)*, *4*(5), 1221–1224.
8. Kiwan, H. and Morgan, Y. L. (2013). Hierarchical networks: Routing and clustering (A concise survey). In *2013 26th IEEE Canadian Conference of Electrical and Computer Engineering (CCECE)*, Regina, SK, Canada, May 5–8, 2013, pp. 1–4.
9. Rathi, N., Saraswat, J., and Bhattacharya, P. P. (2012). A review on routing protocols for application in wireless sensor networks. *International Journal of Distributed and Parallel Systems (IJDPS)*, *3*(5), 39–58.

10. Jain, P. and Chaudhary, A. (2017). The comparison between leach protocol and pegasis protocol based on lifetime of wireless sensor networks. *International Journal of Computer Science and Mobile Computing*, *6*(12), 15–19.
11. Kirichek, R., Paramonov, A., and Koucheryavy, A. (2015). Flying ubiquitous sensor networks as a queening system. In *Proceedings, International Conference on Advanced Communication Technology, ICACT 2015*, Phoenix Park, Korea, July 1–3, 2015, pp. 127–132.
12. Attarzadeh, N. and Mehrani, M. (2011). A new three-dimensional clustering method for wireless sensor networks. *Global Journal of Computer Science and Technology*, *11*(6), version 1.0, pp. 127–132.
13. Hooggar, M., Mehrani, M., Attarzadeh, N., and Azimifar, M. (2013). An energy efficient three-dimensional coverage method for wireless sensor networks. *Journal of Academic and Applied Studies*, *3*(3), 24–33.
14. Abakumov, P. and Koucheryavy, A. (2014). The cluster head selection algorithm in the 3D USN. In *Proceedings, International Conference on Advanced Communication Technology*, Pyeongchang, Korea (South), 2014, pp. 462–466.
15. Hai, D. T., Son, L. H., and Le, V. T. (2017). A novel fuzzy clustering algorithm. In *Proceedings 2003 IEEE International Symposium on Computational Intelligence in Robotics and Automation for the New Millennium*, (Cat. No. 03EX694), Kobe, Japan, Vol. 2, pp. 647–652.
16. Chong, C.-Y. and Kumar, S. (2003). Sensor networks: Evolution, opportunities, and challenges. *Proceedings of the IEEE*, *91*(8), 1247–1256.
17. Akyildiz, I., Su, W., Sankarasubramaniam, Y., and Cayirci, E. (2002). Wireless sensor networks: A survey. *Computer Networks*, *38*(4), 393–422.
18. Ghosal, A., Holder, S., and Das, S. K. (2020). Distributed on-demand clustering algorithm for lifetime optimization in wireless sensor networks. *Journal of Parallel and Distributed Computing*, *141*, 129–142. DOI: https://doi.org/10.1016/j.jpdc.2020.03.014.
19. Xu, L., Delaney, D., O'Hare, G., and Collier, R. (2013). The impact of transmission power control in wireless sensor networks. In *12th IEEE International Symposium on Network Computing and Applications (NCA)*, pp. 255–258.
20. Amis, A. D., Prakash, R., Vuong, T. H., and Huynh, D.T. (2000). — Maxmin d-cluster formation in wireless ad hoc networks. In *Proceedings on IEEE, Nineteenth Annual Joint Conference of the IEEE Computer and Communications Societies, INFOCOM 2000,* Vol. 1, pp. 32–41.
21. Brust, M. R., Frey, H., and Roth kugel, S. (2007). Adaptive multi-hop clustering in mobile networks. In *Proceedings of the 4th International Conference on Mobile Technology, Applications, and Systems and the 1st International Symposium on Computer Human Interaction in Mobile Technology*, pp. 132–138. ACM.

22. Baker, D. and Ephremides, A. (1981). The architectural organization of a mobile radio network via a distributed algorithm. *IEEE Transactions on Communications*, *29*(11), 1694–1701.
23. Arboleda, L. M. C. and Nasser, N. (2006). Comparison of clustering algorithms and protocols for wireless sensor networks. In *2006 Canadian Conference on Electrical and Computer Engineering*, Ottawa Congress Centre, Ottawa, Canada, May 7–10, 2006, pp. 1787–1792.
24. Younis, O., Krunz, M., and Ramasubramanian, S. (2006). Node clustering in wireless sensor networks: Recent developments and deployment challenges. *IEEE Network*, *20*(3), 20–25.
25. Abbasi, A. A. and Younis, M. (2007). A survey on clustering algorithms for wireless sensor networks. *Computer Communications*, *30*, 3–7.
26. Kumarawadu, P., Dechene, D. J., Luccini, M., and Sauer, A. (2008). Algorithms for node clustering in wireless sensor networks: A survey. In *2008 4th International Conference on Information and Automation for Sustainability*, Sri Lanka, December 12–14, 2008, pp. 295–300.
27. Deosarkar, B. P., Yadav, N. S., and Yadav, R. P. (2008). Cluster head selection in clustering algorithms for wireless sensor networks: A survey. In *2008 International Conference on Computing, Communication and Networking*, Karur, Tamil Nadu, India, December 18–20, 2008, pp. 1–8.
28. Zhao, M., Yang, Y., and Wang, C. (2015). Mobile data gathering with load balanced clustering and dual data uploading wireless sensor networks. *IEEE Transactions on Mobile Computing*, *14*(4), 770–785.
29. Nayak, P. and Devulapalli, A. (2016). A fuzzy logic-based clustering algorithm for WSN to extend the network lifetime. *IEEE Sensors Journal*, *16*(1), 137–144.
30. Xu, L., O'Hare, G., and Collier, R. (2014). A Balanced Energy-Efficient Multihop clustering scheme for Wireless Sensor Networks, *2014 7th IFIP Wireless and Mobile Networking Conference (WMNC)*, Vilamoura, Portugal, May 20–22, 2014, pp. 1–8.
31. Sirisha, G., Babu, R. B., and Rao, K. R. (2016). Establishing path quality management in wireless sensor networks through cluster head determination. *Indian Journal of Science and Technology*, *9*(5), 1–9.
32. Lin, C. R. and Gerla, M. (1997). Adaptive clustering for mobile wireless networks. *IEEE Journal on Selected Areas in Communications*, *15*, 1265–1275.
33. Nagpal, R. and Coore, D. (1998). An algorithm for group formation and maximal independent set in an amorphous computer. Technical Report, Cambridge, MA, USA.
34. Heintzelman, W., Chandrakasan, A., and Balakrishnan, H. (2000). Energyefficient communication protocol for wireless microsensor networks.

In *Proceedings of the 33rd Annual Hawaii International Conference on System Sciences,* Maui, Hawaii, January 7, 2000, Vol. 2, p. 10.
35. Banerjee, S. and Khullar, S. (2001). A clustering scheme for hierarchical control in multi-hop wireless networks. In *Twentieth Annual Joint Conference of the IEEE Computer and Communications Societies. Proceedings on INFOCOM 2001*. IEEE, 22–26 April 2001, Anchorage, AK, USA, Vol. 2, pp. 1028–1037.
36. Manjeshwar, A. and Agrawal, D. (2001). Teen: A routing protocol for enhanced efficiency in wireless sensor networks. In *Proceeding's 15th International, Parallel and Distributed Processing Symposium.* IEEE Conference held at San Francisco, CA, USA, pp. 2009–2015.
37. Basu, P., Khan, N., and Little, T. D. (2001). A mobility-based metric for clustering in mobile ad hoc networks. In *2001 International Conference on IEEE, Distributed Computing Systems Workshop*. IEEE Conference held at Mesa, AZ, USA, pp. 413–418.
38. Manjeshwar, A. and Agrawal, D. (2002). Apteen: A hybrid protocol for efficient routing and comprehensive informat retrieval in wireless. In *Proceedings International, Parallel and Distributed Processing Symposium, IPDPS*. IEEE Conference held at Ft. Lauderdale, FL, USA.
39. Hongwei, Z. and Arora, A. (2002). Scalable self-configuration and self-healing in wireless networks. In *Proceedings of the Twenty-First Annual Symposium on Principles of Distributed Computing (PODC'21)*, July 26–30, 2021, pp. 58–67.
40. Sharma, K. K., Patel, R. B., and Singh, H. (2010). A reliable and energy efficient transport protocol for wireless sensor networks. *International Journal of Computer Networks & Communications*, *2*(5), 92–103.
41. Mittal, A., Sharma, K. K., and Dalal, S. (2014). Applying clustering algorithm in case retrieval phase of the case-based reasoning. *International Journal of Research Aspects of Engineering and Management*, *1*(2), 14–16.
42. Hai, D. T., Son, L. H., Le, V. T. Novel fuzzy clustering scheme for 3D wireless sensor networks. *Applied Soft Computing Journal.*
43. Murugesan, S. (2008). Harnessing green it: Principles and practices. *IT Professional*, *10*(1), 24–33.
44. Singh, G. P., Sohi, B. S., and Raj, B. (2017). Material properties analysis of graphene base transistor (GBT) for VLSI analog circuits. *Indian Journal of Pure & Applied Physics (IJPAP)*, *55*, 896–902.
45. Singh, A., Khosla, M., and Raj, B. (2016). Comparative analysis of carbon nanotube field effect transistor and nanowire transistor for low power circuit design. *Journal of Nanoelectronics and Optoelectronics*, *11*, 388–393.

46. Kumar, S. and Raj, B. (2017). Estimation of stability and performance metric for inward access transistor based 6T SRAM cell design using n-type/p-type DMDG-GDOV TFET. *IEEE VLSI Circuits and Systems Letter*, *3*(2), 25–39.
47. Sharma, S., Kumar, A., Pattanaik, M., and Raj, B. (2013). Forward body biased multimode multi-threshold CMOS technique for ground bounce noise reduction in static CMOS adders. *International Journal of Information and Electronics Engineering*, *3*(3), 567–572.
48. Singh, H., Kumar, P., and Raj, B. (2013). Performance analysis of majority gate SET based 1-bit full adder. *International Journal of Computer and Communication Engineering (IJCCE)*, *2*(4), pp. 24–33.
49. Bhardwaj, A. K., Gupta, S., and Raj, B. (2020). Investigation of parameters for Schottky Barrier (SB) height for Schottky barrier based carbon nanotube field effect transistor device. *Journal of Nanoelectronics and Optoelectronics*, *15*, 783–791.
50. Bansal, P. and Raj, B. (2017). Memristor: A versatile nonlinear model for dopant drift and boundary issues. *JCTN*, *14*(5), 2319–2325.
51. Jain, N. and Raj, B. (2016). An analog and digital design perspective comprehensive approach on fin-FET (Fin-Field Effect transistor) technology — A review. *Reviews in Advanced Sciences and Engineering (RASE)*, *5*, 1–14.
52. Sharma, S., Raj, B., and Khosla, M. (2017). Subthreshold performance of $In_{1-x}Ga_xAs$ based dual metal with gate stack cylindrical/surrounding gate nanowire MOSFET for low power analog applications. *Journal of Nanoelectronics and Optoelectronics*, *12*, 171–176.
53. Raj, B., Saxena, A. K., and Dasgupta, S. (2009). Analytical modeling for the estimation of leakage current and subthreshold swing factor of nanoscale double gate FinFET device. *Microelectronics International*, *26*, 53–63.
54. Soniya, S. S., Wadhwa, G., and Raj, B. (2019). An analytical modeling for dual source vertical tunnel field effect transistor. *International Journal of Recent Technology and Engineering (IJRTE)*, *8*(2), 603–608.
55. Singh, S. and Raj, B. (2019). Design and analysis of hetrojunction vertical T-shaped tunnel field effect transistor. *Journal of Electronics Material*, *48*(10), 6253–6260.
56. Goyal, C., Ubhi, J. S., and Raj, B. (2019). A low leakage CNTFET based inexact full adder for low power image processing applications. *International Journal of Circuit Theory and Applications*, *47*(9), 1446–1458.
57. Raj, B., Saxena, A. K., and Dasgupta, S. (2008). A compact drain current and threshold voltage quantum mechanical analytical modeling for

FinFETs. *Journal of Nanoelectronics and Optoelectronics (JNO)*, *3*(2), 163–170.

58. Wadhwa, G. and Raj, B. (2020). An analytical modeling of charge plasma based tunnel field effect transistor with impacts of gate underlap region. *Superlattices and Microstructures*, *142*, 106512.
59. Singh, S. and Raj, B. (2020). Modeling and simulation analysis of SiGe hetrojunction double gatevertical t-shaped tunnel FET. *Superlattices and Microstructures*, *142*, 106496.
60. Singh, A., Saini, D. K., Agarwal, D., Aggarwal, S., Khosla, M., and Raj, B. (2016). Modeling and simulation of carbon nanotube field effect transistor and its circuit application. *Journal of Semiconductors (JoS)*, *37*, 074001-6.
61. Jain, N. and Raj, B. (2016). Device and circuit co-design perspective comprehensive approach on FinFET technology — A review. *Journal of Electron Devices*, *23*(1), 1890–1901.
62. Kumar, S. and Raj, B. (2016). Analysis of I_{ON} and ambipolar current for dual-material gate-drain overlapped DG-TFET. *Journal of Nanoelectronics and Optoelectronics*, *11*, 323–333.
63. Anjum, N., Bali, T., and Raj, B. (2013). Design and simulation of handwritten multiscript character recognition. *International Journal of Advanced Research in Computer and Communication Engineering*, *2*(7), 2544–2549.
64. Sharma, S., Raj, B., and Khosla, M. (2016). A Gaussian approach for analytical subthreshold current model of cylindrical nanowire FET with quantum mechanical effects. *Microelectronics Journal*, *53*, 65–72.
65. Singh, K. and Raj, B. (2015). Performance and analysis of temperature dependent multi-walled carbon nanotubes as global interconnects at different technology nodes. *Journal of Computational Electronics*, *14*(2), 469–476.
66. Kumar, S. and Raj, B. (2015). Compact channel potential analytical modeling of DG-TFET based on evanescent–mode approach. *Journal of Computational Electronics*, *14*(2), 820–827.
67. Singh, K. and Raj, B. (2015). Temperature dependent modeling and performance evaluation of multi-walled CNT and single-walled CNT as global interconnects. *Journal of Electronic Materials*, *44*(12), 4825–4835.
68. Sharma, V. K., Pattanaik, M., and Raj, B. (2014). INDEP approach for leakage reduction in nanoscale CMOS circuits. *International Journal of Electronics*, *102*(2), 200–215.
69. Singh, K. and Raj, B. (2015). Influence of temperature on MWCNT bundle, SWCNT bundle and copper interconnects for nanoscaled technology

nodes. *Journal of Materials Science: Materials in Electronics*, *26*(8), 6134–6142.
70. Anjum, N., Bali, T., and Raj, B. (2013). Design and simulation of handwritten Gurumukhi and Devanagri Numerical recognition. *International Journal of Computer Applications*, *73*(12), 16–21.
71. Khandelwal, S., Gupta, V., Raj, B., and Gupta, R. D. (2015). Process variability aware low leakage reliable nano scale DG-FinFET SRAM cell design technique. *Journal of Nanoelectronics and Optoelectronics*, *10*(6), 810–817.
72. Sharma, V. K., Pattanaik, M., and Raj, B. (2014). ONOFIC approach: Low power high speed nanoscale VLSI circuits design. *International Journal of Electronics*, *101*(1), 61–73.
73. Khandelwal, S., Raj, B., and Gupta, R. D. (2015). FinFET based 6T SRAM cell design: Analysis of performance metric, process variation and temperature effect. *Journal of Computational and Theoretical Nanoscience*, *12*, 2500–2506.
74. Singh, S., Yadav, S., Rahul, J., Srivastava, A., and Raj, B. (2015). Impact of HfO_2 in graded channel dual insulator double gate MOSFET. *Journal of Computational and Theoretical Nanoscience*, *12*(6), 950–953.
75. Sharma, V. K., Pattanaik, M., and Raj, B. (2014). PVT variations aware low leakage INDEP approach for nanoscale CMOS circuits. *Microelectronics Reliability*, *54*, 90–99.
76. Raj, B., Saxena, A. K., and Dasgupta, S. (2013). Quantum mechanical analytical modeling of nanoscale DG FinFET: Evaluation of potential, threshold voltage and source/drain resistance. *Journal of Material Science in Semiconductor Processing*, *16*(4), 1131–1137.
77. Gopal, M., Prasad, S. S. D., and Raj, B. (2013). 8T SRAM cell design for dynamic and leakage power reduction. *International Journal of Computer Applications*, *71*(9), 43–48.
78. Pattanaik, M., Raj, B., Sharma, S., and Kumar, A. (2012). Diode based trimode multi-threshold CMOS technique for ground bounce noise reduction in static CMOS adders. *Advanced Materials Research*, *548*, 885–889.
79. Raj, B., Saxena, A. K., and Dasgupta, S. (2011). Nanoscale FinFET based SRAM cell design: Analysis of performance metric, process variation, underlapped FinFET and temperature effect. *IEEE Circuits and System Magazine*, *11*(2), 38–50.
80. Sharma, V. K., Pattanaik, M., and Raj, B. (2013). Leakage current ONOFIC approach for deep submicron VLSI circuit design. *International Journal of Electrical, Computer, Electronics and Communication Engineering*, *7*(4), 239–244.

81. Chawla, T., Khosla, M., and Raj, B. (2021). Design and simulation of triple metal double-gate germanium on insulator vertical tunnel field effect transistor. *Microelectronics Journal, 114,* 105125.
82. Kaur, P., Gill, S. S., and Raj, B. (2022). Comparative analysis of OFETs materials and devices for sensor applications. *Journal of Silicon, 14,* 4463–4471.
83. Sharma, S. K., Kumar, P., Raj, B., and Raj, B. (2022). $In_{1-x}Ga_xAs$ double metal gate-stacking cylindrical nanowire MOSFET for highly sensitive photo detector. *Journal of Silicon, 14,* 3535–3541.
84. Raj, B., Saxena, A. K., and Dasgupta, S. (2009). Analytical modeling of quasi planar nanoscale double gate FinFET with source/drain resistance and field dependent carrier mobility: A quantum mechanical study. *Journal of Computer (JCP), 4*(9), 1–8.
85. Bhushan, S., Khandelwal, S., and Raj, B. (2012). Analyzing different mode FinFET based memory cell at different power supply for leakage reduction. In *Seventh International Conference on Bio-Inspired Computing: Theories and Application, (BIC-TA 2012)*, 14–16 December 2012, Vol. 202, pp. 89–100.
86. Singh, J. and Raj, B. (2018). Temperature dependent analytical modeling and simulations of nanoscale memristor. *Engineering Science and Technology, 21,* 862–868.
87. Singh, S., Bala, S., Raj, B., and Raj, B. (2020). Improved sensitivity of dielectric modulated junctionless transistor for nanoscale biosensor design. *Sensor Letter, 18,* 328–333.
88. Kumar, V., Vishvakarma, S. K., and Raj, B. (2020). Design and performance analysis of ASIC for IoT applications. *Sensor Letter ASP, 18,* 31–38.
89. Jaiswal, A., Sarin, R. K., Raj, B., and Sukhija, S. (August 2018). A novel circular slotted microstrip-fed patch antenna with three triangle shape defected ground structure for multiband applications. *Advanced Electromagnetic (AEM), 7*(3), 56–63.
90. Wadhwa, G. and Raj, B. (2018). Label free detection of biomolecules using charge-plasma-based gate underlap dielectric modulated junctionless TFET. *Journal of Electronic Materials (JEMS), 47*(8), 4683–4693.
91. Singh, G., Sarin, R. K., and Raj, B. (2017). Design and performance analysis of a new efficient coplanar quantum-dot cellular automata adder. *Indian Journal of Pure & Applied Physics (IJPAP), 55,* 97–103.
92. Singh, A., Khosla, M., and Raj, B. (2017). Design and analysis of electrostatic doped Schottky barrier CNTFET based low power SRAM. *International Journal of Electronics and Communications, 80,* 67–72.
93. Kaur, P., Pandey, V., and Raj, B. (2020). Comparative study of efficient design, control and monitoring of solar power using IoT. *Sensor Letter, 18,* 419–426.

94. Bhardwaj, A. K., Gupta, S., and Raj, B. (2020). Development & analysis of compact model for double gate Schottky barrier CNTFET. *Journal of Nanoelectronics and Optoelectronics*, *15*, 1199–1208.
95. Wadhwa, G., Kamboj, P., Singh, J., and Raj, B. (2020). Design and investigation of junctionless DGTFET for biological molecule recognition. *Transactions on Electrical and Electronic Materials*, *8*, 1–8.
96. Chawla, T., Khosla, M., and Raj, B. (2020). Optimization of double-gate dual material GeOI-vertical TFET for VLSI circuit design. *IEEE VLSI Circuits and Systems Letter*, *6*(2), 13–25.
97. Verma, S. K., Singh, S., Wadhwa, G., and Raj, B. (2020). Detection of biomolecules using charge-plasma based gate underlap dielectric modulated dopingless TFET. *Transactions on Electrical and Electronic Materials (TEEM)*, *21*, 528–535.
98. Jain, N. and Raj, B. (2017). Impact of underlap spacer region variation on electrostatic and analog/RF performance of symmetrical high-k SOI FinFET at 20 nm channel length. *Journal of Semiconductors (JoS)*, *38*(12), 122002.
99. Singh, S. and Raj, B. (2020.). Analytical modeling and simulation analysis of T-shaped III-V heterojunction vertical T-FET. *Superlattices and Microstructures*, *147*, 106717.
100. Singh, G., Sarin, R. K., and Raj, B. (2017). Design and analysis of area efficient QCA based reversible logic gates. *Journal of Microprocessors and Microsystems*, *52*, 59–68.
101. Singh, A., Khosla, M., and Raj, B. (2016). Compact model for ballistic single wall CNTFET under quantum capacitance limit. *Journal of Semiconductors (JoS)*, *37*, 104001-8.
102. Singh, S., Khosla, M., Wadhwa, G., and Raj, B. (2021). Design and analysis of double-gate junctionless vertical TFET for gas sensing applications. *Applied Physics A*, *127*(16), 2.
103. Singh, I., Raj, B., Khosla, M., and Kaushik, B. K. (2020). Potential MRAM technologies for low power SoCs. *SPIN*, *10*(4), 2050027.
104. Stergiou, C. L., Psannis, K. E., and Gupta, B. B. (2020). IoT-based big data secure management in the fog over a 6G wireless network. *IEEE Internet of Things Journal*, *8*(7), 5164–5171.
105. Alsmirat, M. A., Jararweh, Y., Obaidat, I., and Gupta, B. B. (2017). Internet of surveillance: A cloud supported large-scale wireless surveillance system. *The Journal of Supercomputing*, *73*, 973–992.
106. Alsmirat, M. A., Jararweh, Y., Obaidat, I., and Gupta, B. B. (2017). Automated wireless video surveillance: an evaluation framework. *Journal of Real-Time Image Processing*, *13*, 527–546.

107. Amphawan, A., Chaudhary, S., and Gupta, B. (2015). Secure MDM-OFDM-Ro-FSO system using HG modes. *International Journal of Sensors Wireless Communications and Control*, *5*(1), 13–18.
108. Cao, N., Liu, P., Li, G., Zhang, C., Cao, S., Cao, G., Yan, M., and Gupta, B. B. (2018). Evaluation models for the nearest closer routing protocol in wireless sensor networks. *IEEE Access*, *6*, 77043–77054.

https://doi.org/10.1142/9789811279539_0009

Chapter 9

Development of Autonomous Vehicles: Perspectives and Challenges

Md. Hashim[*,†,§], Vivek Kumar[†,¶], Garima Jain[†,||], Hitesh Singh[†,], K. Prabhanjan Kumar[†,††], and Shyam Akashe[‡,‡‡]**

[†]*Noida Institute of Engineering and Technology, Noida, Uttar Pradesh, India*

[‡]*ITM University Gwalior, Madhya Pradesh, India*

[§]*md.hashim127@gmail.com*
[¶]*vivek.bansal1977@niet.co.in*
[||]*garimajain@niet.co.in*
[**]*hsingh12@gmail.com*
[††]*kpkumar.it@niet.co.in*
[‡‡]*shyam.akashe@itmuniversity.ac.in*

Abstract

Although research in autonomous vehicles had started in the early 20th century, due to a lack of efficient and highly powerful computational devices, they were only controlled by radio wave technologies, widely known as remotes, despite rapid developments in the related technologies. Then, in the 1950s, research and development activities picked up pace with the development of computers. With the introduction and

[*]Corresponding author.

advancement of artificial intelligence (AI) and computing hardware technologies, autonomous vehicles have become a primary field of research and development for researchers seeking to address various day-to-day problems faced by drivers, traffic controllers, and human transportation as a whole, such as traffic jams, road accidents, expensive infrastructure maintenance, and the engagement of human resources (drivers) in inefficient and repetitive work. Considering these factors, automobile and tech giants have identified a new market opportunity with the potential for huge profits. This has resulted in rapid research and development as companies strive to be the first to capture the market. As we progressed with new developments, various technical, social, and administrative problems and challenges started arising which need to be resolved. This chapter contains a review and analysis of various studies carried out in this field and proposes alternative approaches wherever needed. The main technical challenge is to design and develop a superfast and energy-efficient AI model which must be capable of performing millions of calculations in microseconds, which would require a very powerful GPU and a highly efficient system architecture design. As infrastructure is designed and developed solely for human drivers everywhere, it becomes a challenge for the physical testing and deployment of autonomous vehicles. As such, there are either no clear administrative rules and regulations or no rules at all for such vehicles. Even if all these technical and administrative challenges of autonomous vehicles are overcome, social challenges may remain since the extent to which customers and passengers accept and adopt them will usher in completely new changes to conventional transportation.

Keywords: Autonomous vehicles, computer vision, artificial intelligence, machine learning, deep learning

9.1. Introduction

With the rapid research and development in artificial intelligence (AI) and sensor technologies, researchers are provided with opportunities for a broader and more efficient harnessing of the benefits of these technologies to achieve accurate decision-making in computerized driving of vehicles. After an intensive analysis and research of customers' needs and behavior, researchers concluded that autonomous vehicles might be one of the best solutions to satisfy people's transportation needs and requirements while tackling other major issues parallelly, such as traffic jams and accidents.

In particular, accidents mainly occur because of wrong or delayed decisions made by human drivers, most of which are very common and repetitive mistakes that could be prevented to some extent through the implementation of AI for vehicle driving, replacing human drivers. A number of surveys and studies claim that traffic jams and accidents can be drastically reduced, and even completely prevented, through accurate and timely decisions made by AI models. The AI models, which will be used in the vehicles, are trained on thousands of hours of driving data collected from well-trained and professional drivers. The key components for hardware sensing used in autonomous vehicles are RGB cameras, LIDAR, RADAR, and ultrasonic sensors. Autonomous vehicles (AVs) are much more secure in many aspects where humans are vulnerable to mistakes, which can lead to distrust of drivers among co-passengers, especially when using cab services provided by companies. Another key benefit is the lower maintenance costs of AVs because the chances of misinterpretation of a vehicle's control commands are significantly reduced, which results in a drastic fall in the wear and tear of the vehicle. This will be a boon for cab companies in terms of their profits since, besides drivers' salaries, a major portion of their revenues are spent on vehicle maintenance, which could be minimized by avoiding wear and tear caused by wrong decision-making while driving. So, after considering the many advantages of AVs, the governments of many countries have made proposals for the identification and development of AV-enabled zones or city infrastructure to support and ease the growth of AVs. Therefore, in the near future, AVs are expected to become a necessity rather than a luxury.

Along with the social and infrastructural factors, the design and development of AVs also include the implementation of autonomic computing, in which a system is designed to be capable of self-management, meaning that a vehicle operated by such a system will autonomically manage its internal processing with nominal human intervention. Autonomic computation technologies are a combination of four major self-adapting features: self-configurable, self-healing, self-protective, and self-optimizing. Among these, in AVs, the development processes are focused mainly on self-configuration and self-optimization features, in which the former handles the system's configuration-related processes and the latter handles the optimization of resources available to the AV.

With the development of highly efficient and effective computing hardware and AI algorithms, the research and development of AVs have received a major boost in recent decades. Their development will make driving and commuting highly efficient and safe while consuming fewer human resources for repetitive work. In addition to AVs, many countries are planning to develop AV-enabled cities and infrastructures, which could nullify human error while driving and hence prevent traffic jams and road accidents.

This chapter contains a detailed survey and analysis of different technical, social, infrastructural, and social challenges that have emerged or are likely to emerge due to the introduction of AVs into heterogeneous traffic or of pure connected AV (CAV) traffic infrastructure. It also reviews and proposes potential solutions to encounter those challenges.

9.2. State of the Art

A significant amount of research work has been done on the various aspects of the design and development of AVs. Here, we briefly discuss a few relevant studies which have been previously performed by researchers or organizations. The key contributions, merits, and demerits of the studies are summarized in Tables 9.1–9.4. Table 9.1 contains an outline of the previous research surveys conducted in the field. Table 9.2 represents the different technical implementations used in past developments of AVs. Table 9.3 presents the infrastructural reforms needed for the most efficient use of AVs in the future, as suggested by different studies. Table 9.4 outlines the research done on social acceptance and reforms toward the development of AVs.

One of the main challenges, as reported in Ref. [1], is that many autonomous driving systems consist of machine learning components, such as deep neural networks, for which formal properties are difficult to characterize. To overcome this common challenge of characterization, the system components, such as machine learning, deep learning, and neural networks, must be used according to the need. This research paper demonstrates test case generation and automatic falsification to evaluate closed-loop properties of the system and identify problem scenarios, which results in greater reliability of the AV systems. The authors of [2] describe the validation of an autonomous driving architecture in a simulated environment, considering some challenging driving scenarios and focusing on the decision-making layer, based on hierarchical interpreted binary petri nets (HIBPN). Another study [3] describes a method of efficient and effective sensor calibration, which is a fundamental part of any AV. This research evaluates the capabilities and technical performance of sensors commonly used in AVs, including LIDAR, RGB cameras, RADAR, and GPS, in various weather conditions, such as sunny, rainy, windy, and fog. This paper presents three main approaches to sensor fusion and reviews multi-fusion technologies and algorithms for object detection and ranging for AVs. A discussion of efforts and progress in various aspects of the ethical challenges in the development of AVs is taken up in Ref. [4]. This critical review reports that, currently, a consistent and effective solution for ethical challenges does not exist; thus, various recommendations are provided in order to potentially encourage the players involved to collaborate to tackle the various challenges [29–33].

An overview of the key aspects of autonomous driving is provided in Ref. [5], taking various social, geographical, and political perspectives, and based on the information received from top research centers, it highlights the most important and advanced findings so far. The authors also discuss the various approaches to traffic control and various ethical perspectives that could be considered further in the development of AVs. The authors of [6] describe in five stages the various barriers encountered in the adoption of AVs, including government policies, infrastructure developments, and public acceptance [34–39]. In the first and second stages, the authors review the literature, followed by an industry and academic expert discussion to find out the interrelationship between these barriers. In the third and fourth stages, the authors present various technical and lab-based analyses to determine the interconnections among the barriers. In the final stage, the authors depict the key results in a causal loop diagram (CLD) for cause-and-effect relationships and highlight two main reasons for the lack of customer acceptance: the lack of industry standards and the absence of regulations and certifications. Another study [7] provides an insight into the adoption of AVs in smart cities considering various parameters, such as safety, security, efficiency, and various social factors. The authors of [8] analyze previously collected AV crash test statistics and develop a dynamic runtime safety assessment system to simulate the case study using various AI techniques, such as OpenCV, for lane detection. This system predicts possible crashes using machine learning and neural network models and provides a response mechanism for the avoidance of crashes. Additionally, it provides techniques and parameters for the development and enhancement of AVs. A survey [9] analyzes the potential impact and performance of AVs after their deployment on heterogeneous motorways (current traffic infrastructure) using microscopic traffic simulation by taking various real/hypothetical scenarios and parameters into consideration. Then, various governance laws and traffic infrastructural and social reforms are suggested by the authors [40–44].

Table 9.1. Survey of Research

Related Works	Year	Key Contributions	Merits	Demerits
[4]	2022	It proposes effective solutions to tackle different ethical challenges associated with the development of AVs	Proposes different solutions to overcome ethical challenges	No clear outcomes are described after implementation of those recommendations
[16]	2019	Presents a review of the principles, perception systems, physical fundamentals, simulators, and sensor technologies used in AVs	Reviews almost every aspect of the development of AVs	Doesn't consider social, governance, and infrastructural reforms needed for the adoption of AVs
[17]	2018	It presents state-of-the-art results, applications, and issues that hinder the large-scale development of AVs	Presents results, applications, and challenges faced during development	Lacks in providing statistical/ data-based overview and comparisons
[18]	2016	Presents the research and developments and the potential advantages of connected AVs (CAVs)	Summarizes current and future scope and the advantages of CAVs	Lacks in providing statistical or data-driven insights of the advantages of CAVs
[19]	2018	This survey of people depicts the impact of AVs on the labor marketplace, such as losses of drivers' jobs, along with creation of new jobs and safety of commuters	Provides people's perspectives toward the impact of AVs in terms of job changes and safety	Its outcome is solely based on people's perspectives/ interviews; no statistical or data-driven illustration is given

In Ref. [10], a conceptual framework is given for a holistic approach to the design and development of an efficient infrastructure planning method for AVs. It proposes a stepwise approach, starting with a literature review for the identification of infrastructure options, followed by a geographical analysis to prioritize the best-suited roadway development [45–52]. In the third step, the authors develop a bass diffusion model to forecast the adoption of AVs in the short, mid, and long terms. Subsequently, a traffic micro-simulation model is employed to predict the efficiency and performance of infrastructure. The authors of [11] use a deep reinforcement learning (DRL) algorithm called deep *Q*-network in the control layer of an AV and test it in a simulated environment. The aim is to obtain a trained model that can send control commands to the vehicle to navigate it properly. Another study [12] proposes a DRL-based algorithm for collision avoidance between pedestrians and vehicles in diverse environmental conditions. It uses a deep *Q*-network in which the learning agent is deployed and trained using a reward-based technique [53–58].

In another study [13], the authors propose an extension of the popular state–action–reward–state–action (SARSA) RL technique, named "Harmonic SK Deep SARSA," which uses the stability technique of the SARSA algorithm. It uses similar notations and cumulative states provided by SARSA, saves in an alternative memory to enhance the stability of the algorithm, and achieves remarkable performance. In Ref. [14], a safe motion planning for AVs with a stability guarantee is proposed by integrating the risk identification method and the Lyapunov function with a well-known soft actor-critic (SAC) algorithm [59–67]. To overcome the accuracy challenges faced by various RL algorithms, a study [15] proposes an

Table 9.2. Technical Implementation

Related Works	Year	Key Contributions	Merits	Demerits
[3]	2022	Describes different efficient and effective calibrations of sensors commonly used by AVs, evaluates their capabilities and technical performance	Proposes different ways of sensor calibrations for more effective object detection and ranging	It only provides approaches and review of different sensor fusion algorithms and techniques but lacks implementations
[11]	2022	Uses deep reinforcement learning (DRL) at control layer to send control commands to navigate the vehicle properly	Provides control commands to the vehicle to navigate and move it properly	Only DRL algorithm is used. Training of learning agents took longer

(*Continued*)

Table 9.2. (*Continued*)

Related Works	Year	Key Contributions	Merits	Demerits
[12]	2022	Proposes a new intelligent approach for pedestrian collision avoidance using a deep Q-network algorithm	This learning algorithm reduces pedestrian collision by up to 64%	Uses only DRL, so results might vary dramatically with environments and scenarios
[13]	2022	Proposes an extension of the RL algorithm (SARSA), "Harmonic SK Deep SARSA" algorithm, for complex decision-making for Avs	This algorithm performs very well while lane changing, merging, and navigating dynamic obstacles	It uses notations of similar and cumulative states saved in alternative memory, so it is less space- and energy-efficient
[14]	2021	Safe motion planning for AVs with stability guarantees using safe actor critic (SAC) algorithm	This experiment shows that the proposed method achieves a better success rate than SAC	This method is limited to a specific size and power of vehicles, and the experiment was done in a simulated environment
[15]	2022	It proposes an "Alternative Bidirectional Q Network" for better stability and accuracy of control commands to the vehicle under different scenarios	It provides more accurate control commands than the Q-Network algorithm in different environments	As it is also an extension of one of the RL algorithms, the training of learning agents needs much time to be acceptably accurate

Table 9.3. Infrastructural and Traffic Reforms

Related Works	Year	Key Contributions	Merits	Demerits
[20]	2018	Briefs on the potential changes and challenges due to the deployment of AVs in current cities and traffic infrastructure	Shows physical and habitual changes to city infrastructure and human lifestyle	It solely depends on the prediction based on considering a number of current scenarios, not on data
[21]	2016	Represents the effect of AVs on the alertness and behavior of human drivers, along with travel time differences due to them	Test was taken in a microscopic traffic simulator environment to gather data	Data presented are solely based on tests in a simulated environment and not on physical road tests
[22]	2018	It determines the smallest number of AVs required to stabilize a single-lane traffic system of human-driven vehicles	Determines that only 6% uniform penetration of AVs is required to stabilize traffic	This research is done on a small scale and doesn't discuss the impact of the deployment of a large number of Avs
[23]	2020	It proposes a traffic model for trajectory planning for Avs, considering multiple mandatory lane change demands and traffic scenes in the most efficient way	It proposes a solution to two major vehicular traffic issues: vehicle grouping and motion planning for AVs	Validation of the performance of the model is tested only in simulated environments, no physical implementation results are given
[24]	2018	It presents driving strategies for CAVs to stabilize heterogeneous traffic by controlling vehicles using vehicle-to-infrastructure communications	Presents a traffic model which is V2I communication enabled for efficient commute	This proposed model will be applicable solely for traffic systems with AVs and can't be implemented for heterogeneous traffic

Table 9.4. Governance and Social Acceptance of AVs

Related Works	Year	Key Contributions	Merits	Demerits
[6]	2021	Represents the five-stage method to describe the barriers associated with the adoption of AVs	Presents a lack of governance, social, infrastructural, and industry readiness	Doesn't propose any effective and efficient solution to overcome these boundaries in a practical way
[25]	2019	Presents psychological research to show the acceptance of AVs socially based on three measures: general acceptance, willingness to pay, and behavioral intention	Illustrates how people would respond to assess their acceptance of AVs	This research was mostly theoretical; data and statistical parameters were not used
[26]	2019	This survey research focuses on the public's acceptance and social perception toward different levels of AVs	It presents social perceptions and responses toward AVs of different levels of autonomy	This research is solely based on a survey, so no statistical or data-based analysis was available
[27]	2018	This reports a balanced overview of the pros and cons of AV from the perspectives of society, social governance, and governance regulatory bodies	It illustrates the impact of AVs on society and its governance using some famous examples	This illustration could be more detailed and to the point by considering a few more cases of AV crashes and their judgments
[28]	2019	It presents the association between different risk factors and proposes strategies to encounter those, along with emerging government responses toward risks such as crashes, privacy concerns, and cybersecurity risks	It also illustrates different legal reforms in some countries to address liabilities as well as environmental and employment issues	Consideration of some additional parameters and case studies of real issues, such as crashes, could be better

algorithm called "Alternative Bi-directional Q Network," an extension of the deep Q-network algorithm, to overcome the limitations, such as stability, convergence rate, the handling of dynamic weather, and the complexity of traditional algorithms [68–79].

The analysis of various research papers and research and development scenarios shows that there are several problems and challenges which still need rigorous investigation, along with innovations and reforms in conventional infrastructure, governance rules, social mindset, and perceptions [109–112]. Despite various challenges, considerable progress has been made in the development of AVs, with a few companies have successfully tested their AVs up to the third level of autonomy. Therefore, AVs have the potential to overcome these challenges and become the future mode of commute for everyone [80–92].

9.3. Conclusion

Based on the detailed analysis and survey of a significant number of research papers and real development scenarios, we can conclude that through advancements in AI, sensor technologies, and machine learning algorithms, the research and development of AVs is reaching great heights at a rapid pace, as a result of which AVs equipped with new advancements and improvements are becoming smarter every day. These advancements are going to transform our means of transportation in the near future and will become an irreplaceable part of our daily commute [93–98]. Despite several challenges, the current research and development, along with its outcomes, is providing promising indications toward a safer, hassle-free, and economical mode of commute. So, considering the potential for rapid advancement in transport technologies, it becomes a necessity and priority for automobile companies to invest in the research and development of autonomous vehicles to maintain a sustainable and profitable business in the industry [99–108]. Above all, the potential commuters and society also have to adopt the changes and shift their travel preferences from the conventional way of human driving to AI-enabled autonomous driving of vehicles. The governing bodies also need to reform transportation infrastructures and regulations applicable to automobiles, which will support the smooth transition toward using autonomous vehicles and thus harnessing their advantages.

References

1. Tuncali, C. E., Fainekos, G., Ito, H., and Kapinski, J. (2018). Simulation-based adversarial test generation for autonomous vehicles with machine learning components. In *2018 IEEE Intelligent Vehicles Symposium (IV)*. IEEE.
2. Rosique, F., Navarro, P. J., Fernández, C., and Padilla, A. (2019). A systematic review of perception system and simulators for autonomous vehicles research. Sensors, *19*(3), 648.
3. Yeong, D. J., Velasco-Hernandez, G., Barry, J., and Walsh, J. (2021). Sensor and sensor fusion technology in autonomous vehicles: A review. *Sensors, 21*(6), 2140.
4. Wang, H., Khajepour, A., Cao, D., and Liu, T. (2020). Ethical decision making in autonomous vehicles: Challenges and research progress. *IEEE Intelligent Transportation Systems Magazine*, *14*(1), 6–17.
5. Martínez-Díaz, M. and Soriguera, F. (2018). Autonomous vehicles: Theoretical and practical challenges. *Transportation Research Procedia, 33*, 275–282.
6. Raj, A., Ajith Kumar, J., and Bansal, P. (2020). A multicriteria decision making approach to study barriers to the adoption of autonomous vehicles. *Transportation research Part A: Policy and Practice*, *133*, 122–137.
7. Manfreda, A., Ljubi, K., and Groznik, A. (2021). Autonomous vehicles in the smart city era: An empirical study of adoption factors important for millennials. *International Journal of Information Management*, *58*, 102050.
8. Dixit, A., Chidambaram, R. K., and Allam, Z. (2021). Safety and risk analysis of autonomous vehicles using computer vision and neural networks. *Vehicles*, *3*(3), 595–617.
9. Desta, R. and Tóth, J. (2022). Impacts of autonomous vehicle driving logics on heterogenous traffic and evaluating transport interventions with microsimulation experiments. In *International Conference on Human-Computer Interaction*, June 16, 2022, pp. 355–370. Springer International Publishing: Cham.
10. Sanusi, F., Choi, J., Kim Y.H., and Moses, R. (2022). Development of a knowledge base for multiyear infrastructure planning for connected and automated vehicles. *Journal of Transportation Engineering, Part A: Systems*. *48*(4), 03122001.
11. Pérez-Gil, Ó., Barea, R., López-Guillén, E., Bergasa, L. M., Gomez-Huelamo, C., Gutiérrez, R., and Diaz-Diaz, A. (2022). Deep reinforcement learning based control for Autonomous Vehicles in CARLA. *Multimedia Tools and Applications*, *81*(3), 3553–3576.
12. Rafiei, A., Fasakhodi, A. O., and Hajati, F. (2022). Pedestrian collision avoidance using deep reinforcement learning. *International Journal of Automotive Technology*, *23*(3), 613–622.

13. Rais, M. S., Boudour, R., Zouaidia, K., and Bougueroua, L. (2022). Decision making for autonomous vehicles in highway scenarios using Harmonic SK Deep SARSA. *Applied Intelligence, 53*(3), 2488–2505.
14. Zhang, L., Zhang, R., Wu, T., Weng, R., Han, M., and Zhao, Y. (2021). Safe reinforcement learning with stability guarantee for motion planning of autonomous vehicles. *IEEE Transactions on Neural Networks and Learning Systems, 32*(12), 5435–5444.
15. Rais, M. S., Zouaidia, K., and Boudour, R. (2022). Rais MS, Zouaidia K, Boudour R. Enhanced decision making in multi-scenarios for autonomous vehicles using alternative bidirectional Q network. *Neural Computing and Applications, 34*(18), 15981–15996.
16. Rosique, F., Navarro, P. J., Fernández, C., and Padilla, A. (2019). A systematic review of perception system and simulators for autonomous vehicles research. *Sensors, 19*(3), 648.
17. Hussain, R. and Zeadally, S. (2018). Autonomous cars: Research results, issues, and future challenges. *IEEE Communications Surveys & Tutorials, 21*(2), 1275–1313.
18. Rios-Torres, J. and Malikopoulos, A. A. (2016). A survey on the coordination of connected and automated vehicles at intersections and merging at highway on-ramps. *IEEE Transactions on Intelligent Transportation Systems, 18*(5), 1066–1077.
19. Pettigrew, S., Fritschi, L., and Norman, R. (2018). The potential implications of autonomous vehicles in and around the workplace. *International Journal of Environmental Research and Public Health, 15*(9), 1876.
20. Duarte, F. and Ratti, C. (2018). The impact of autonomous vehicles on cities: A review. *Journal of Urban Technology, 25*(4), 3–18.
21. Aria, E., Olstam, J., and Schwietering, C. (2016). Investigation of automated vehicle effects on driver's behavior and traffic performance. *Transportation Research Procedia, 15*, 761–770.
22. Wu, C., Bayen, A. M., and Mehta, A. (2018). Stabilizing traffic with autonomous vehicles. In *2018 IEEE International Conference on Robotics and Automation (ICRA)*, May 21, 2018, pp. 6012–6018. IEEE.
23. Li, T., Wu, J., Chan, C.-Y., Liu, M., Zhu, C., Lu, W., and Hu, K. (2020). A cooperative lane change model for connected and automated vehicles. *IEEE Access, 8*, 54940–54951.
24. Wang, M. (2018). Infrastructure assisted adaptive driving to stabilise heterogeneous vehicle strings. *Transportation Research Part C: Emerging Technologies, 91*, 276–295.
25. Liu, P., Yang, R., and Xu, Z. (2019). Public acceptance of fully automated driving: Effects of social trust and risk/benefit perceptions. *Risk Analysis, 39*(2), 326–341.

26. Liu, H., Run Yang, R., Wang, L., and Liu, P. (2019). Evaluating initial public acceptance of highly and fully autonomous vehicles. *International Journal of Human-Computer Interaction, 35*(11), 919–931.
27. Stilgoe, J. (2018). Machine learning, social learning and the governance of self-driving cars. *Social Studies of Science, 48*(1), 25–56.
28. Taeihagh, A. and Lim, H. S. M. (2019). Governing autonomous vehicles: Emerging responses for safety, liability, privacy, cybersecurity, and industry risks. *Transport Reviews, 39*(1), 103–128.
29. Jain, N. and Raj, B. (2018). Parasitic capacitance and resistance model development and optimization of raised source/drain SOI FinFET structure for analog circuit applications. *Journal of Nanoelectronics and Optoelectronins, 13*, 531–539.
30. Singh, S., Vishvakarma, S. K., and Raj, B. (2018). Analytical modeling of split-gate junction-less transistor for a biosensor application. *Sensing and Bio-sensing, 18*, 31–36.
31. Gopal, M. and Raj, B. (2013). Low power 8T SRAM cell design for high stability video applications, *ITSI Transaction on Electrical and Electronics Engineering, 1*(5), 91–97.
32. Raj, B., Mitra, J., Bihani, D. K., Rangharajan, V., Saxena, A. K., and Dasgupta, S. (2010). Analysis of noise margin, power and process variation for 32 nm FinFET based 6T SRAM cell. *Journal of Computer (JCP), 5*(6), 1–8.
33. Sharma, D., Mehra, R., and Raj, B. (2021). Comparative analysis of photovoltaic technologies for high efficiency solar cell design. *Superlattices and Microstructures, 153*, 106861.
34. Kaur, P., Buttar, A. S., and Raj, B. (2021). A comprehensive analysis of nanoscale transistor based biosensor: A review. *Indian Journal of Pure and Applied Physics, 59*, 304–318.
35. Yadav, D., Raj, B., and Raj, B. (2020). Design and simulation of low power microcontroller for IoT applications. *Journal of Sensor Letters, 18*, 401–409.
36. Singh, S. and Raj, B. (2020). A 2-D analytical surface potential and drain current modeling of double-gate vertical t-shaped tunnel FET. *Journal of Computational Electronics, 19*, 1154–1163.
37. Singh, J. and Raj, B. (2019). An accurate and generic window function for non-linear memristor model. *Journal of Computational Electronics, 18*(2), 640–647.
38. Kaur, M., Gupta, N., Kumar, S., Raj, B., and Singh, A. K. (2021). Comparative RF and crosstalk analysis of carbon based nano interconnects. *IET Circuits, Devices & Systems, 15*(6), 493–503.
39. Kandasamy, N., Ahmad, F., Ajitha, D., Raj, B., and Telagam, N. (2020). Quantum dot cellular automata based scan flip flop and boundary scan register. *IETE Journal of Research, 66*, 535–548.

40. Sharma, S. K., Raj, B., and Khosla, M. (2019). Enhanced photosensivity of highly spectrum selective cylindrical gate $In_{1-x}Ga_xAs$ nanowire MOSFET photodetector. *Modern Physics letter-B*, *33*(12), 1950144.
41. Singh, J. and Raj, B. (2019). Design and investigation of 7T2M NVSARM with enhanced stability and temperature impact on store/restore energy. *IEEE Transactions on Very Large Scale Integration Systems*, *27*(6), 1322–1328.
42. Bhardwaj, A. K., Gupta, S., Raj, B., and Singh, A. (2019). Impact of double gate geometry on the performance of carbon nanotube field effect transistor structures for low power digital design. *Computational and Theoretical Nanoscience*, *16*, 1813–1820.
43. Jain, N. and Raj, B. (2019). Thermal stability analysis and performance exploration of asymmetrical dual-k underlap spacer (ADKUS) SOI FinFET for security and privacy applications. *Indian Journal of Pure & Applied Physics (IJPAP)*, *57*, 352–360.
44. Singh, A., Khosla, M., and Raj, B. (2019). Design and analysis of dynamically configurable electrostatic doped carbon nanotube tunnel FET. *Microelectronics Journal*, *85*, 17–24.
45. Jain, N. and Raj, B. (2019). Dual-k spacer region variation at the drain side of asymmetric SOI FinFET structure: Performance analysis towards the analog/RF design applications. *Journal of Nanoelectronics and Optoelectronics*. *14*, 349–359.
46. Singh, J., Sharma, S., Raj, B., and Khosla, M. (2018). Analysis of barrier layer thickness on performance of $In_{1-x}Ga_xAs$ based gate stack cylindrical gate nanowire MOSFET. *JNO*, *13*, 1473–1477.
47. Jain, N. and Raj, B. (2018). Analysis and performance exploration of high-k SOI FinFETs over the conventional low-k SOI FinFET toward analog/RF design. *Journal of Semiconductors (JoS)*, *39*(12), 124002.
48. Goyal, C., Ubhi, J. S., and Raj, B. (2018). A reliable leakage reduction technique for approximate full adder with reduced ground bounce noise. *Journal of Mathematical Problems in Engineering*, *2018*, 16. Article ID 3501041.
49. Anuradha, J. S., Raj, B., and Khosla, M. (2018). Design and performance analysis of nano-scale memristor-based nonvolatile SRAM. *Journal of Sensor Letter*, *16*, 798–805.
50. Wadhwa, G. and Raj, B. (2018). Parametric variation analysis of charge-plasma-based dielectric modulated JLTFET for biosensor application. *IEEE Sensor Journal*, *18*(15), 6070–6077.
51. Singh, J. and Raj, B. (2018). Comparative analysis of memristor models for memories design. *JoS*, *39*(7), 074006.
52. Yadav, D., Chouhan, S. S., Vishvakarma, S. K., and Raj, B. (2018). Application Specific Microcontroller Design for IoT based WSN. *Sensor Letter*, *16*, 374–385.

53. Singh, G., Sarin, R. K., and Raj, B. (2018). Fault-tolerant design and analysis of quantum-dot cellular automata based circuits. *IEEE/IET Circuits, Devices & Systems*, *12*, 638–64.
54. Singh, J. and Raj, B. (2018). Modeling of mean barrier height levying various image forces of metal insulator metal structure to enhance the performance of conductive filament based memristor model. *IEEE Nanotechnology*, *17*(2), 268–267.
55. Jain, A., Sharma, S., and Raj, B. (2018). Analysis of triple metal surrounding gate (TM-SG) III-V nanowire MOSFET for photosensing application. *Opto-electronics Journal*, *26*(2), 141–148.
56. Jain, A., Sharma, S., and Raj, B. (2016). Design and analysis of high sensitivity photosensor using cylindrical surrounding gate MOSFET for low power sensor applications. *Engineering Science and Technology*, *19*(4), 1864–1870.
57. Singh, A., Khosla, M., and Raj, B. (2017). Analysis of electrostatic doped Schottky barrier carbon nanotube FET for low power applications. *Journal of Materials Science: Materials in Electronics*, *28*, 1762–1768.
58. Kumar, G. S., Singh, A., and Raj, B. (2018). Design and analysis of gate all around CNTFET based SRAM cell design. *Journal of Computational Electronics*, *17*(1), 138–145.
59. Singh, G. P., Sohi, B. S., and Raj, B. (2017). Material properties analysis of graphene base transistor (GBT) for VLSI analog circuits. *Indian Journal of Pure & Applied Physics (IJPAP)*, *55*, 896–902.
60. Singh, A., Khosla, M., and Raj, B. (2016). Comparative analysis of carbon nanotube field effect transistor and nanowire transistor for low power circuit design. *Journal of Nanoelectronics and Optoelectronics*, *11*, 388–393.
61. Kumar, S. and Raj, B. (2017). Estimation of stability and performance metric for inward access transistor based 6T SRAM cell design using *n*-type/*p*-type DMDG-GDOV TFET. *IEEE VLSI Circuits and Systems Letter*, *3*(2), 25–39.
62. Sharma, S., Kumar, A., Pattanaik, M., and Raj, B. (2013). Forward body biased multimode multi-threshold CMOS technique for ground bounce noise reduction in static CMOS adders. *International Journal of Information and Electronics Engineering*, *3*(3), 567–572.
63. Singh, H., Kumar, P., and Raj, B. (2013). Performance analysis of majority gate SET based 1-bit full adder. *International Journal of Computer and Communication Engineering (IJCCE)*, *2*(4), ISSN: 2010–3743.
64. Bhardwaj, A. K., Gupta, S., and Raj, B. (2020). Investigation of parameters for Schottky barrier (SB) height for Schottky barrier based carbon nanotube field effect transistor device. *Journal of Nanoelectronics and Optoelectronics*, *15*, 783–791.
65. Bansal, P. and Raj, B. (2017). Memristor: A versatile nonlinear model for dopant drift and boundary issues. *JCTN*, *14(*5), 2319–2325.

66. Jain, N. and Raj, B. (2016). An analog and digital design perspective comprehensive approach on Fin-FET (Fin-Field Effect transistor) technology — A review. *Reviews in Advanced Sciences and Engineering (RASE)*, *5*, 1–14.
67. Sharma, S., Raj, B., and Khosla, M. (2017). Subthreshold performance of $In_{1-x}Ga_xAs$ based dual metal with gate stack cylindrical/surrounding gate nanowire MOSFET for low power analog applications. *Journal of Nanoelectronics and Optoelectronics*, *12*, 171–176.
68. Raj, B., Saxena, A. K., and Dasgupta, S. (2009). Analytical modeling for the estimation of leakage current and subthreshold swing factor of nanoscale double gate FinFET device. *Microelectronics International*, *26*, 53–63.
69. Soniya, S. S., Wadhwa, G., and Raj, B. (2019). An analytical modeling for dual source vertical tunnel field effect transistor. *International Journal of Recent Technology and Engineering (IJRTE)*, *8*(2), 603–608.
70. Singh, S. and Raj, B. (2019). Design and analysis of hetrojunction vertical T-shaped tunnel field effect transistor. *Journal of Electronics Material*, *48*(10), 6253–6260.
71. Goyal, C., Ubhi, J. S., and Raj, B. (2019). A low leakage CNTFET based inexact full adder for low power image processing applications. *International Journal of Circuit Theory and Applications*, *47*(9), 1446–1458.
72. Raj, B., Saxena, A. K., and Dasgupta, S. (2008). A compact drain current and threshold voltage quantum mechanical analytical modeling for FinFETs. *Journal of Nanoelectronics and Optoelectronics (JNO)*, *3*(2), 163–170.
73. Wadhwa, G. and Raj, B. (2020). An analytical modeling of charge plasma based tunnel field effect transistor with impacts of gate underlap region. *Superlattices and Microstructures*, *142*, 106512.
74. Singh, S. and Raj, B. (2020). Modeling and simulation analysis of SiGe hetrojunction double gate vertical *T*-shaped tunnel FET. *Superlattices and Microstructures*, *142*, 106496.
75. Singh, A., Saini, D. K., Agarwal, D., Aggarwal, S., Khosla, M., and Raj, B. (2016). Modeling and simulation of carbon nanotube field effect transistor and its circuit application. *Journal of Semiconductors (JoS)*, *37*, 074001-6.
76. Jain, N. and Raj, B. (2016). Device and circuit co-design perspective comprehensive approach on FinFET technology — A review. *Journal of Electron Devices*, *23*(1), 1890–1901.
77. Kumar, S. and Raj, B. (2016). Analysis of ION and ambipolar current for dual-material gate-drain overlapped DG-TFET. *Journal of Nanoelectronics and Optoelectronics*, *11*, 323–333.
78. Anjum, N., Bali, T., and Raj, B. (2013). Design and simulation of handwritten multiscript character recognition. *International Journal of Advanced Research in Computer and Communication Engineering*, *2*(7), 2544–2549.

79. Sharma, S., Raj, B., and Khosla, M. (2016). A Gaussian approach for analytical subthreshold current model of cylindrical nanowire FET with quantum mechanical effects. *Microelectronics Journal*, *53*, 65–72.
80. Singh, K. and Raj, B. (2015). Performance and analysis of temperature dependent multi-walled carbon nanotubes as global interconnects at different technology nodes. *Journal of Computational Electronics*, *14*(2), 469–476.
81. Kumar, S. and Raj, B. (2015). Compact channel potential analytical modeling of DG-TFET based on evanescent–mode approach. *Journal of Computational Electronics*, *14*(2), 820–827.
82. Singh, K. and Raj, B. (2015). Temperature dependent modeling and performance evaluation of multi-walled CNT and single-walled CNT as global interconnects. *Journal of Electronic Materials*, *44*(12), 4825–4835.
83. Sharma, V. K., Pattanaik, M., and Raj, B. (2014). INDEP approach for leakage reduction in nanoscale CMOS circuits. *International Journal of Electronics*, *102*(2), 200–215.
84. Singh, K. and Raj, B. (2015). Influence of temperature on MWCNT bundle, SWCNT bundle and copper interconnects for nanoscaled technology nodes. *Journal of Materials Science: Materials in Electronics*, *26*(8), 6134–6142.
85. Anjum, N., Bali, T., and Raj, B. (2013). Design and simulation of handwritten Gurumukhi and Devanagri numerical recognition. *International Journal of Computer Applications*, *73*(12), 16–21.
86. Khandelwal, S., Gupta, V., Raj, B., and Gupta, R D. (2015). Process variability aware low leakage reliable nano scale DG-FinFET SRAM cell design technique. *Journal of Nanoelectronics and Optoelectronics*, *10*(6), 810–817.
87. Sharma, V. K., Pattanaik, M., and Raj, B. (2014). ONOFIC approach: Low power high speed nanoscale VLSI circuits design. *International Journal of Electronics*, *101*(1), 61–73.
88. Khandelwal, S., Raj, B., and Gupta, R D. (2015). FinFET based 6T SRAM cell design: Analysis of performance metric, process variation and temperature effect. *Journal of Computational and Theoretical Nanoscience*, *12*, 2500–2506.
89. Singh, S., Yadav, S., Rahul, J., Srivastava, A., and Raj, B. (2015). Impact of HfO_2 in graded channel dual insulator double gate MOSFET. *Journal of Computational and Theoretical Nanoscience*, *12*(6), 950–953.
90. Sharma, V. K., Pattanaik, M., and Raj, B. (2014). PVT variations aware low leakage INDEP approach for nanoscale CMOS circuits. *Microelectronics Reliability*, *54*, 90–99.
91. Raj, B., Saxena, A. K., and Dasgupta, S. (2013). Quantum mechanical analytical modeling of nanoscale DG FinFET: Evaluation of potential, threshold voltage and source/drain resistance. *Journal of Material Science in Semiconductor Processing*, *16*(4), 1131–1137.
92. Gopal, M., Prasad, S. S. D., and Raj, B. (2013). 8T SRAM cell design for dynamic and leakage power reduction. *International Journal of Computer Applications*, *71*(9), 43–48.

93. Pattanaik, M., Raj, B., Sharma, S., and Kumar, A. (2012). Diode based trimode multi-threshold CMOS technique for ground bounce noise reduction in static CMOS adders. *Advanced Materials Research, 548*, 885–889.
94. Raj, B., Saxena, A. K., and Dasgupta, S. (2011). Nanoscale FinFET based SRAM cell design: Analysis of performance metric, process variation, underlapped FinFET and temperature effect. *IEEE Circuits and System Magazine, 11*(2), 38–50.
95. Sharma, V. K., Pattanaik, M., and Raj, B. (2013). Leakage current ONOFIC approach for deep submicron VLSI circuit design. *International Journal of Electrical, Computer, Electronics and Communication Engineering, 7*(4), 239–244.
96. Chawla, T., Khosla, M., and Raj, B. (2021). Design and simulation of triple metal double-gate germanium on insulator vertical tunnel field effect transistor. *Microelectronics Journal, 114*, 105125.
97. Kaur, P., Gill, S. S., and Raj, B. (2022). Comparative analysis of OFETs materials and devices for sensor applications. *Journal of Silicon, 14*, 4463–4471.
98. Sharma, S. K., Kumar, P., Raj, B., and Raj, B. (2022). $In_{1-x}Ga_xAs$ double metal gate-stacking cylindrical nanowire MOSFET for highly sensitive photo detector. *Journal of Silicon, 14*, 3535–3541.
99. Raj, B., Saxena, A. K., and Dasgupta, S. (2009). Analytical modeling of quasi planar nanoscale double gate FinFET with source/drain resistance and field dependent carrier mobility: A quantum mechanical study. *Journal of Computer (JCP), 4*(9), 1–8.
100. Bhushan, S., Khandelwal, S., and Raj, B. (2012). Analyzing different mode FinFET based memory cell at different power supply for leakage reduction. In *Proceedings of Seventh International Conference on Bio-Inspired Computing: Theories and Applications (BIC-TA 2012)*, Vol. 2, pp. 89–100. Springer: India.
101. Singh, J. and Raj, B. (2018). Temperature dependent analytical modeling and simulations of nanoscale memristor. *Engineering Science and Technology, 21*, 862–868.
102. Singh, S., Bala, S., Raj, B., and Raj, B. (2020). Improved sensitivity of dielectric modulated junctionless transistor for nanoscale biosensor design. *Sensor Letter, 18*, 328–333.
103. Kumar, V., Vishvakarma, S. K., and Raj, B. (2020). Design and performance analysis of ASIC for IoT applications. *Sensor Letter, 18*, 31–38.
104. Jaiswal, A., Sarin, R. K., Raj, B., and Sukhija, S. (2018). A novel circular slotted microstrip-fed patch antenna with three triangle shape defected ground structure for multiband applications. *Advanced Electromagnetic (AEM), 7*(3), 56–63.
105. Wadhwa, G. and Raj, B. (2018). Label free detection of biomolecules using charge-plasma-based gate underlap dielectric modulated junctionless TFET. *Journal of Electronic Materials (JEMS), 47*(8), 4683–4693.

106. Singh, G., Sarin, R. K., and Raj, B. (2017). Design and performance analysis of a new efficient coplanar quantum-dot cellular automata adder. *Indian Journal of Pure & Applied Physics (IJPAP)*, *55*, 97–103.
107. Singh, A., Khosla, M., and Raj, B. (2017). Design and analysis of electrostatic doped Schottky barrier CNTFET based low power SRAM. *International Journal of Electronics and Communications*, *80*, 67–72.
108. Kaur, P., Pandey, V., and Raj, B. (2020). Comparative study of efficient design, control and monitoring of solar power using IoT. *Sensor Letter*, *18*, 419–426.
109. Gaurav, A., Gupta, B. B., Peñalvo, F. J. G., Nedjah, N., and Psannis, K. (2022). Ddos attack detection in vehicular ad-hoc network (vanet) for 5g networks. *Security and Privacy Preserving for IoT and 5G Networks: Techniques, Challenges, and New Directions*, 263–278.
110. Gaurav, A., Gupta, B. B., Castiglione, A., Psannis, K., and Choi, C. (2020). A novel approach for fake news detection in vehicular ad-hoc network (VANET). In *Proceedings of Computational Data and Social Networks: 9th International Conference, (CSoNet 2020)*, Dallas, TX, USA, December 11–13, 2020, pp. 386–397. Springer International Publishing.
111. Liang, Y., Liu, Y., and Gupta, B. B. (2022). PPRP: Preserving-privacy route planning scheme in VANETs. *ACM Transactions on Internet Technology*, *22*(4), 1–18.
112. Chui, K. T., Kochhar, T. S., Chhabra, A., Singh, S. K., Singh, D., Peraković, D., Almomani, A., and Arya, V. (2022). Traffic accident prevention in low visibility conditions using vanets cloud environment. *International Journal of Cloud Applications and Computing (IJCAC)*, *12*(1), 1–21.

https://doi.org/10.1142/9789811279539_0010

Chapter 10

Review of Time Domain Signal Processing Using Time-to-Digital Converter

Pooja Saxena[*,†,§] **and K. Hari Prasad**[‡,¶]

[†]*Department of Electronics & Communication Engineering, Greater Noida Institute of Technology, Greater Noida, Uttar Pradesh, India*

[‡]*Homi Bhabha National Institute, Anushakti Nagar, Mumbai, Maharashtra, India*

[§]*saxena06pooja@gmail.com*

[¶]*hari.barc@gmail.com*

Abstract

In the current scenario, technology scaling is being implemented in digital integrated circuits to improve their power, speed, size, and cost-effectiveness. On the other hand, scaling induces reductions in supply voltage and intrinsic transistor gain, because of which analog circuit designers face various design challenges. This paved the way for using a time domain approach to analog designs, which is based on the representation of signals in time domain parameters, such as time period, transition time, and time delays. These signals are processed using time

[*]Corresponding author.

domain signal processing implemented by time-mode circuits, such as voltage-to-time converters, time amplifiers, time registers, and time-to-digital converters (TDCs). This chapter incorporates various reported time-mode circuits but emphasis is placed on TDCs. It covers the details of a wide range of TDC applications and performance metrics. Further, a classified review of various time interval measurement techniques used to design TDCs is given. The key design concepts of each technique are comprehensively discussed, along with their merits, demerits, salient features, and applications. It concludes with a timeline of the evolution of various TDC techniques, right from the 1960s until today, using a flow diagram, and provides a comprehensive bibliography.

Keywords: Time-to-digital converter, time interval, dynamic range, delay lock loop, phase lock loop, time of flight

10.1. Introduction

The rapid advancement in CMOS technology paved the way for the achievement of optimal performance of digital circuits in terms of area and power efficiency and high speed. However, because of side effects such as worsening MOS device mismatch and reduced voltage headroom, analogue circuits are unable to make use of specialized and process-controlled components. The device mismatch issue leads to a reduction in IC performance and its reliability. The reduction in voltage headroom is due to a decline in device threshold and supply voltage. This leads to degradation in the signal-to-noise ratio and dynamic range [1]. The current mode approach provides an alternative means to cope with losses in voltage headroom. However, a large and steady current results in high power consumption. Therefore, this approach is preferred for applications where speed is preferred at the expense of high power consumption.

The speed improvement achieved in MOS technology is due to the following two reasons: (a) a reduction in intrinsic logic gate delay attributed to a short transit time and less parasitic capacitance of the MOS device; (b) improved switching characteristics of the MOS device. These features have prominently improved the timing resolution in digital circuits as compared to the voltage resolution achieved in analog circuits. The improvement in timing resolution offers good timing accuracy for systems, thereby resulting in a new design paradigm: the so-called time-mode signal processing. Here, the information is represented in terms of timing parameters, such as pulse duration, delay, rise and fall times, and time periods of the clock. These timing signals are processed using time-mode blocks, such as timing adders, amplifiers, time shifters (time delays), time-based analog-to-digital converters (ADCs), and time-to-digital converters (TDCs).

In this chapter, Section 10.2 covers the various time-mode blocks reported in the literature, including voltage-to-time converters (VTCs), timing registers, timing adders, and amplifiers. Section 10.3 covers the comprehensive details of TDCs, a main pillar of time-mode signal processing due to their role in digitizing the time domain intermediate signal.

10.2. Time-Mode Blocks

10.2.1. *VTC Based on Current Starved Inverter*

A VTC is one of the key building blocks of time-mode ADCs. It converts an analog input voltage into a pulse signal with a linear relation between the pulse width and the analog voltage. This pulse is digitized with the help of TDCs.

The basic structure of a VTC is based on the simple current starved inverter (CSI) [2], as shown in Figures 10.1(a) and 10.1(b). It is realized by adding an input-voltage-controlled NMOS or PMOS transistor (M_3) in series with the pull-up or pull-down paths of a basic CMOS inverter. The choice of added transistors depends on the requirement of delay introduced in either the falling-edge or rising-edge transition of the applied signal "In." The input voltage causes a change in current through the added transistor M_3, which leads to a change in the rate of discharging (or charging) of a load capacitor, thereby introducing variation in the falling- (or rising-) edge delay. As shown in Figure 10.2, the rising-edge delay across the design process corners is proportional to the input voltage; therefore, the pulse width of the output is inversely proportional to V_{in}.

An advantage of a CSI-based VTC is its small area consumption due to less number of required transistors. In addition, it exhibits a high speed as its working depends on discharging (or charging) a small capacitor on the application of an input signal. However, the VTC circuit exhibits a trade-off between linearity and dynamic range. The increment in dynamic range results in a higher nonlinearity effect, and vice versa. In addition, it has limited sensitivity, defined as the slope of the delay versus V_{in}

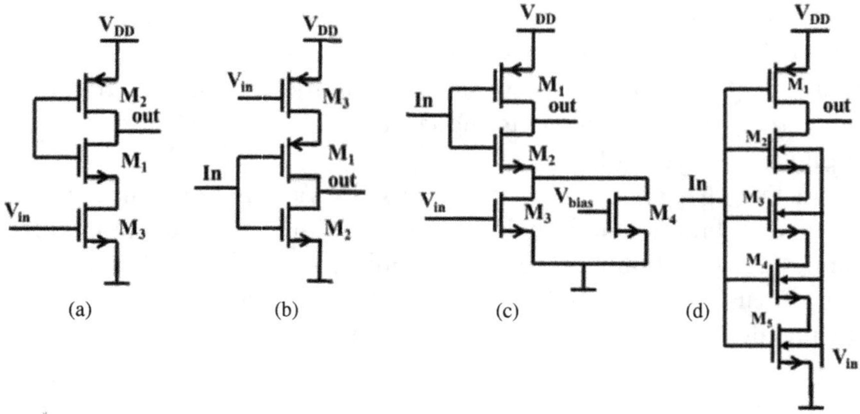

Figure 10.1. Schematic Diagram of Current Starved Inverter: (a) CSI-1 for Falling-Edge Delay, (b) CSI-2 for Rising-Edge Delay, (c) Addition of Transistor M_4 to Improve Sensitivity, (d) CSI Bbased on Body Bias Technique

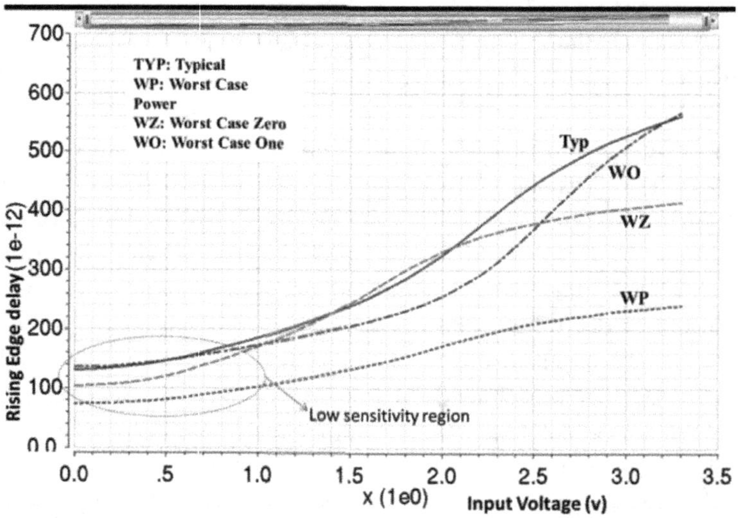

Figure 10.2. Rising-Edge Delay vs. Input Characteristics of Current Starved Inverter

characteristic curve, as shown in Figure 10.2. In order to address these issues, a VTC with improved linearity and sensitivity by using parallel current starving devices, M_5, M_6, and M_9–M_{14}, and source degeneration techniques has been reported, as shown in Figure 10.3. The parallel-connected MOS transistors (M_{11}–M_{14}) improve sensitivity with respect to the

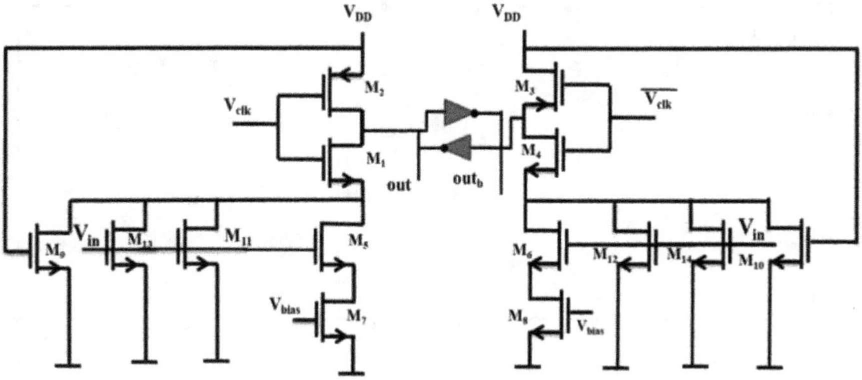

Figure 10.3. CSI-Based VTC with Improved Linear Characteristics and Sensitivity

input control voltage. The MOS transistors M_7 and M_8 implement the source degeneration for the current starving devices M_5 and M_6, which results in linear delay versus control voltage characteristics. The devices M_9 and M_{10} are weak devices with small aspect ratios and low g_m, which are used to ensure the operation of the inverter at low input voltages. The cross-coupled inverters are used to provide a small transition time at the output.

In Ref. [4], a VTC with linear delay versus voltage characteristics is reported, where a control voltage is applied to the substrate terminal of the current starving transistor instead of its gate terminal, as shown in Figure 10.1(d). This substrate biasing technique provides a linear drain current with respect to the substrate-to-source voltage of the current starved transistors, as given by Equation (10.1). This technique can be applied to both NMOS and PMOS current starved transistors, as shown in Figures 10.1(a) and 10.1(b), respectively:

$$\Delta T = \frac{C * dV_{\text{in}}}{(K_1 + K_2 V_{\text{bs}})} \quad \text{where } K_1 = \frac{K_n}{2}\left(V_{\text{gs}} - V_{\text{to}}\right)^2$$

and

$$K_2 = \frac{K_{n\gamma}}{2\sqrt{2(\varphi F)}}\left(V_{\text{gs}} - V_{\text{to}}\right). \tag{10.1}$$

Moreover, the addition of series transistors, M_4 and M_5, improves the resolution and linearity of the VTC circuit. The resolution improvement is attributed to the decrement in the equivalent aspect ratio of series NMOS

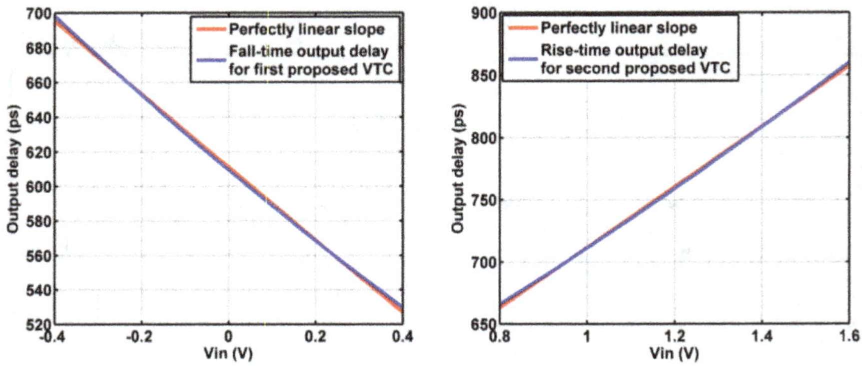

(a) (b)

Figure 10.4. CSI-Based VTC Implemented Using Body Bias Technique [3]

transistors ($W/L_{eq} = W/(L \times N)$), resulting in a low discharge current. This manifests as an increased delay in the fixed control voltage step. Moreover, the addition of series NMOS transistors decreases the gate-to-source voltage of the driver transistor M_1. This ensures the saturation operating region for driving the transistor to maintain linearity during discharging of the load capacitor. Figure 10.4 shows the achieved VTC characteristics corresponding to both the rising and falling-edge delays.

In Refs. [2] and [5], a technique based on the difference between the rising- and falling-edge delays is used to cancel the nonlinearity errors. This method provides better linearity over the same dynamic range as compared to the CSI-based VTC. As shown in Figure 10.5, CSI-1 consists of an *n*-channel current starved transistor in the pull-down path. Therefore, the required time (falling-edge delay, T_f) to discharge the output capacitor is inversely proportional to V_{in}. In CSI-2, a *p*-channel current starved transistor is used in the pull-up path; therefore, the charging time (rising-edge delay, T_r) of the output capacitor is directly proportional to V_{in}. An inverted delayed version of the clock signal is applied to CSI-2. The outputs of CSI-1 and CSI-2 are applied to an exclusive-OR (XOR) gate to generate a pulse whose width is $\Delta + T_r - T_f$, where Δ is the delay introduced by the delay line. Here, the difference in the T_r and T_f delays versus voltage characteristics provides the nonlinearity error cancellation, thereby showing improved linearity.

The difference between T_r and T_f is required to be at its maximum to improve the dynamic range. Therefore, in CSI-1, the pull-down network is designed to be faster (less T_f) by choosing a large width for the current starved transistor. Similarly, in CSI-2, the current starved transistor in the pull-up path is designed with low strength. Also, in CSI-1 and CSI-2,

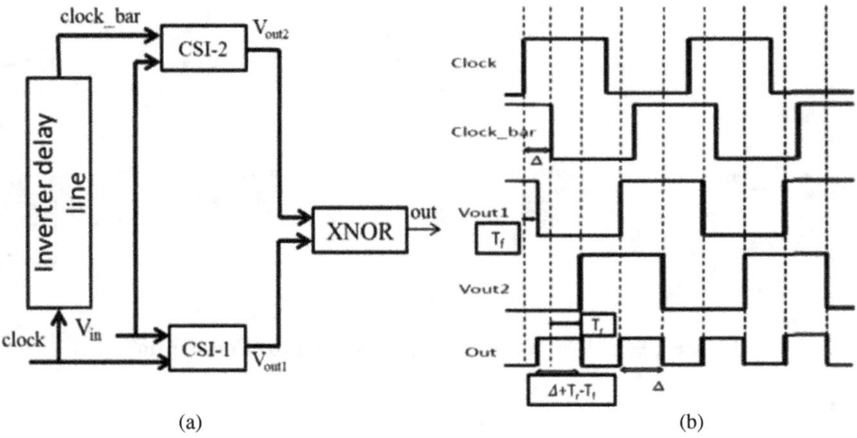

(a) (b)

Figure 10.5. (a) CSI-Based VTC with Improved Linear Characteristics by Taking the Difference of Rising and Falling Delays (b) (CSI-1 and CSI-2 are Shown in Figure 10.1)

NMOS and PMOS transistors, respectively, are added in parallel to the current starved device (as in Figure 10.1(c)) to limit the maximum discharging (or charging) time when V_{GS} is lower than the threshold voltage of the transistor.

10.2.2. *VTC Based on Ramp and Comparator*

It is used in time-based ADCs [2], where the input voltage is sampled and held across a capacitor *C*, as shown in Figure 10.6(a). The "start" signal connected to the switch initiates the generation of a ramp voltage by discharging the capacitor using a constant current. The ramp voltage is compared with the ground voltage using a comparator. When discharging is completed, it generates the output "stop" signal. The time interval between "start" and "stop" is proportional to the applied sample of the input voltage. This scheme can also be used in another way [6], as shown in Figure 10.6(b), where a capacitor is charged using a constant current source during a low clock period. A high clock duration enables quick discharging of the capacitor. This ramp voltage across the capacitor is compared with the applied input signal using the comparator. It generates an output pulse when the ramp crosses the input voltage signal. The benefit of employing a ramp-based VTC is that the ramp is created linearly, and the input range is constrained by the ramp's "bias compliance." Its limitations are a large conversion time and high power consumption as

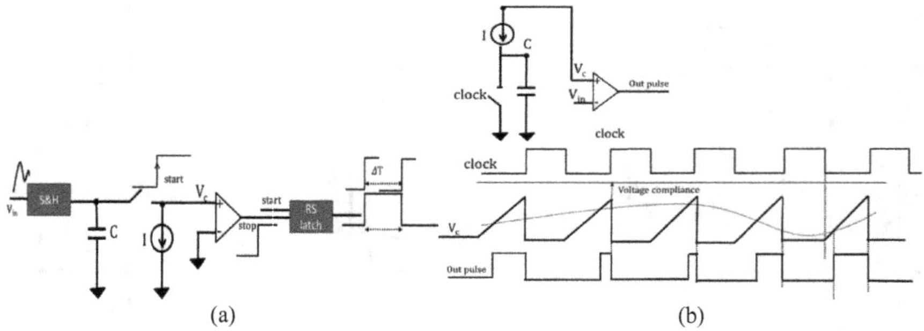

Figure 10.6. VTC Based on (a) Ramp and (b) Comparator

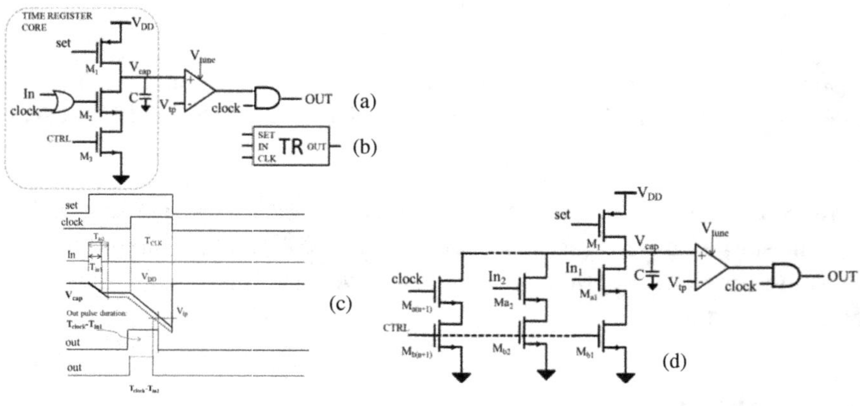

Figure 10.7. Time Register: (a) Circuit, (b) Block Diagram, (c) Timing Diagram, (d) Time Adder

compared to a CSI-based VTC. Moreover, in order to prevent nonlinearity, the comparator must also be constructed with sufficient common mode rejection.

10.2.3. *Time-Based Register*

A time-based register (TR) which stores a pulse of a given time interval T_{in} is reported in Ref. [7]. As shown in Figure 10.7(a and b), when the set signal is at logic 0, the transistor M_1 is turned on. It provides a charging path for the capacitor C up to the supply voltage V_{DD}. The capacitor discharges through the transistor M_2, which is turned on during the high duration of either the clock or the input pulse T_{in}, as shown in Figure 10.7(c). The voltage across the capacitor remains fixed when both T_{in} and the clock are at logic “0.” The pulse width of the clock signal is fixed with a 25% duty cycle. As the discharging slope of the capacitor and the discharging time during clock high duration are fixed, the output pulse will have a width equal to $T_{clock} - T_{in}$. This results in registering the value of T_{in} with the synchronization of the output pulse with the clock signal.

10.2.4. *Time Adder*

A time adder circuit derived from the time register is shown in Figure 10.7(d). It adds n number of input pulses, T_{in1}, T_{in2}, …, T_{inn}. As the aspect ratios of transistors M_{b1}, M_{b2}, …, M_{bn}, M_{bn+1} are chosen to be equal, the discharging slopes caused by T_{in1}, T_{in2}, …, T_{inn} will be given by slope$_{total}$ = slope$_{in1}$ + slope$_{in2}$ + ⋯ + slope$_{inn}$. Therefore, the output pulse width will be $T_{out} = T_{clock} - (T_{in1} + T_{in2} + \cdots + T_{inn})$.

10.2.5. *Time Interval Amplification*

The primary component of a time amplifier is a set–reset (SR) latch followed by an XOR gate, as shown in Figure 10.8. In an SR-latch, if inputs are close to each other, there will be a setup time violation. Therefore, the latch will enter a metastable state, i.e., the output voltage will be stuck at some intermediate voltage level. The inbuilt regenerative feedback in the latch leads the NAND gate outputs to attain complementary logic levels, which gives the XOR gate output an amplified logic pulse. The regeneration time is the logarithmic function of the initial time interval, as shown in Figure 10.8(d). Also, introducing a delay, T_{off}, to the S and R inputs causes a shift in the metastability point of the SR latch, as shown in Figure 10.8(c). The gain is expressed as $2C/(g_m T_{off})$, where g_m is the transistor’s transconductance gain in the SR latch at metastability and C is the load capacitor. The drawbacks of time amplifiers include a small input range, poor gain linearity, and significant sensitivity to PVT variations.

Another time amplifier based on delay lock loop (DLL) has been reported [84], which uses a closed-loop approach to minimize the effect of PVT. As shown in Figure 10.9, the “start” and “stop” signals trigger two identical ring oscillators to provide clocks, allowing the DLL to maintain the loop in a locked condition via iterative corrections. The respective rising edges of these clocks have a delay equal to the applied time interval.

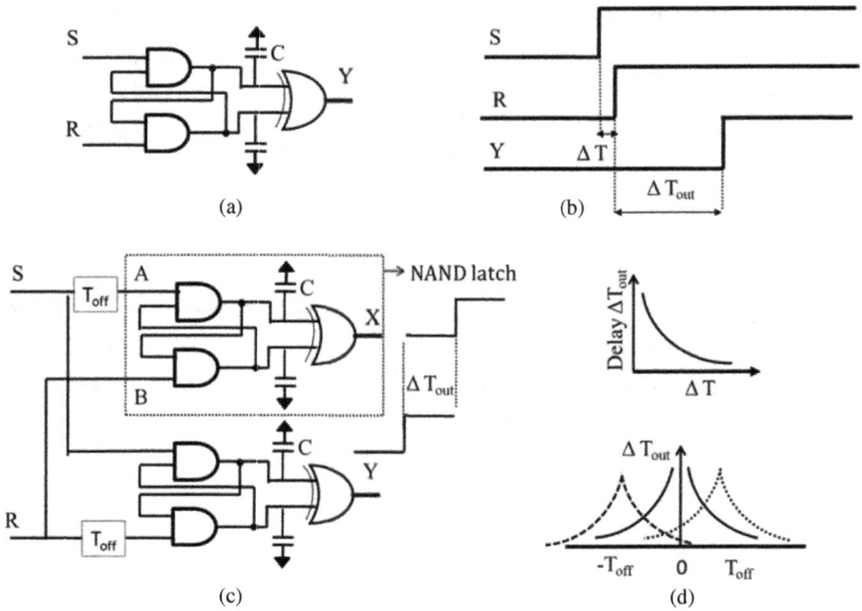

Figure 10.8. Time Amplifier Based on Metastability in SR Latches

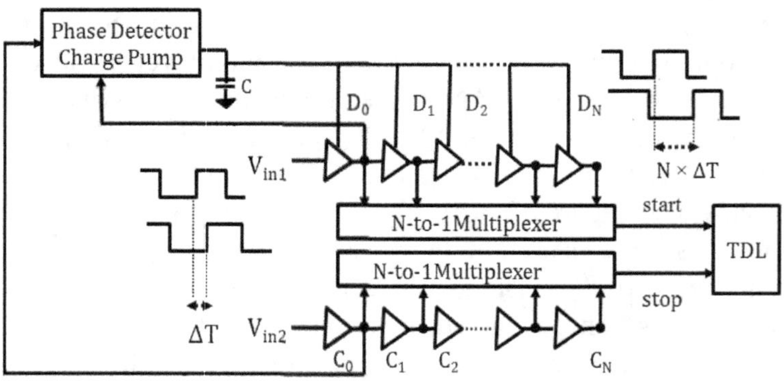

Figure 10.9. Block Diagram of Time Amplifier Using DLL

These clocks are then applied to the delay elements "D_0" and "C_0," which are identical in structure and sizing. The phase detector (PD) compares the rising edges of the delayed outputs. The feedback loop adjusts the delay of "D_0" using a control voltage to ensure output phase coincidence. At coincidence, the difference in the delays of "C_0" and "D_0" is equal to the input time interval ΔT expressed by Equation 10.2(a). To produce amplification, these delay cells are replicated N times, so that the output delay difference is equal to N times of the input time interval. The amplified output is given by Equation 10.2(b).

$$\Delta T = (T_{c0} - T_{D0}), \tag{10.2(a)}$$

$$\text{Output time interval } \Delta T_{\text{out}} = N \times \Delta T. \tag{10.2(b)}$$

The dynamic range of the time amplifier is limited as it depends on the delay regulation range of the voltage-controlled delay element (such as a CSI) "D_0." Therefore, it can be further enhanced by expanding the delay element count of the feedback loop. Additionally, a multiplexer has been used in Ref. [85] in order to achieve the amplification to any integral value from 1 to N.

10.3. Time-to-digital Converter

10.3.1. *Applications*

TDC is used to digitize the time domain information represented by either pulse duration or time interval (TI) between two designated physical events: "start" and "stop." It finds applications in various fields, including medical, consumer, industrial, and research instrumentation, as described in the following.

10.3.1.1. *High-energy physics and nuclear experiments*

In high-energy physics (HEP) and nuclear experiments, corresponding to a physical event, an analog signal is generated using a relevant radiation detector. When this analog signal crosses the set thresholds of a discriminator, the logical transition signals "start" and "top" are generated. A TDC measures the TI between "start" and "stop," and this timing information is used to measure the time of flight (TOF) of particles [8–16], particle track reconstruction, and particle identification. The TI measurement is also used in TOF mass spectrometry (TOFMS) to determine the mass-to-charge (m/e) ratio of charged particles. In these measurements, the particles are equally accelerated, but they gain different velocities due to their different m/e ratios. A TDC is therefore required to measure the difference in the arrival times of various particles.

10.3.1.2. *TOF laser range finders*

TDC is a key building block of the TOF laser range finder instruments [17, 18]. Laser range finding has a variety of applications, such as profiling of hot surfaces [19], the industrial method of surface analysis [20], traffic condition monitoring [21], height measurement [22], and 3D imaging. Here, the laser range is determined by using the expression $1/2 \times c \times \Delta T$, where c is the velocity of light and ΔT is the time interval between the pulse sent from the transmitter and received back by the receiver. Here, a TDC with a high resolution (of the order of a few ps) and a wide dynamic range (of approximately hundreds of ns) is used in order to achieve mm-level accuracy over a measurement range of several hundreds of meters.

10.3.1.3. *Distance measurement based on sonar*

A TDC finds application in sonar and pulse echo technique to determine distances. Here, distance $d = 1/2 \times v \times \Delta T$ is evaluated by measuring the time interval ΔT between the transmitted and reflected ultrasound waves from the target object. Here, v is the velocity of the wave in the target material. This information is used for medical imaging, the thickness measurement of metal layers [23], and to locate underwater objects. In such applications, TDCs featuring moderate resolutions (~300 ps) with large dynamic ranges are used, as the thickness in cm is equivalent to a time interval in the order of tens of μs.

10.3.1.4. *Ultrasonic density meter*

In this application [23], a TDC is used to measure the time interval that is equivalent to the change in the velocity of sound while passing through a fluid of different concentrations. Here, the velocity of sound in the fluid directly varies with its concentration at a fixed temperature. A TDC with a resolution of ~500 ps is usually sufficient for this instrument.

10.3.1.5. *Positron emission tomography medical imaging*

Positron emission tomography (PET) images [24] are constructed using radiographic tracers injected into a patient's body. These tracers decay into positrons that further annihilate with the electrons present in the body, thereby producing a pair of two 511 keV photons traveling in opposite directions. These photons are detected by a scintillator-PMT detection ring of the PET system, where a TDC determines the time coincidence (the time interval between arrivals of photons at the detector) of these gamma-ray photons to find their line of response (LOR). With the help of various measured LORs using parallel processing channels, a tomography image is reconstructed. In such applications, a low-cost, low-power, and area-efficient multi-channel TDC is required. The other medical applications of TDC are ultrasound imaging diagnostic equipment and X-ray imaging detectors [25].

10.3.1.6. *Time-based ADC*

A standard voltage-mode ADC implements the function of digitization with the help of voltage amplifiers and comparators, and switch capacitor networks. These analog circuit blocks operate at relatively slow data processing speeds and are power and area inefficient as compared to digital circuitry. TDCs are traditionally applied in the development of high-precision and low-power ADCs, such as dual-slope ADCs. Another method of power- and area-efficient ADC implementation using TDC is reported in Refs. [2–5, 26]. Here, the VTC is used to convert the analog voltage to a corresponding pulse delay time, which is digitized using a TDC.

10.3.1.7. *Frequency syntheses in RF communications systems*

With the advancements in CMOS technologies in the sub-micron domain, TDCs have replaced the analog blocks such as phase detectors and charge pumps in the design of all digital PLLs for modern RF communication systems [27, 28]. Here, the TDC output controls the frequency of a digitally controlled oscillator (DCO).

10.3.1.8. *Test and measurement instrumentation*

A high-resolution TDC is required in test and measurement instruments [29] that are used to characterize timing performance parameters, such as skew, jitter, clock-to-output delay, and setup and hold times of high-speed circuits. The other applications of TDCs are in electronic test equipment, such as digital storage oscilloscopes [30, 31], logic analyzers, CMOS process corner/strength estimators of analog circuits [32], single photon counting [33], PM demodulators [34], temperature sensing [35], and the optical characterization of MOS circuits [36].

Considering the wide spectrum of applications of TDC, it is important to study various TI measurement methods and TDC implementation techniques based on performance parameters. Initially, TDCs were implemented using ECL technology and discrete devices [18, 37, 38] that have the merits of high precision and accuracy. However, they were limited by high power consumption and complex circuitry. Later, an ECL gate array was preferred, which reduced power and area consumption [39].

Contemporary TDC solutions are based on FPGA and CMOS ASIC. The FPGA-based implementation offers a fast development cycle and reconfiguration. However, issues such as high power consumption, unpredictable P&R delays, logic resource bottlenecks, and delay sensitivity toward temperature and voltage have to be taken into account during design and implementation. In addition, for multi-channel TDC, the FPGA has constraints of non-identical placement and routing of channels. This leads to a higher channel-to-channel specification spread. In contrast, the ASIC-based solutions exhibit better power and area efficiency and are preferred for bulk requirements for multi-channel TDCs. It also provides the design flexibility to achieve the high performance of TDC. A review of TDC techniques based on the above-discussed implementation platforms was reported earlier in 1984 [40] and 2004 [41].

10.3.2. *Performance Parameters of TDC*

10.3.2.1. *Resolution (LSB)*

This parameter is defined as the least value of time interval that can be measured distinguishably and is represented in terms of LSB, as shown in Figure 10.10. The count (digital word) given by the TDC is multiplied by the LSB value in order to calculate the measured time interval.

10.3.2.2. *Precision (RMS resolution)*

This parameter characterizes TDC for repetitive fixed time interval measurement. Due to the effect of various random noise sources, the repeatedly measured fixed time interval tends to vary around the mean value with a standard deviation σ, known as the precision of measurement. It is a performance parameter used to compare different TDCs.

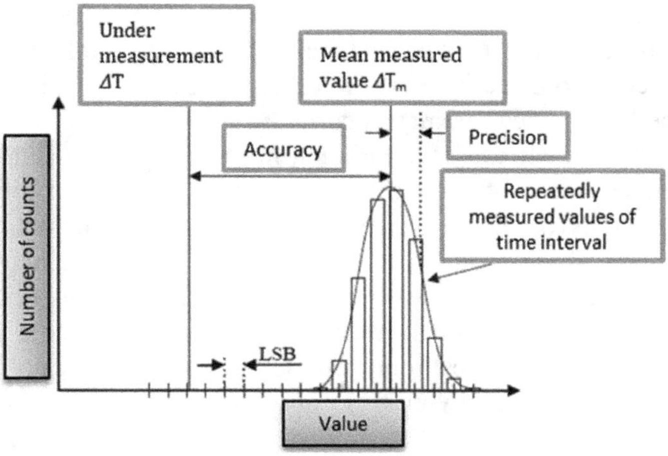

Figure 10.10. Visualization of Performance Parameters of TDC

10.3.2.3. *Accuracy*

The degree of matching between a measured TI and its actual value is referred to as accuracy. The main sources of inaccuracy are systematic error and linearity error encountered in the time interval measurement. The systematic error manifests as a fixed constant offset that is simple to remove from the outcome. However, as the linearity error might vary with the measured time interval, a look-up table (LUT) is the only way to eliminate it. The sources of error, which contribute to the degradation of accuracy are as follows.

Quantization error Q_e: It occurs due to the finite amplitude of the TI that can be digitized by TDC. It is measured in terms of the LSB of the time interval measurement. The quantization is assumed as a random variable with a uniform distribution in the bounded range and is independent of the input time interval to formulate its impact on the performance of TDC. For time interval measurement, where an asynchronous input signal "start" or "stop" arrives at any time within the LSB of time measurement, the quantization error varies from 0 to LSB, and its root mean square (RMS) value is LSB/√12. In TI measurement techniques, where both "star" and "stop" are asynchronous to the LSB of measurement, Q_e varies from −LSB to + LSB and its RMS value is given by LSB/√6.

Nonlinearity error: Nonlinearity error appears as a deviation of the TDC transfer characteristic (output versus input) from the ideal straight line. The sources of nonlinearity error are asymmetries in circuit layout, non-homogeneity of device parameters during device fabrication, and systematic noise sources. It causes a variation in the unit step of quantization from the ideal value of one LSB, appearing in the form of differential nonlinearity (DNL). The DNL errors accumulate over a long time interval measurements. The accumulated error appears as integral nonlinearity (INL) over a range of time interval measurements. For instance, in tapped delay line (TDL)-based time measurement techniques, the static temporal variation in the signal edges from their ideal positions causes nonlinearity error. The non-homogeneity in device parameters is the dominant source of nonlinearity in modern CMOS processes because, with the reduction in the physical dimensions of the device, the mismatch in device parameters increases [42]. The choice of large device dimensions and careful layout design can reduce the degree of mismatch and layout asymmetries.

10.3.2.4. *Dynamic ranges*

It defines the longest time interval that may be digitalized by TDC.

10.3.2.5. *Dead times*

Dead time represents the time gap between the end of a measurement and the start of the next one. It depends on the conversion time of the measurement technique and the readout speed. It defines the measurement rate of TDC.

10.3.3. *Types of Time Measurement*

Time measurement can be generalized into two categories: time interval measurement and time stamping. Time interval measurement is for the time gap between "start" and "stop" signals. Time stamping is used to "stamp" the time of occurrence of logical events with respect to the reference clock. This method can also be used to indirectly measure the time interval between "start" and "stop" by subtracting their corresponding time-stamped values with respect to the reference clock. It is further sub-classified into single-hit and multi-hit time stamping, as shown in Figure 10.11. In single-hit time stamping, the time of occurrence of a single transition (hit) is stamped. In multi-hit time stamping, the time of

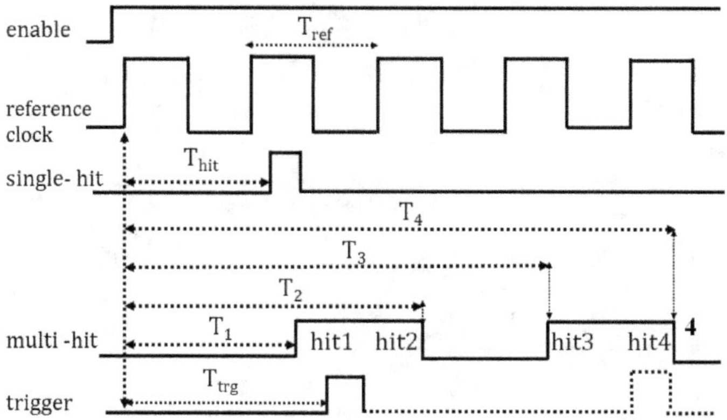

Figure 10.11. Timing Diagram of Single-Hit and Multi-Hit Event Time Stampings

occurrence of multiple rising and falling transitions (multi-hit) is stamped. Further, the relative time of hits is measured with respect to the trigger (reference hit), whose time is also stamped with respect to the reference clock.

10.3.4. *TDC Implementation Techniques*

The various TDC implementation techniques, as illustrated in Figure 10.12, exhibit different performance metrics. Detailed discussions on each technique while incorporating their merits and demerits are given in the following section.

10.3.4.1. *Analog techniques*

10.3.4.1.1. TAC with ADC

The analog TI measurement techniques are primarily based on the use of a time-to-amplitude (TAC) converter [39], where a capacitor (C) is linearly charged using a constant current source (I) for the time interval (ΔT) to be measured. The charging of the capacitor is controlled by a switch, as shown in Figure 10.13(a). Here, the voltage ΔV across the capacitor is proportional to the time interval ΔT and is given by Equation (10.3). This voltage can be converted into digital code by using an ADC:

$$\Delta V = \frac{I}{C}(T_2 - T_1) = \frac{I}{C}\Delta T. \tag{10.3}$$

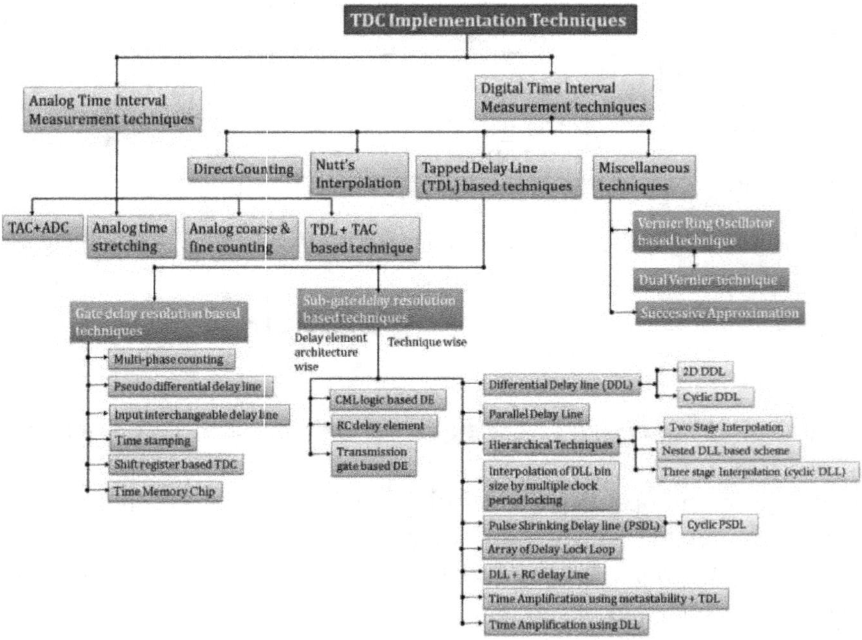

Figure 10.12. Classification of Time Interval Measurement Techniques

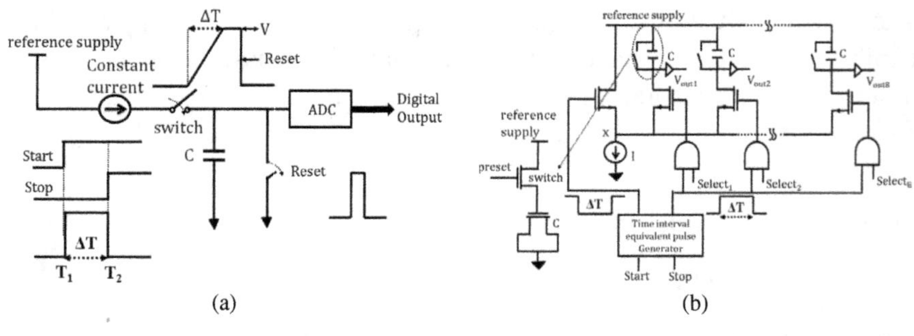

Figure 10.13. (a) Schematic Diagram of TAC Method. (b) Schematic Diagram of 8-channel TAC-Based Analog Memory

Ideally, a high time resolution (order of 1 ps) that depends on the LSB of the ADC and the choice of the constant of proportionality (C/I) can be attained through this technique. However, various factors such as charge injection, clock feed through in switch and leakage & memory effects of the capacitor result in a variable offset in the voltage across the capacitor, leading to a deterioration of performance. In addition, the stability and linearity of current source and the accuracy of the ADC are other factors that affect the accuracy and linearity of the measurement. Moreover, variations in supply and reference voltages (common mode noise) degrade the accuracy of measurement.

The dead time of TAC-based current integration techniques is on the order of a few microseconds, which depends on the conversion time of the ADC and the resetting time of logic circuits. The conversion time of the ADC can be reduced by using a high-speed flash ADC.

In multi-channel TACs, the dead time can be reduced by sharing the SAR ADC among multiple multiplexed channels [8], as shown in Figure 10.13(b). Here, the number of TAC channels that can be integrated is limited by the amount of parasitic capacitor at node "X." This parasitic capacitance increases with the number of channels, and its nonlinear charging during transients degrades the measurement linearity. The mismatch in capacitors causing variations in the specifications of different TAC channels is minimized by using the common centroid layout technique described in Ref. [8].

10.3.4.1.2. Differential TAC-based TDC

To lessen the effect of common mode noise, differential TAC [43] is preferred. It uses two identical TACs to convert the time intervals from "start" and "stop" to the end of a "gate" window into voltages independently, as shown in Figure 10.14. The differential amplifier amplifies the difference of the output voltages of two TACs (V_{start} and V_{stop}), which is equivalent to the time interval between "start" and "stop" (ΔT) and is given by Equation (10.4):

$$\Delta T = t_{start} - t_{stop} = \frac{c(V_{start} - V_{stop})}{I}. \tag{10.4}$$

10.3.4.1.3. Dual-slope or analog time stretching technique

In the above-discussed TAC-based TI measurement techniques, separate ADCs are used, which increases circuit complexity as well as power and area consumption. The dual-slope or analog time stretching technique [44] based on the principle of time-to-time-to-digital conversion avoids the need for a separate ADC. In this technique, the time interval under measurement is extended by the stretching factor (S), independent of circuit parameters. The extended time interval window thus obtained is digitized by counting the number of reference clock periods (T_{ref}) using a counter.

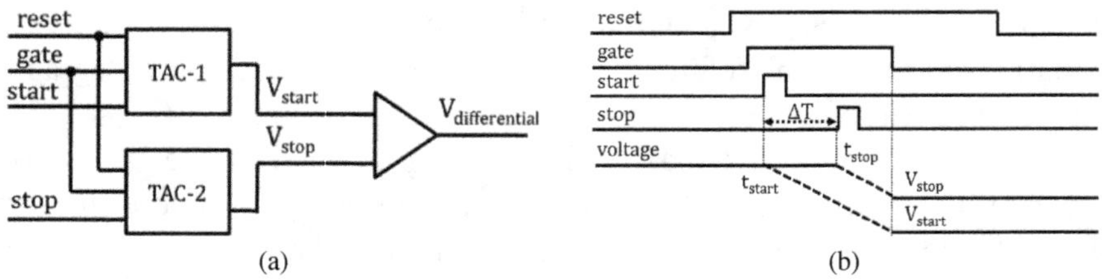

Figure 10.14. Differential TAC: (a) Schematic Diagram, (b) Timing Diagram

To implement this scheme, the Wilkinson ADC technique, which has the salient features of high resolution and linearity, is used. As shown in Figure 10.15(a), it charges a capacitor C linearly through a large constant current (I_1) within the applied time interval. Subsequently, this capacitor is discharged through a small constant current ($I_2 = I_1/N$) until it attains 0 V. As $I_2 < I_1$, the capacitor discharges slowly and takes a longer time $T > \Delta T$ to fully discharge. The voltage across this capacitor is compared to 0 V with the help of a comparator to obtain a stretched time interval window. The stretching factor S is defined as the ratio of the discharging time (T) to the charging time (ΔT) of the capacitor. For a stored voltage ΔV across the capacitor C, the charging and discharging durations are given by

$$\Delta T = \frac{C}{I_1}\Delta V,$$

$$T = \frac{C}{I_2}\Delta V,$$

$$S = \frac{T}{\Delta T} = \frac{C/I_2}{C/I_1} = N. \tag{10.5}$$

Thus, the stretching factor is independent of the absolute values of the circuit parameters (current and capacitor). Further, the extended duration T is digitized by a counter. For the count n, the conversion expression is

$$\Delta T = n \times \frac{T_{\text{ref}}}{S}. \tag{10.6}$$

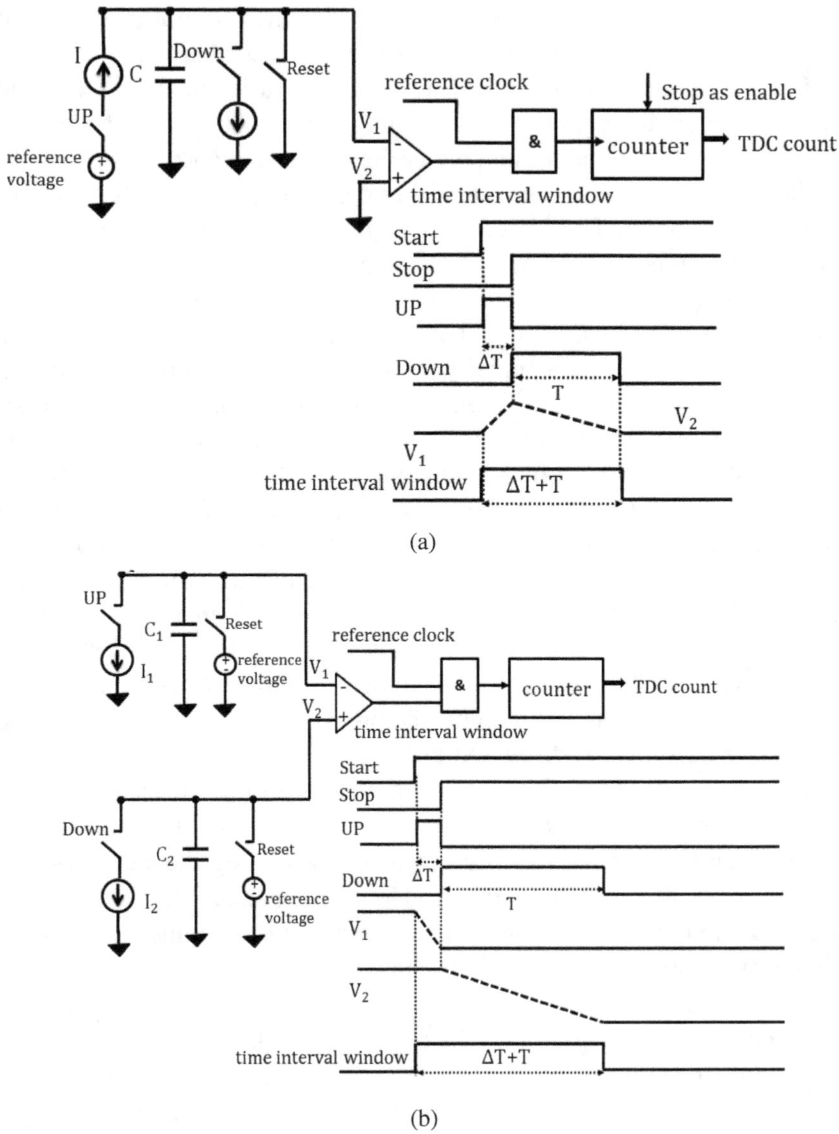

Figure 10.15. Dual-Slope Techniques Using Wilkinson Principle: (a) With Single Capacitor and Current Source, (b) With Separate Capacitors and Current Sources

The measured time interval (ΔT) using the dual-slope technique is given by Equation (10.6). Here, the resolution T_{ref}/S is independent of the absolute values of current and capacitor. This technique is therefore independent of process variations. However, the inaccuracies of switches (due to charge injection and clock feed-through) and switching noise coupled to the substrate contribute to the DNL error of TI measurement.

From Equation (10.6), the LSB (resolution) varies inversely with the stretching factor and directly with the time period of the clock. Therefore, it can be improved by increasing either the clock frequency or the stretching factor. However, the use of a higher clock frequency leads to higher power consumption. The increment in stretching factor is therefore preferred in the revised versions of this technique at the cost of conversion time. In Ref. [44], the stretching factor is increased by using two separate capacitors C_1 and C_2 ($C_1 < C_2$) that are discharged at different rates, as shown in Figure 10.15(b). Initially, both capacitors are pre-charged to a reference voltage. The "start" signal enables the discharging of capacitor C_1 through a large constant current I_1. After a time interval ΔT, the "stop" signal disables the discharging. Simultaneously, it enables the discharging of the big capacitor C_2 using a small constant current I_2 until it attains a voltage similar to that across C_1. The modified stretching factor is $S = M \times N$, where $C_2 = M \times C_1$ and M is a constant. Here, a significant stretching can be achieved if there is a large difference in both capacitor sizes. However, the utilization of large capacitors is avoided in CMOS ASIC designs due to die area and cost considerations. Other area-efficient techniques of increasing the stretching factor are synchronized discharging and Vernier charging, as discussed in the following.

In synchronized discharging [45], the capacitor discharges on a high duration of the clock with a duty cycle (<50%), as shown in Figure 10.16(a). This reduces the speed of discharging the capacitor and thus improves the stretching factor. The Vernier charging method [46] improves stretching factor through successive approximation of a target voltage by various iterations of alternate charging and discharging of a capacitor for durations, T_D and T_E ($T_D > T_E$), respectively. It also avoids the dependency of resolution on the reference clock frequency.

As shown in Figure 10.16(b), the time interval under measurement is divided into three parts such that $\Delta T = T_C + T_{fA} - T_{fB}$. The time T_C equivalent to full clock periods is measured by the counter. To measure the difference between fractional time intervals T_{fA} and T_{fB}, the capacitor C_1 is charged by using a large constant current I_1 during T_{fA}. Then, it is discharged during T_{fB} using the same current and attains an intermediate voltage level. This voltage is compared to the voltage V_2 across a larger capacitor C_2 using a comparator. The voltage V_2 approaches this intermediate voltage successively through the Vernier charging method. In this

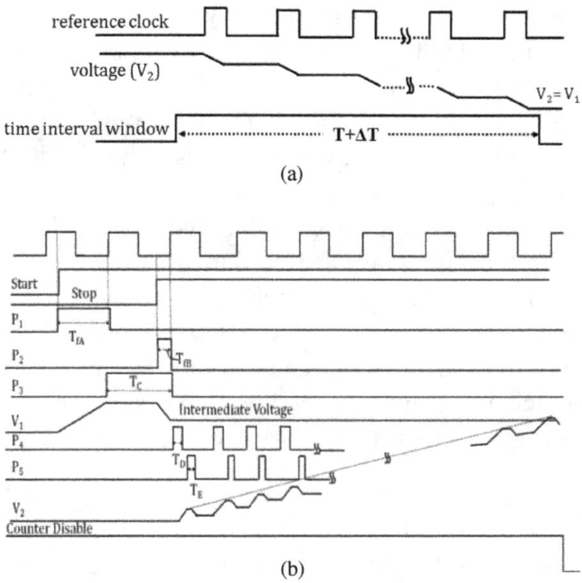

Figure 10.16. (a) Dual Slope Using Clock-Synchronized Discharging of Capacitor. (b) Waveform Diagram in Vernier Charging Method

method, the durations of non-overlapping time interval windows T_D and T_E are chosen such that both lie in each half of the reference clock cycle and are counted by the counter. When the voltages across both the capacitors become equal, the comparator toggles its output and disables the counter.

For n number of counted cycles to achieve the equal voltage across both the capacitors,

$$\frac{I_1 \times (T_{fA} - T_{fB})}{C_1} = 2n\frac{I_2 \times (T_D - T_E)}{C_2},$$

$$=> \frac{N \times I_2 \times (T_{fA} - T_{fB})}{C_1} = 2n\frac{I_2 \times (T_D - T_E)}{M \times C_1},$$

$$=> (T_{fA} - T_{fB}) = n \times \text{LSB where LSB} = 2 \times (T_D - T_E)/(M \times N). \quad (10.7)$$

The limitation of the dual-slope method is the input-dependent large conversion time expressed by $\Delta T \times S$. In addition, it has an inherent trade-off between resolution and the measurement rate of the TDC. For instance, a resolution of 1 ps with a 100 MHz (10 ns) clock requires a stretch ratio of $S = 10{,}000$ and implies a conversion time of 150 μs for a 15 ns of the time interval measurement. The TAC-based techniques are efficient for time interval measurements with high resolution but are limited by dynamic range. Time interval measurement with high resolution and a large dynamic range requires measurement in two parts. First is coarse time measurement by counting the full cycles of reference clock within the time interval. The second is the fine measurement of the fractional time (within one clock period) by the TAC or time stretcher. Based on this scheme, a TDC using a time-to-charge converter with an on-chip Wilkinson ADC and counter has been reported [47]. Another approach to implementing the above scheme is the time-to-amplitude conversion chain (TACC), based on analog coarse and fine counting.

10.3.4.1.4. Time-to-amplitude conversion chain (analog coarse and fine counting)

This technique [48] uses periodic charging and discharging of a capacitor triggered by "start" and ended by "stop." The number of cycles of the triangular clock thus obtained until the arrival of "stop" gives the coarse count, as shown in Figure 10.17(a). The voltage across the capacitor is sampled by "stop." It is digitized by an ADC to provide the fine count. This scheme is implemented using an operational-amplifier-based charge integrator, two comparators, an ADC, two flip-flops, and a counter. Here, "start" enables the charge integrator to produce a voltage ramp-up at its

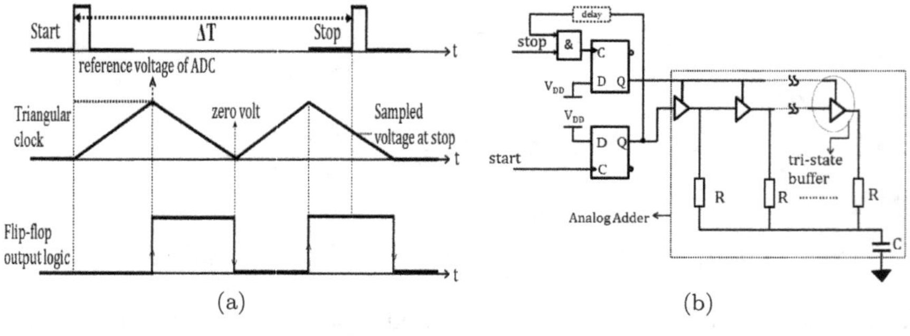

Figure 10.17. (a) Timing Diagram of TAC Conversion Chain. (b) Time Interval Measurement Technique Using TAC with TDL

output until it becomes equal to the reference voltage of the ADC. This is detected by a comparator, which toggles its output to set a flip-flop. The output of this flip-flop enables the voltage ramp down until it becomes 0 V. This is also detected by another comparator, which resets the flip-flop to further enable the voltage ramp-up. Thus, the cycle of subsequent voltage ramp-up and -down provides a triangular clock. The "stop" signal disables the integrator and enables the sample and hold circuit with the help of switches. The final voltage across the capacitor in the integrator is sampled and digitized by the ADC. This analog scheme is used to improve the dynamic range of TDC without sacrificing the resolution.

10.3.4.1.5. TDC using TAC and TDL

In this technique [9, 10], the charging of capacitor C is regulated by a TDL that is realized by a cascaded chain of tri-state buffers, as shown in Figure 10.17(b). The delay line taps are connected to a ladder of resistor R. The other end of the resistor ladder is connected together to the capacitor C at the summing point. This structure forms an analog adder as each delayed replica of "start" passing through TDL provides an additional unit of charging current to the capacitor C through resistor R. The "stop" signal disconnects the resistor ladder from the delay line by disabling the tri-state buffers. The voltage thus obtained across the capacitor C is equivalent to the time interval between "start" and "stop" and is digitized by an ADC.

The merits of this technique include high resolution beyond the intrinsic buffer delay and lower static power consumption. In Ref. [11], to achieve the accuracy of the measured time interval, the delay of TDL is regulated against PVT variations using DLL.

10.3.4.1.6. Cyclic TDC using triple-slope interpolation with time amplification

This technique [49] is based on triple-slope analog interpolation with a time amplification technique for digitizing the time interval ΔT. Here, in the first step, as shown in Figure 10.18, a capacitor C is charged linearly to voltage V_{in} such that

$$V_{in} = \frac{I_{ref} \times \Delta T}{C}.$$

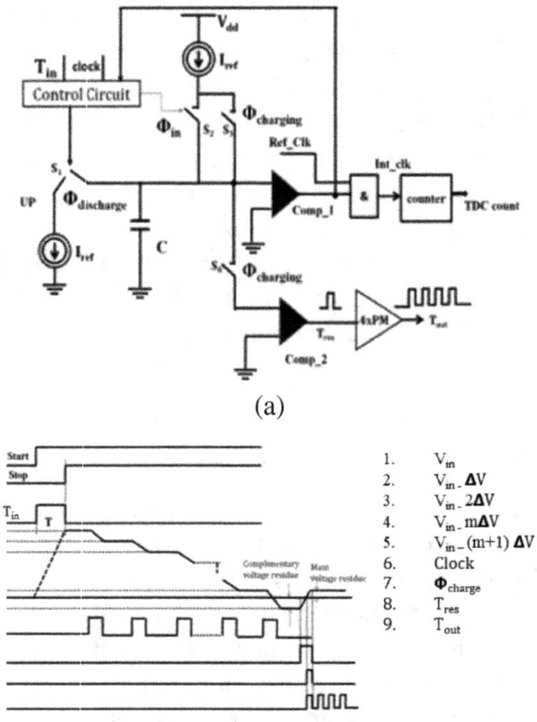

Figure 10.18. (a) Block Diagram and (b) Timing Diagram of Cyclic TDC

In step two, through a constant current source I_{ref}, the capacitor C is slowly discharged to generate the complement voltage residue. The slow discharging is achieved by synchronization with a clock of pulse width T_w in each period. The drop in voltage ΔV across C in each period of the clock is calculated as

$$\Delta V = \frac{I_{\text{ref}} \times T_w}{C}.$$

A three-bit digital counter is used to count the number of discharging steps. If T is the total discharging time and m $(= \Delta T/T_{\text{ref}})$ is the number of clock cycles required to discharge the capacitor, then

$$T = (m + 1)\, T_{\text{ref}}.$$

The voltage across capacitor at the end of second step is given as

$$V_{\text{res}} = V_{\text{in}} - (m+1)\Delta V = \frac{I_{\text{ref}} \times \Delta T}{C} - (m+1)\frac{I_{\text{ref}} \times T_w}{C}. \qquad (10.8)$$

From Equation (10.8), it is clear that V_{res} has a negative value, which is proportional to the complimentary time residue. A comparator (comp_1) is used to compare this negative voltage with zero, leading to toggling its output to zero from one. This stops further counting of the number of clock cycles required for the discharging the capacitor. A register is used to save the count provided by the counter. In the third step, through a digital logic correction module, a time residue (T_{res}) corresponding to the main voltage residue is generated, as shown in Figure 10.18(b). This is carried out by further charging of the capacitor, synchronized with a clock for a single cycle. The comparator (comp_2) compares the voltage level $V_{\text{in}} - m\Delta V$ to 0 V and thereby generates a pulse whose width is proportional to the main time residue.

In the fourth step, the main time residue (T_{res}) is amplified by a four-pulse multiplier (PM) circuit. The amplified output $T_{\text{out}} = 4 \times T_{\text{res}}$ is used as an input in the next cyclic step. In this way, this process is repeated a total of three times. The register is used to record the output of the digital counter in a total of four cyclic steps. Finally, the four generated digital output codes are combined in the digital error correction logic to produce a nine-bit output.

10.3.4.2. *Digital techniques*

The TAC-based analog time interval measurement techniques are very efficient in providing high time resolutions (approximately tens of ps), but this merit is counteracted by a reduction in the supply voltages and hence reduced noise margins in the current submicron semiconductor technologies. On the contrary, the advancements in submicron semiconductor technologies have enabled the digital methods of time interval measurement to achieve resolutions comparable to the analog techniques, along with large scales of integration with optimum power consumption. Digital TDC is therefore gaining attention as an alternative to data converters in various applications, such as temperature sensors, all-digital PLLs, and ADCs. In this section, TDCs implemented using digital topologies are presented. A few TDCs use small analog blocks, such as DLL/PLL, to provide delay variation immunity against PVT variations, thus making it a mixed signal architecture.

10.3.4.2.1. Direct-counting-based TDC

This technique has been used in digital stopwatches and timers [50] for time interval measurements with a large dynamic range. Here, the number of elapsed cycles of the stable reference clock between "start" and "stop" is proportional to the time interval between them. This number is determined using a counter by applying a clock for a given time interval duration. The period of this clock defines the time resolution, so the maximum quantization error is within one count. There are two ways of applying the clock to the counter for the time interval duration: One is direct gating, and another is synchronized gating. The direct gating [51] is simple to construct but may result in more than one count error due to the truncation of the clock pulse (at the moment of gate opening and closing), as shown in Figure 10.19(a). In order to minimize this error, synchronized gating is

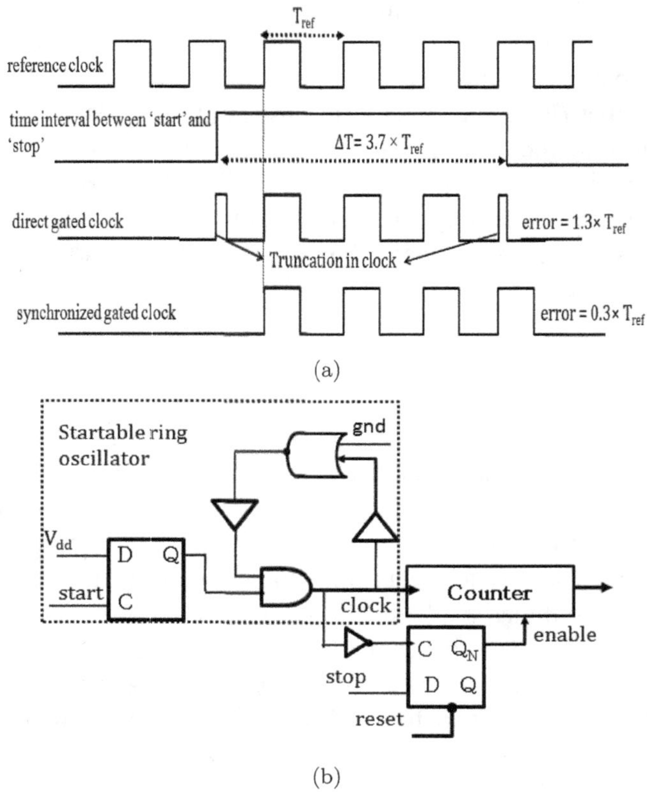

Figure 10.19. (a) Timing Diagram of Direct Counting Method. (b) Time Interval Measurement Using Startable Ring Oscillator

used that operates on the edges of the clock instead of level. In the case of synchronized gating, the maximum error would be one clock cycle at the time of gate opening as well as closing (refer to Figure 10.19(a)). The total error in time interval measurement is the difference between these two that is within one count. Further, to avoid count error at the beginning of the time interval measurement, a startable ring oscillator is used, as shown in Figure 10.19(b). Here, the "start" signal triggers an oscillator, and then, its cycles are counted until the arrival of "stop." The end of conversion (eoc) signal, which stops the counting, is obtained by synchronizing the "stop" at the falling edge of the clock. To limit the measurement error within one count at the time of "stop," one count is subtracted from the result if stop occurs in the falling edge duration of the clock.

The gated clock method is not efficient for multi-stop or multi-channel time interval measurement as it requires replication of counters as per the number of stops or measurement channels. Therefore, a free-running counter, shared between all measurement channels, is used until its rollover. It is initialized by an "enable" signal. The inputs "start" and "stop" individually sample the state of the counter and latch the corresponding counts into registers. The subtraction of both counts gives the time interval. Here, if "start" or "stop" occurs during toggling of the counter, the metastability of registers causes unpredictable outputs. Therefore, a synchronizer is used between input (start/stop) and counter to sample its state on the safe edge of the clock.

The merit of direct counting is its small conversion time, as the settling time of registers is hundreds of picoseconds. In addition, good linearity in time interval measurement can be obtained by ensuring the stability of the clock. However, the accuracy of measurement in direct counting technique is limited due to poor resolution (one clock period). Higher resolution could be achieved by using a high-frequency clock, either from an external source or from an on-chip PLL. However, the maximum operating frequency is limited by the choice of technology. Also, the on-chip PLL-based solution turns out to be power consuming and complex due to the issue of phase and frequency locking.

An improvement in accuracy of the direct counting method has been reported [51] while using a low-frequency clock by averaging the large number of measurements. It is based on the statistical reduction in quantization and random errors if the time interval is asynchronous to the reference clock and the inputs are repetitive. The averaging reduces the measurement error by taking the square root of the number of measurements, but it limits the measurement rate of TDC.

10.3.4.2.2. Nutt's interpolation method

Nutt's interpolation technique (developed in the 1960s) was used to improve the resolution of the direct counting method. In this technique, the fractional time bound between consecutive states of a counter is measured using techniques featuring high resolution and is named a fine interpolator. Thus, the entire time interval measurement is carried out in two parts: one is coarse and the other is fine, as shown in Figure 10.20.

The coarse time T_C is equivalent to full reference clock cycles within the time interval window and is counted by the counter (Figure 10.20(a)). The fractional time intervals T_A and T_B at the beginning and end of the time interval window are measured by the fine interpolator. The fine interpolator is designed for the dynamic range of one coarse clock period to synchronize the coarse and fine counts. The measured time interval is given by

$$\Delta T = T_C + T_A - T_B. \tag{10.9}$$

Figure 10.20(b) shows the block diagram of the interpolation technique, where a synchronizer is used to obtain time interval windows T_A, T_B, and

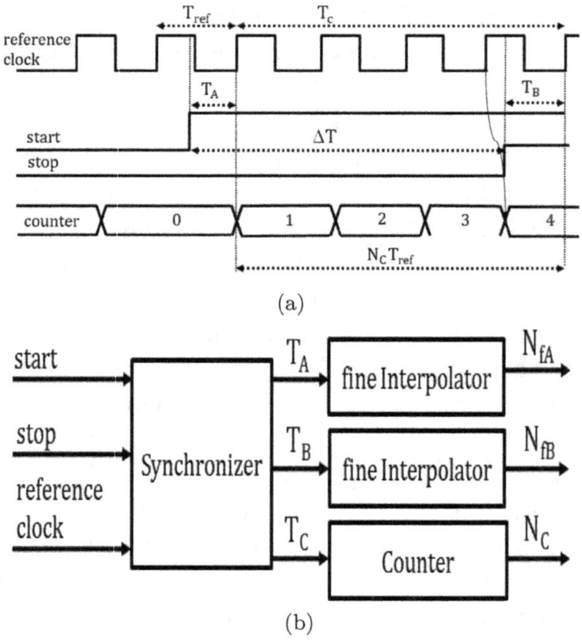

Figure 10.20. Nutt's Interpolation Technique: (a) Timing Diagram, (b) Block Diagram

T_C. It is also used to mask the "stop" signal from occurring prior to "start." The merit of this interpolation technique is the reduction in nonlinearity of the fine interpolator by averaging several measurements of the fixed time interval. Here, the values of T_A and T_B vary randomly, but the absolute value of $T_A - T_B$ remains constant. This averages out the nonlinearity of the interpolator. Nutt's technique has been implemented using TAC [17], analog time stretcher [44, 52], and TDL [39, 53] in several TDC prototypes.

In Nutt's interpolation technique with a TDL-based fine interpolator; a DLL is used to provide delay variation immunity in PVT variations. However, the DLL is power and area inefficient in view of the requirement of a large number of delay elements to achieve the sub-nanosecond LSB using a low-frequency clock (hundreds of MHz). The requirement of a DLL has been avoided by using an inbuilt digital calibrator to calibrate the unit delay (LSB) [54]. It is based on measuring the time period (T_{osc}) of a ring oscillator derived from TDL. This time period, when divided by the number of stages in TDL, gives the LSB of the time interval measurement.

To implement this calibration scheme, a multiphase ring oscillator is designed using an even number of cascade differential inverters [55]. The oscillations are maintained by cross-coupled feedback of outputs, as shown in Figure 10.21. This type of structure of ring oscillator avoids the coupling stage and intrinsically realizes the interpolation of clock period by using its intermediate phases.

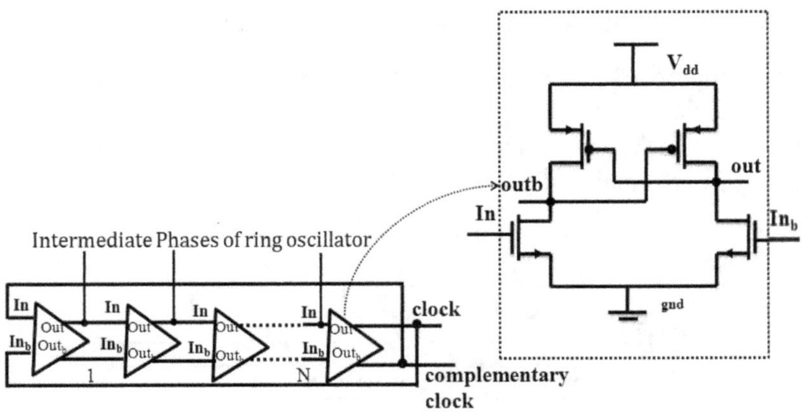

Figure 10.21. Use of Multiphases of Ring Oscillator with Calibration in Nutt's Interpolation Technique

To calibrate the LSB, the measurement circuitry is used in the calibration mode for a known time interval (ΔT). An on-chip calibration counter generates the "start" and "stop" signals with a known time interval using a reference clock provided by an off-chip crystal oscillator. Using conversion equation (10.10), of Nutt's interpolation technique, the LSB = T_{osc}/N can be calibrated:

$$\Delta T = N_C \times T_{osc} + \left(N_{fA} - N_{fB}\right) \times \frac{T_{osc}}{N}. \tag{10.10}$$

The other variant of interpolation reported is an asynchronous interpolation technique [22], where an event-triggerable oscillator with a time period (T_{osc}) triggered by "start" is used for coarse counting T_C instead of using a reference clock, as shown in Figure 10.22. The fractional time interval T_F in between "stop" and the next rising edge of the clock is measured using a fine interpolator realized by the time interval measurement technique with a resolution (T_d) higher than that of direct counting.

The conversion expression for this technique is given by Equation (10.11), where N_C and N_F are the counts provided by the counter and fine interpolator, respectively. Here, the commencement of time measurement is asynchronous to the system clock. This technique is used for TDC applications in space exploratory missions and HEP experiments with random events:

$$\Delta T = T_C - T_F = T_{osc} \times N_c - N_f \times T_d. \tag{10.11}$$

10.3.4.2.3. TDL-based time interval measurement

A TDL comprises a cascaded chain of delay elements that generate delayed replicas of either "start" or "stop' or the "reference clock." These delayed replicas are then tapped and used in numerous schemes to implement high-performance TDCs.

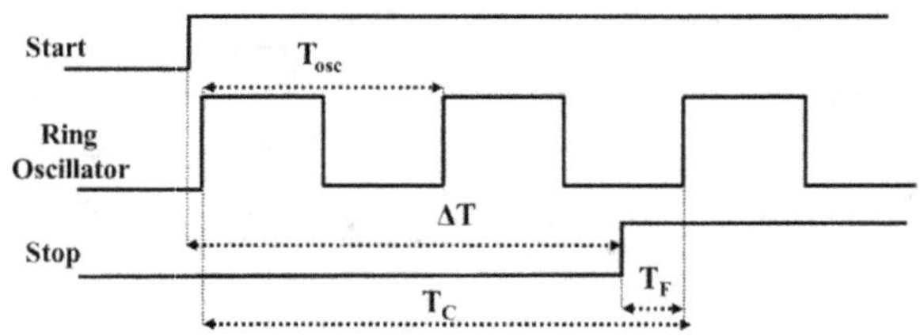

Figure 10.22. Timing Diagram of Reference Clock Asynchronous Time Interpolation Technique

The simplest form of TDL-based time interval measurement technique uses the flash conversion principle, where, as shown in Figure 10.23, the delayed taps of the "start" signal are concurrently sampled and latched into a register on the arrival of the "stop" signal, as reported in Refs. [39, 41, 56–58]. Here, the delay (T_d) provided by each delay element defines the LSB or resolution of the time interval measurement. The number of elapsed delayed "start" until "stop" is proportional to the time interval between them. The latched register code is in the form of thermometer code, where the thermometer provides the first logic transition (0 to 1 or 1 to 0) count to the binary encoder. This count multiplied by the unit delay (T_d) gives the time interval.

The thermometer-to-binary encoder is a vital component of delay-line-based TDCs. The performance of TDC is affected if there is an invalid transition in the thermometer code, known as bubble error. The sources of bubble error are the metastability of flip-flops, clock jitter, cross-talk, and timing uncertainty in the sampling moment. In order to overcome this issue, in Ref. [105], three types of encoding methods, namely Wallace tree, fat tree, and MUX-based, are reported. These encoders are efficient in eliminating bubble error up to the third order.

It is prudent to discuss various other aspects of TDL-based measurement before discussing the architectures of TDL-based TDCs:

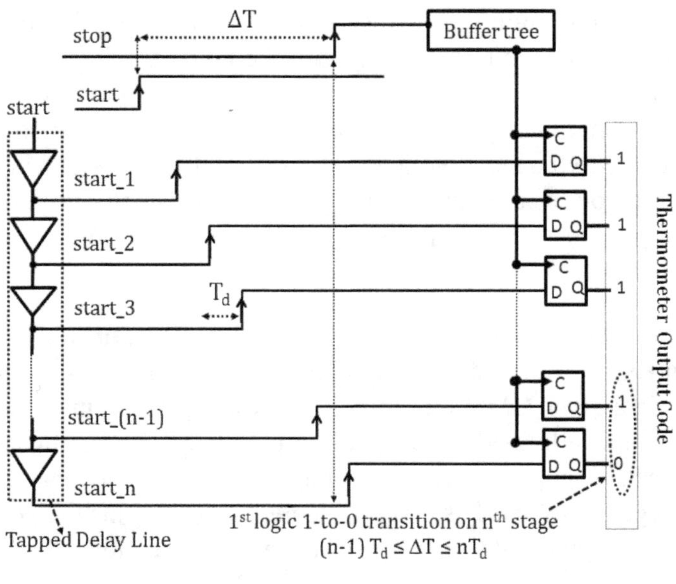

Figure 10.23. Time Interval Measurement Using TDL

(a) **Unit delay:** The LSB or resolution of TDC depends on the smallest attainable delay from the delay element. This makes the choice of delay element a crucial aspect of TDC design.
(b) **Matching among tapped delays:** The linearity of TDL-based TDCs depends on the matching in tapped delays. The sources of delay mismatch across the taps are systematic errors and random variation in device parameters [59, 60]. As the random variation in device parameters is inversely proportional to the device area [42], it can be reduced by choosing a large aspect ratio of transistors in the design of delay elements.
(c) **Length of TDL:** The dynamic range of TDC is given by the product of the unit delay and the number of delay elements in the delay line. However, the use of a long delay line in order to achieve a large dynamic range is not preferred due to the following two reasons: (1) It is power and area inefficient; (2) the differential nonlinearity error due to mismatch in delays among consecutive taps of the delay line accumulates over the delay line. This accumulated error appears in the form of integral nonlinearity, which increases with the length of the delay line.
(d) **Accuracy and stability of TDL:** Moreover, the stability of unit delay across process and operating condition variations determines the accuracy of time interval measurements. The CMOS buffer delay is sensitive to PVT variations. Therefore, a delay variation compensation circuit DLL [59] is used. Here, the delay of TDL is referenced to an external clock with the help of a feedback loop. It measures the total delay of the delay line and compares it with an external reference clock. The timing error after comparison is corrected due to the negative feedback of the DLL in various iterations. This method ensures the accuracy of measurement. However, DLL turns out to be a complex and power- and area-consuming solution.

(i) Techniques using TDL based on gate delay resolution

An inverter is the basic delay element that provides the minimum delay in a given CMOS technology. However, due to the asymmetric rising- and falling-edge delays of an inverter, there is a mismatch in consecutive tapped delays. Therefore, a buffer (two cascaded inverters) is generally used as a delay element. It maintains the polarity of the input signal with uniform delays among the taps. However, the attainable delay of the buffer is twice that provided by an inverter. Further, the minimum delay achievable is also limited by the parasitic capacitance of a given technology, thereby limiting the resolution of the time interval measurement in that technology. The TDC topologies reported to exhibit gate-delay resolution are as follows:

- *Direct counting using TDL*: In this technique, a TDL provides delayed copies of the applied reference clock. Corresponding to each delayed clock, a counter is used to count the number of cycles between "start" and "stop." All the counters are simultaneously enabled by "start" and disabled by "stop." The addition of counts provided by the counters multiplied by the resolution gives the time interval. Here, the resolution is defined by the unit delay of TDL. This technique achieves an improvement in the resolution of the direct counting method but at the cost of using multiple numbers of counters that are proportional to the number of delayed clocks.
- *Latch-based TDL*: In this technique, the TDL itself is realized by using a chain of latches that are transparent for the propagation of "start" until "stop" occurs. The D-to-Q delay of the latch defines the resolution of time interval measurement. This architecture consumes relatively less power and area as it does not require a separate chain of delay elements.
- *TDL using inverter as a delay element (pseudo-differential delay line)*: The resolution of TDL can be improved by a factor of two if an inverter is used instead of a buffer as a delay element. However, due to the contribution of both the rising- and falling-edge delays of the inverter in the time interval measurement, the linearity of tapped delays is poor as compared to that in buffer-based TDL. By adjusting the aspect ratio of MOS transistors, both edge delays of the inverter can be made nearly symmetric. However, the symmetry in delays is not maintained across process variations. In addition, the setup time of the flip-flop used in TDL is not symmetric for input rising- and falling-edge transitions.

The nonlinearity of inverter-based TDL is circumvented in Ref. [32], where a pseudo-differential delay line with sense-amplifier-based flip-flops (SAFF) is used. The SAFF consists of a sense amplifier followed by an R-S latch. The complementary data signals are applied to the differential inputs of the sense amplifier. It produces a high-to-low transition at one of the outputs corresponding to the leading input on the leading edge of the clock. The S-R latch records each transition and holds the state until the next clock edge arrives, thereby making the whole structure work as a flip-flop. As shown in Figure 10.24(a), "start," and its complementary signal pass through two identical delay lines of cascaded inverters. The "stop" signal samples and latches the status of delayed versions of "start" and its complementary signal using SAFF. In the latch register chain, the inputs of alternate flip-flops are interchanged between delayed start and corresponding delayed complementary to avoid the impact of asymmetric delays of the inverter. The sense-amplifier-based flip-flop is designed with a small and symmetric setup time window for both rising- and falling-edge input transitions. Due to the small setup time window of flip-flops, the resolution of time interval measurement is limited by the inverter delay. As shown in Figure 10.24(b), the combined output of flip-flops is a thermometer code, as in buffer-based TDL, where the

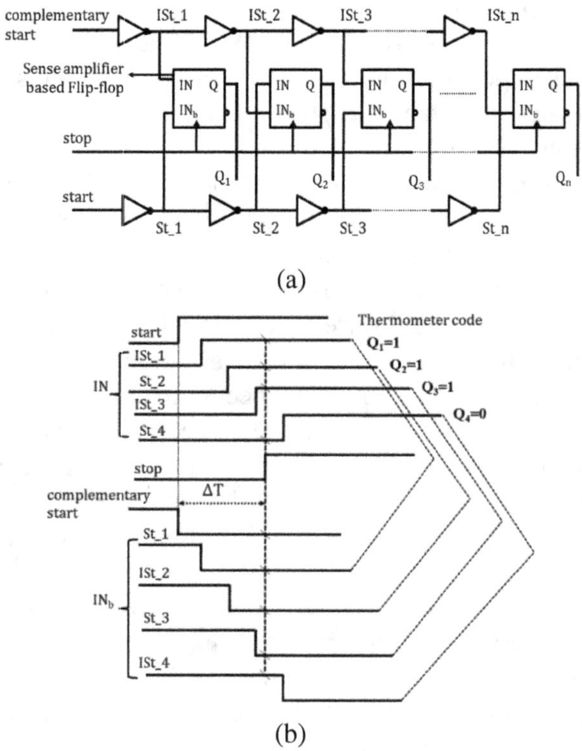

Figure 10.24. Pseudo-Differential Delay Line: (a) Block Diagram, (b) Timing Diagram

first logic 1 to 0 transition count multiplied by inverter delay (average of rising and falling delays) gives the measured time interval. This technique is used to design TDCs with 20 ps resolution in the 90 nm CMOS process for phase error detection in all-digital PLLs [32].

- *Input interchangeable TDL*: This time interval measurement scheme [61] is capable of measuring negative time intervals, i.e., when "stop" occurs before "start." This type of measurement is similar to the sound direction sensitivity in the nerve system of barn owls. Here, the direction of sound is detected by comparing the timing of sound coming from a source to its ears. For time interval measurement, the "start" and "stop" pass through identical TDLs in opposite directions, as shown in Figure 10.25(a). The timings of the delayed versions of "start" and

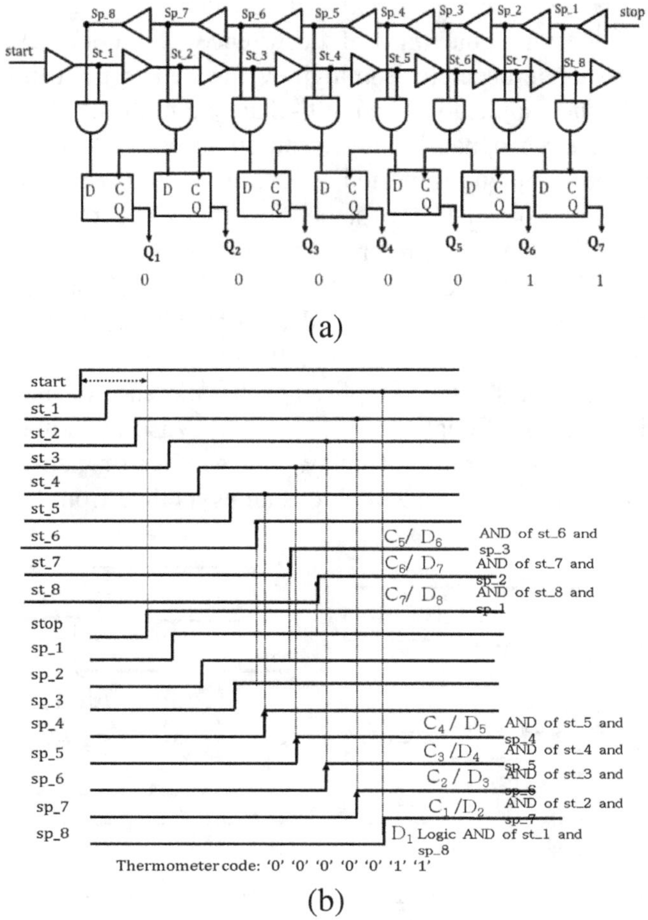

Figure 10.25. Input Interchangeable TDL: (a) Block Diagram, (b) Timing Diagram

"stop" are compared in the reverse order, i.e., the first delayed version of "start" is compared with the last delayed version of "stop." In this order of comparison, the time interval between "start" and "stop" reduces on each tap, which leads to coincidence, as shown in Figure 10.25(b). The AND logic gate is used to detect coincidence and the leading of "stop" from "start" by setting the data input to logic 1 before the clock input of the flip-flop. The number of logic 1 in the output code multiplied by the unit delay gives the time interval to be measured. To determine the total number of logic 1 in the code, each AND gate output signal samples the status of its previous gate output. In this architecture, the roles of "start" and "stop" are interchangeable due to the structure and functional symmetry of the circuit. Hence, it fulfills the requirement of negative time interval measurement.

- *TDL-based time-stamping technique*: In many HEP experiments, it is required to stamp the time of occurrence of random events with respect to a reference clock in high event rate and multi-hit environments. In such applications, a multi-channel, multi-hit TDC with high resolution, large dynamic range, and small conversion time is preferred. The techniques reported in Refs. [15, 60, 62, 63] have implemented a DLL-based TDL to stamp the time of occurrence of random events with respect to a reference clock. The total delay of TDL is continuously calibrated using the feedback loop of DLL to maintain it equal to one reference clock period, as shown in Figure 10.26(a). The DLL ensures the accuracy of time stamping across PVT variations. The TDL provides delayed replicas of the reference clock with time-period T_{ref}, as per Equation (10.12), where N is the number of delay elements and T_d is the unit delay. The number of elapsed delayed clocks until the occurrence of "hit" stamps its time from the previous

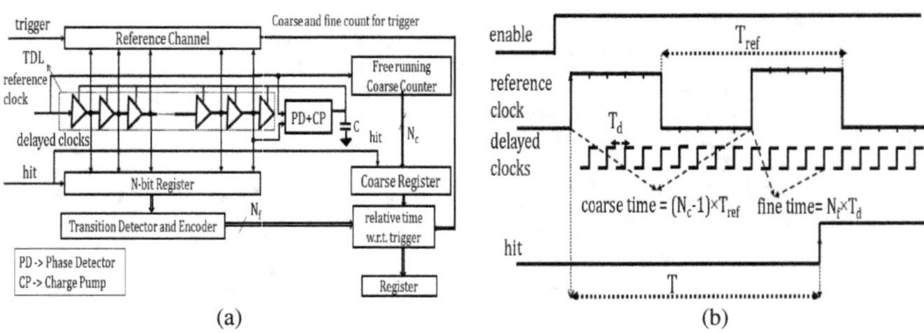

Figure 10.26. Time Stamping Using TDL: (a) Block Diagram, (b) Timing Diagram

rising edge of the reference clock. Here, the range of time stamping is confined within one clock period (T_{ref}). To improve the range, the frequency of clock is required to be reduced, which further demands an increase in the length of delay line (refer to Equation (10.12)). Alternatively, a counter synchronized to the reference clock is used to improve the range of time stamping. Thus, time stamping of hit is carried out in two parts: coarse measurement (full clock cycles) using counter and fine measurement (fractional time within one clock period) using TDL, as shown in Figure 10.26 (b):

$$T_{\text{ref}} = N \times T_d. \tag{10.12}$$

The "hit" on its detection samples the status of the counter and TDL in the coarse and fine registers, respectively. The first logic 1 to 0 transition count in the fine register code gives a a fine count (N_f) that is equivalent to the number of elapsed delayed clocks before "hit."

For accurate coarse measurement, it is required to ensure that the latching of coarse counter occurs on the safe edge of the reference clock when the counter is in stable state. If latching occurs on the same clock edge when the counter is in switching state, the coarse register may be in the metastable state, resulting in an ambiguous coarse count. However, latching on the safe edge will cause an error of one coarse count if the hit comes after the safe clock edge during one clock cycle. A dual-edge synchronization [57] and dual-counter method [53] are utilized to avoid metastability and their respective coarse count errors.

If N_c and N_f are coarse and fine register binary values, respectively, the arrival time of "hit" is given by Equation (10.13). Here, as the coarse count is added to the fine count, the accounted value of the coarse count is $N_c - 1$:

$$T = (N_c - 1) \times T_{\text{ref}} + N_f \times T_d. \tag{10.13}$$

In the time stamping of multi-hit events, the double hit resolution is a crucial design aspect. The TDL with a single fine register cannot resolve two consecutive hits occurring within one clock cycle. Therefore, in the reported multi-hit TDCs [65, 66], a single transition is allowed to be detected in the range of TDL (one clock period).

Further, time stamping can be utilized in triggered as well as non-triggered applications. In non-triggered applications, the time of hit is given relative to the time of the reference clock. In triggered applications, the time of trigger is also stamped in the reference channel so that the relative time of hits is given with respect to trigger.

This technique is efficient for multi-channel TDCs as the free-running counter and TDL can be shared between multiple channels of TDC, thereby reducing the power consumption per channel. However, the memory access time for each channel and readout speed add significant dead time to the measurements. This limits the utilization of this technique in HEP experiments, where dead-time-less measurement of events along with data buffering capability is required.

- *Shift-register-based timestamping*: In this technique [12], the hit line is applied to the pipeline of registers synchronized to the clock. On each clock cycle, the hit line is sampled, and the sampled data are shifted subsequently. Thus, the time period of the clock defines the resolution of hit line sampling. Further, the amount of shifting needed depends on the trigger latency, as the relative time of sampled hits is measured with respect to the time of trigger. Therefore, the depth of the shift register is chosen as per the specified trigger latency.

  This technique requires a high clock frequency of the order of 1 GHz to achieve a resolution of 1 ns during hit sampling. In addition, power consumption is high (nearly 1.2 W at the rate of 1 GHz) due to the shifting of sampled data on each cycle. This limits the application of shift-register-based TDC in the HEP experiment with millions of channels. To reduce power consumption, a time memory chip (TMC)-based TDC architecture [13–15] has been reported and is discussed as follows.
- *Time memory chip*: This technique uses a low-frequency clock to achieve the same resolution for sampling the hit line (with multiple events), as in shift-register-based TDC [12]. Here, the delayed replicas of the clock within one clock period are used as memory write signals to sample the hit line in the low-power CMOS static memories. The time interval (T_d) between consecutive delayed clock signals defines the resolution of sampling. A TDL is used to obtain the delayed clocks with a time interval T_d, as shown in Figure 10.27. Here, a reduction in power from 1.2 W for 256-bit memory [12] to 7 mW/channel for four-channel 1 K memory "TMC40040" [16] has been achieved as follows: The required clock frequency is lowered by the length of the delay line to achieve the same resolution (1 ns). This leads to an increment in the word size of memory in the same proportion. (2) This scheme does not involve the shifting of sampled data.

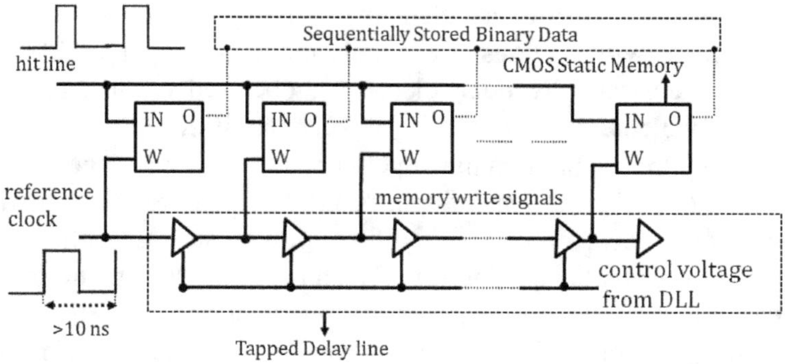

Figure 10.27. Schematic Diagram of Digital-Memory-Based Time Interval Measurement

In pipeline architectures, to increase the dynamic range of measurement, an expansion in memory size is required, which is cost-effective. Therefore, a low-cost gate array version of TMC with separate dual-port memory (24 × 128) for data buffering and four channels of 32-bit memory for hit-line sampling is reported. There are some major differences in time stamping and pipeline techniques. In pipeline, the storage depth of memory is chosen as per the trigger latency. While in time stamping, it is as per the expected hit rates in the experiments.

In the above-discussed digital techniques, irrespective of time interval measurement or time stamping, the best attainable resolution is defined by the smallest delay provided by the used delay element. The other approach to achieving high resolution is to use the sub-gate delay resolution techniques, discussed in the following section.

(ii) Sub-gate delay resolution based TDC techniques

The sub-gate delay resolution can be achieved by using high-speed delay elements in TDL or using TI measurement techniques that do not depend on the absolute delay offered by delay elements. These techniques include differential delay lines, pulse shrinking delay line, gate delay interpolation, parallel delay line and time amplification.

- *Use of high-speed delay element in TDL*: A TDL utilizing the high-speed differential delay element using a current mode logic (CML)-based delay element is reported in Ref. [61]. It provided a measurement resolution of 25.5 ps with the 0.35 μm CMOS technology, where the intrinsic buffer delay is 150 ps. However, it had high static power consumption (1 mA/delay element), making it unsuitable for HEP experiments with millions of channels and low-power portable instruments. The passive RC delay element can be used as a low-power and fast delay element. In addition, the RC delay is stable due to a lower sensitivity to the variation in temperature and supply voltage (typically, 500 ppm/V and a few tens of ppm/°C) [60]. However, it is sensitive to the technological process. Therefore, only a start-up calibration is needed to calibrate its delay for a given process. Despite these advantages, the utilization of these delay elements is constrained due to the following reasons: (1) The propagation delay of the RC delay line is a quadratic function of the number of stages, according to the Elmore delay equation [67]. This causes nonlinearity in the tapped delays. (2) The RC delay line is not able to produce uniform delayed replicas of the applied "clock" or "start" signal due to successive attenuation in the amplitude of clock on the taps. This is attributed to the low-pass filter characteristic of the RC delay element. The number of RC elements (N) in the TDL defines the order of the filter. The gain (V_{out}/V_{in}) and the −3 db frequency of the N-order filter reduce with its order, as per Equations (10.14) and (10.15), respectively, where f_c is the calculated corner frequency. In Ref. [68], buffers are used in between the taps of the long RC delay line in order to minimize the above-discussed issues to some extent:

$$V_{out} = \left(\frac{1}{\sqrt{2}}\right)^N V_{in}, \tag{10.14}$$

$$f_{(-3db)} = f_c\sqrt{2^{\frac{1}{N}} - 1}. \tag{10.15}$$

Transmission lines [69], such as coaxial cables and microstrips, on a PCB of a certain length are used as high-speed TDLs. Here, the resolution is equal to the length of unit tap, which is in the order of ps. The propagation delay for a unit length is given by $\sqrt{(L \times C)}$, where L and C are the inductance and capacitance per unit length of the TM line, respectively. However, the nonlinearity in tapped delay, temperature, and resistive and dielectric losses affect the delay characteristics and hence the TDC precision. A transmission gate serves as a high-speed, low-power, low-parasitic, and area-efficient delay element. However, similar to RC delay lines, cascading of transmission gates also causes attenuation of the signal amplitude and delay nonlinearity due to the quadratic dependency of delay on the number of stages. In

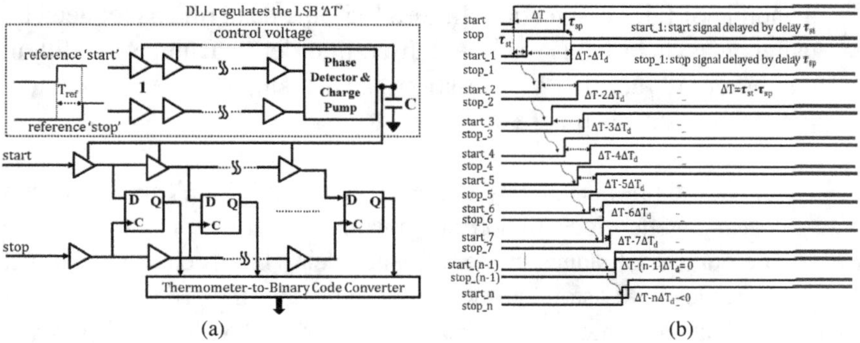

Figure 10.28. Differential Delay Line Technique: (a) Block Diagram Including (LSB) Stabilization using DLL, (b) Timing Diagram

addition, due to switching inaccuracies, such as charge injection, the quality of the waveform degrades at successive stages of TDL.

- *Differential (Vernier) delay line*: This technique, as reported in Refs. [70–72], uses two TDLs with slightly different unit delays in order to achieve sub-gate delay resolution in time interval measurement. The inputs "start" and "stop" are applied to the "slow" and "fast" TDLs with unit delays τ_{st} and τ_{sp} ($\tau_{sp} < \tau_{st}$), respectively, as shown in Figure 10.28(a). Due to the difference in speed, on each tap, the "stop" signal approaches the "start" by the step size of $\Delta T_d = \tau_{st} - \tau_{sp}$, as shown in Figure 10.28(b). Eventually, the "stop" signal coincides with and subsequently leads the "start" signal. The number (n) of stages in the delay lines that are crossed by both signals until the coincidence occurs is used to find the time interval between them. In order to determine this number, the "stop" signal samples and latches the status of "start" on each tap. Finally, the register code gives the value of n in the form of a thermometer code, where the position of logic 1 to 0 transition gives the point of coincidence. The conversion expression is given by Equation (10.16), where the difference of unit delays ($\tau_{sp} - \tau_{st}$) provides the resolution of time interval measurement:

$$T = n\ (\tau_{st} - \tau_{sp}) = n \times \Delta T_d. \tag{10.16}$$

The dynamic range of time interval measurement is the product of the number of delay stages (N) and resolution (ΔT_d). Here, the resolution is higher than that achieved in the single TDL method. However, differential delay line exhibits a longer conversion time ($N \times \tau_{sp}$) than the single TDL-based technique, which works on the principle of flash conversion. In Ref. [65], the dead time is reduced by using an asynchronous read-out buffer. In order to achieve stability in resolution ΔT_d across PVT variations, a DLL architecture, as shown in Figure 10.28(a), is reported.

Here, for the reference delay line, the "start" and "stop" signals with a time interval of one clock period are obtained on-chip using a reference clock. The number of stages in the reference delay line is chosen such that "start" coincides with "stop" at the end of the delay line. The DLL adjusts the control voltage to modify the delay of the slow delay line in PVT variations, so that "start" and "stop" have coincidence at the end of the delay line. This scheme regulates the difference between unit delays and hence the resolution of the differential delay line. Theoretically, a very small resolution (hundreds of ps) can be achieved using this technique. However, with resolutions of tens of ps over dynamic ranges of hundreds of ns, the length (N = 10,000) of delay lines is long enough to have significant systematic and random errors. This may nullify the resolution gain achieved by the differential method. To reduce the requirement of long delay lines for reasonable dynamic ranges, two other architectures of the differential scheme, namely two-dimensional (2D) differential delay line and cyclic differential delay line, have been reported, which are discussed in the following.

- *Two-dimensional differential delay line*: The 2D differential delay line requires fewer stages as compared to the linear differential delay line for the same dynamic range. This reduces power consumption as well as the integral nonlinearity error prevalent in long delay lines.

  In the linear differential delay line, the delays are compared between the taps located in the same position. If all combinations of both delay lines are chosen, a 2D plane of time reference can be implemented, as shown in Figure 10.29. Here, the range of the plane with uniform quantization defines the dynamic range of time interval measurement. For instance, for delay lines with five stages, the range of the linear differential delay line is from 0 to $5\Delta T_d$, while for the 2D (5×5) Vernier plane (shaded area), the dynamic range is from $-3\Delta T_d$ to $9\Delta T_d$.

  Moreover, in 2D Vernier plane, by extending the length of only one delay line, the dynamic range can be extended effectively. For instance, if the number of slow delay elements is extended from five to eight, the modified dynamic range is from $-3\Delta T_d$ to $24\Delta T_d$. Based on this concept, TDC prototypes have been reported [73], which find application as phase comparators in all-digital PLLs.

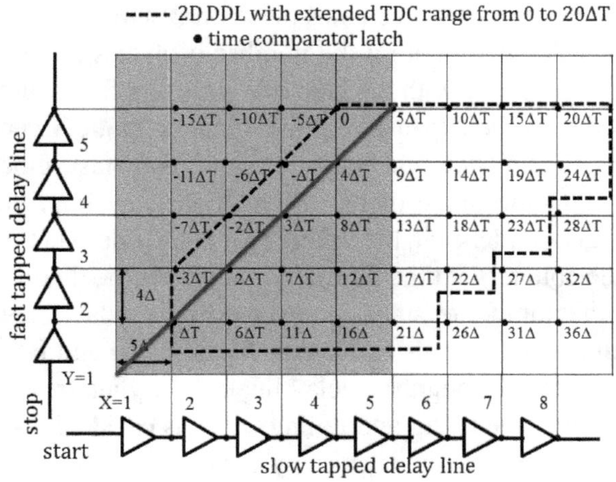

Figure 10.29. 2D Differential Delay Line Technique for TDC

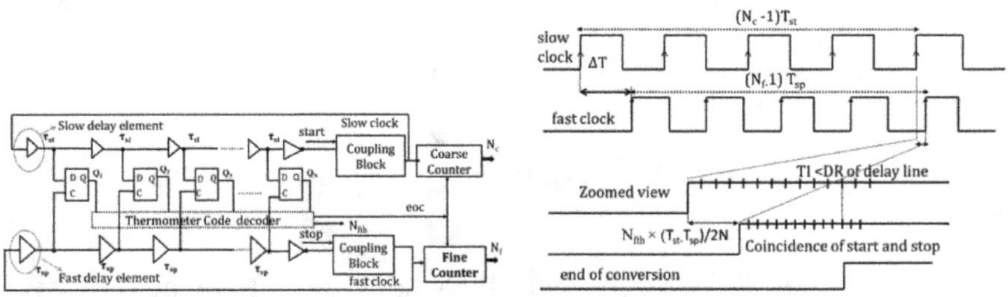

(a) (b)

Figure 10.30. Cyclic Differential Delay Line Implemented Using Buffers: (a) Schematic Diagram, (b) Timing Diagram

- *Cyclic differential delay line*: The cyclic differential delay line provides reusability of the delay stages for a theoretically infinite dynamic range. Here, "start"- and "stop"-triggerable ring oscillators are derived from the differential delay line, as shown in Figure 10.30(a). To establish the oscillations, when "start" and "stop" complete a span of the respective delay line, their inverted outputs are coupled back to the input of the delay line with the help of coupling blocks. This gives "slow" and "fast" clocks corresponding to "start" and "stop" delay lines, respectively, as shown in Figure 10.30(b).

As the delay of the "start" delay line is more than the "stop" one, the time interval between "slow" and "fast" clocks reduces on each rotation. Thus, if this time interval is more than the dynamic range of the differential delay line, no phase coincidence takes place on its taps. So, this time interval is accounted for by counting the number of cycles of the "slow" and "fast" clocks individually with the help of coarse and fine counters, respectively. This provides the expansion of the dynamic range. Eventually, after a few rotations, the time interval between "slow" and "fast" clocks becomes less than the dynamic range of the differential delay line. Subsequently, there is a phase coincidence in the differential delay line. This asserts "eoc," which stops the counters. At the time of coincidence, the thermometer code, equivalent to the number of stages n covered in the delay line until coincidence, is produced in the register. Further, as the rising and falling edges of oscillator clocks pass through the delay line in every alternate rotation, the coincidence of both the rising and falling edges is checked across the taps.

The conversion expression is given by Equation (10.17), where N_c and N_f are the number of cycles provided by coarse and fine counters, respectively:

$$\boldsymbol{T} = (N_c - 1)T_{\text{st}} + N_{\text{fth}} \times (T_{\text{st}} - T_{\text{sp}}) - (N_f - 1)T_{\text{sp}}. \tag{10.17}$$

The above equation can be rewritten as

$$T = (N_c - 1)T_{\text{st}} + N_{\text{fth}}\left(\frac{T_{\text{st}} - T_{\text{sp}}}{2 \times N}\right) - (N_f - 1)T_{\text{sp}}, \tag{10.18}$$

where $\Delta T_d = (T_{\text{st}} - T_{\text{sp}})/2N$ represents the resolution of time measurement for the N-stage cyclic delay line. However, if coincidence takes place at the falling edge, the conversion expression is given by

$$T = \left(N_c - \frac{3}{2}\right)T_{\text{st}} + N_{\text{fth}}\left(\frac{T_{\text{st}} - T_{\text{sp}}}{2 \times N}\right) - \left(N_f - \frac{3}{2}\right)T_{\text{sp}}. \tag{10.19}$$

Here, there are two conversion expressions depending on the type of edge coincidence. The other approach, which has a single conversion expression for the cyclic scheme, is used in Ref. [74]. Here, the delay lines are realized by using an odd number of NAND gates, with each one functioning like a basic inverter. The NAND gate is chosen instead of the inverter to provide trigger ability to the inputs (start/stop) as well as to keep all delay elements identical. The oscillations are maintained by simple feedback owing to the odd number of stages. Here, there are alternate signal-edge transitions on each tap. So, to detect the coincidence, two arbiters corresponding to rising and falling edge coincidence are used on each tap.

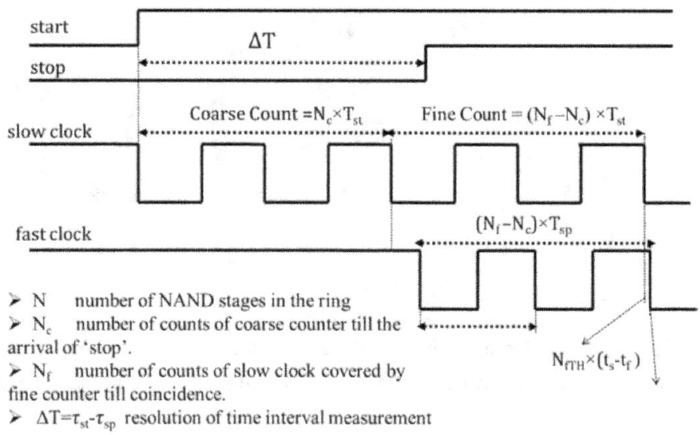

Figure 10.31. Cyclic Differential Delay Line Implemented using Odd Number of NAND Gates

As shown in Figure 10.31, the coarse and fine counters are clocked by a slow oscillator clock. The coarse counter counts full cycles (N_c) of the clock until the occurrence of the "stop" signal. The fine counter counts the number of times (N_f) the ring core is reused before the coincidence of the taps of the delay line. The outputs of arbiters are combined into thermometer code, where the first logic transition gives "eoc," which stops the fine counter and oscillators. In this scheme of coarse and fine counting, the conversion expression is independent of the type of edge coincidence and is given by

$$\Delta T = N_c T_{\text{st}} + (N_f - N_c)T_{\text{st}} + N_{\text{fth}} \times \Delta T_d - (N_f - N_c)T_{\text{sp}}. \qquad (10.20)$$

- *Parallel delay line*: It consists of delay elements with capacitive load scaling, as shown in Figure 10.32(a). Each successive delay element is loaded with a linearly scaled capacitor at node "X." When a rising edge of a signal is applied to a parallel delay line, the time required to discharge the load capacitor is proportional to the scaling factor. Therefore, the delay versus scaling factor characteristic is linear, as shown in Figure 10.32(b).

The delay T_p of the delay element loaded with a capacitor of scaling factor N is given by Equation (10.21), where T_d is the intrinsic propagation delay and ΔT_d is the delay due to unit load ($N = 1C$). The delay

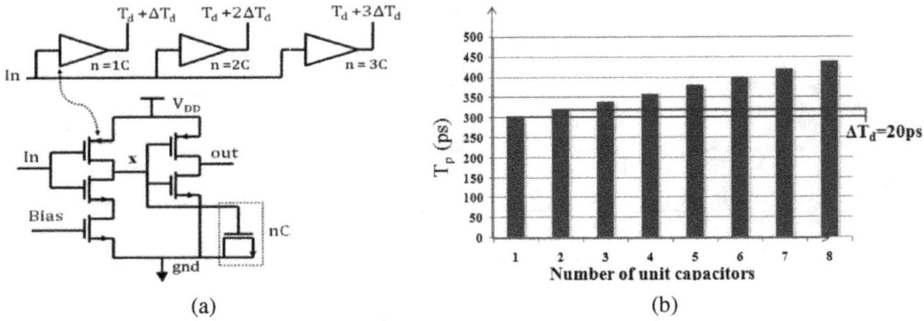

Figure 10.32. Parallel Delay Line with Load Capacitor Scaling-Based Delay Element: (a) Schematic Diagram, (b) Delay vs. Scaling Factor Characteristic

between consecutive delay elements is ΔT_d if their intrinsic propagation delays are identical:

$$T_p = T_d + N \times \Delta T_d. \tag{10.21}$$

For time interval measurement, "start" is applied to the parallel delay line to obtain its delayed replicas with time interval ΔT_d. The "stop" signal samples and latches the status of the delayed replicas of "start" to determine their number just before "stop." The number thus obtained multiplied by ΔT_d gives the time interval measurement.

The salient features of this technique are its small conversion time and sub-gate delay (ΔT_d) resolution in time interval measurement. In addition, the measurement has less nonlinearity errors, as the DNL error does not accumulate to create a high INL as in cascaded delay lines. On the other hand, this technique has large current spikes as all stages change their states at the same time. In addition, the dynamic range is limited by the number of capacitors $N \times (N + 1)/2$ required for N stages. With the limitations of a small dynamic range and high power, this technique is suitable as a fine time interpolator [75]. The key design challenges are to reduce the fan-out delay of the "start" signal and layout complexity in view of the matching of the intrinsic delays of delay elements and matching of capacitors.

- *Hierarchical TI measurement techniques (reduction in number of delay elements)*: Hierarchical techniques (two- and three-stage interpolation and nested DLLs) use the hierarchy of interpolators to reduce the length of the delay line by achieving high resolution. It leads to a reduction in nonlinearity error as well as power and area consumption due to less number of delay elements. These techniques are discussed in the following sections.

- *Two-stage interpolation*: In two-stage interpolation [76–78], the coarse interpolator is implemented by a TDL providing delayed replicas of reference clock for a measurement range of one clock period. It measures the time intervals T_{fst1} and T_{fsp1}, which are equal to the delayed clock edge (next to the "start"/"stop" signal) to the next rising edge of the reference clock, as shown in Figure 10.33. The resolution (T_d) of a coarse interpolator is determined by gate propagation delay. The sub-gate delay-resolution-based time interval measurement techniques are used as fine interpolators over a range that is equal to the resolution of the coarse interpolator. Two fine interpolators (corresponding to "start" and "stop") working in parallel are used to measure the time intervals T_{fst2} and T_{fsp2}. These time intervals are in between the "start" and "stop" of their adjacent delayed clocks, respectively.

The measurement error due to metastability in the coarse register is avoided by using fine synchronizers for "start" and "stop" signals at the second interpolation level. It provides sufficient time for the coarse register to attain the stable state of metastability. The scheme of synchronization is based on generating a signal "syn" corresponding to the delayed clock, adjacent to input (start/stop) with a synchronization delay. This delay is in multiples of the coarse interpolator resolution. The input (start/stop) is also delayed by the same amount to obtain an "asyn" signal, thereby maintaining the time intervals T_{fst2} and T_{fsp2} intact. The conversion time is given by

$$\Delta T = T_c + T_{\text{fst1}} + T_{\text{fst2}} - (T_{\text{fsp1}} + T_{\text{fsp2}}). \tag{10.22}$$

In the reported architectures [77, 78], to provide immunity to the LSBs of coarse and fine interpolators across PVT variations, different schemes of

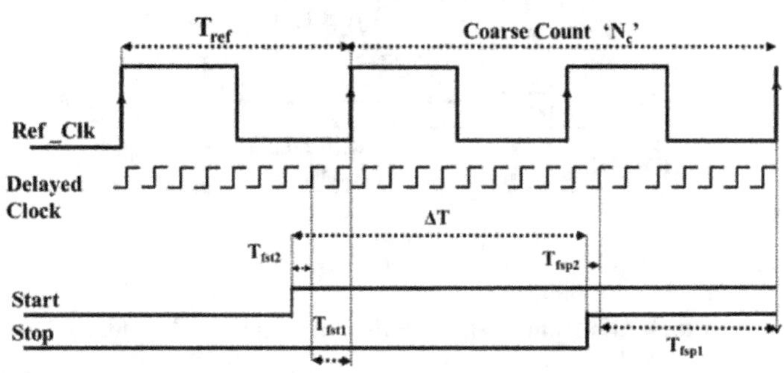

Figure 10.33. Timing Diagram for Hierarchical Technique

DLL have been used. In Ref. [77], a dual-loop dual DLL is used, as shown in Figure 10.34(a). It consists of slow and fast DLLs with identical delay elements. The "fast" DLL locks the delay of the *N*-stage delay line to the time period of the clock, as per Equation (10.23a). In the slow DLL, the delay of the *N*th delay element is matched with the $(N + 1)$th delay element of the fast DLL, as per Equation (10.23b). By virtue of the loop correction mechanism, the control voltages V_{ctrlf} and V_{ctrls} are different, providing smaller unit delay τ_{sp} in the fast DLL than τ_{st} in the slow DLL. The delay τ_{sp} is the LSB of the coarse interpolator, and ΔT_d is for the fine

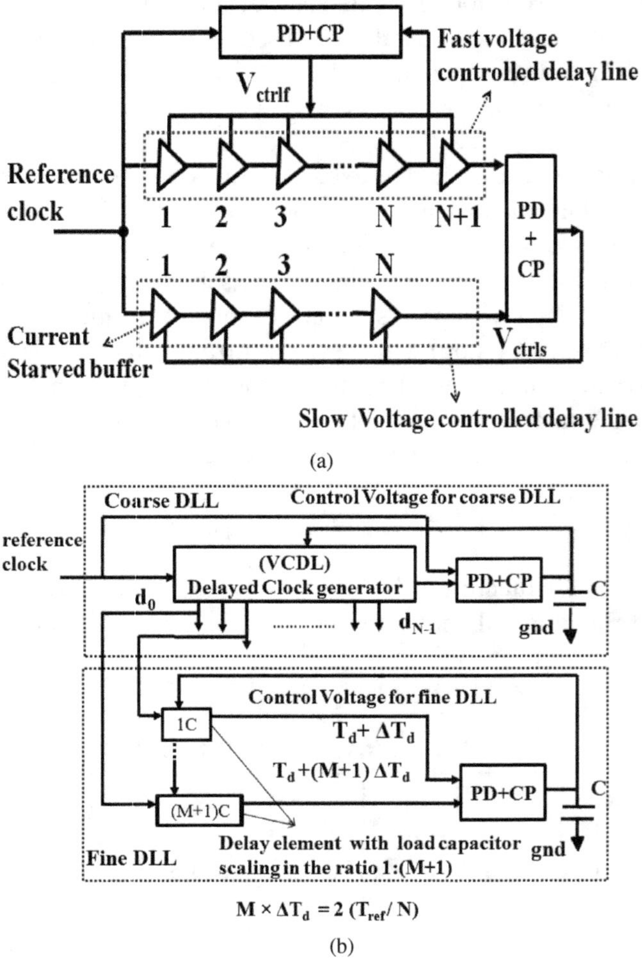

Figure 10.34. Block Diagram of (a) Dual-Loop Dual DLL and (b) Nested DLL Scheme

interpolator. The required number of stages N for a clock frequency T_{ref} and resolution ΔT_d is calculated from Equation (10.23c):

$$\tau_{\text{sp}} = \frac{T_{\text{ref}}}{N}, \tag{10.23a}$$

$$N \times T_{\text{st}} = (N+1) \times \tau_{\text{sp}}, \tag{10.23b}$$

$$\Delta T_d = \frac{T_{\text{ref}}}{N^2}. \tag{10.23c}$$

In Ref. [78], a nested DLL scheme has been used, as shown in Figure 10.34(b), where the coarse DLL locks the delay of the N-stage delay line to one clock period. The fine DLL locks the Mth output delay of the parallel delay line to the two times of bins T_{ref}/N_f of the coarse DLL. The LSB of the fine interpolator is $2 \times T_{\text{ref}}/(M \times N)$.

- *Three-stage interpolation using reference recycling with low-frequency clock (high-precision time interval measurement)*: A low-frequency reference clock, due to its stability and small power requirement, improves the performance of TDCs. On the other hand, it requires a large number of delay elements for delay-line-based fine interpolators. In order to further reduce the number of delay elements, a reference-recycling-based DLL [79–81] architecture with three-stage interpolation has been reported. It reduces the length of the delay line with a low-frequency clock while maintaining high resolution. This results in a reduction of nonlinearity error, which improves the precision of TDC. The scheme is based on recycling the reference clock edge several times in the delay line until one reference clock period is covered. After that, a new reference edge is applied to the delay line. The required number of recycling rounds of the N-stage delay line to complete one reference clock period is defined by the recycling factor R.

The recycling in the delay line is carried out by using a differential-multiplexer-based delay element [81] with a selection line (Sel), as shown in Figure 10.35(a). Initially, “clock” is selected to enter the delay line. When “clock” completes a span of delay line, a feedback loop is selected by cross-coupling between the last element and the second channel of the first element to enable recycling. The selection line of the delay element is controlled by a recycling counter, which toggles when the count is equal

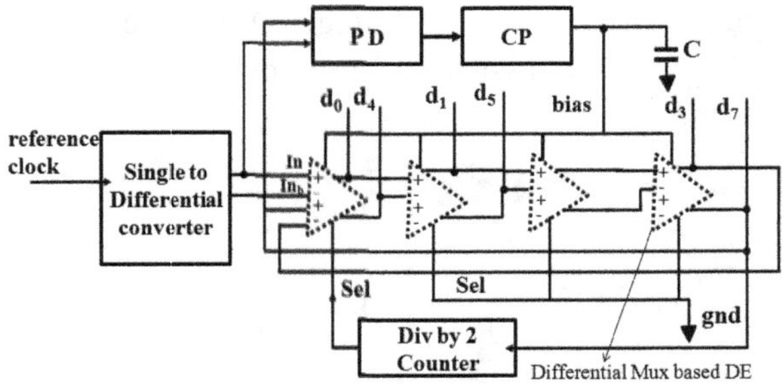

(a)

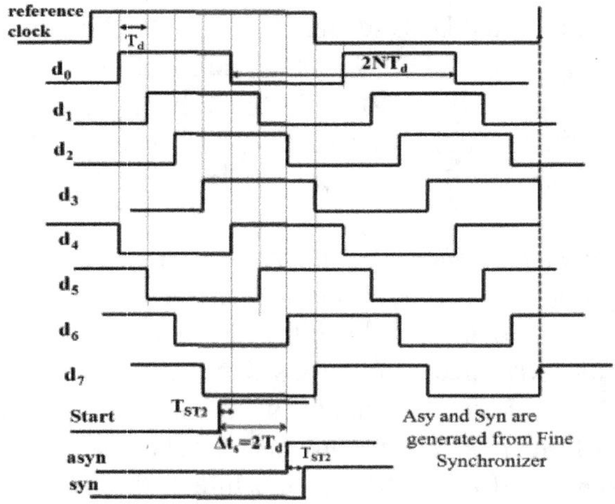

(b)

Figure 10.35. Cyclic-DLL-Based Coarse Interpolator: (a) Block Diagram, (b) Timing Diagram

to the recycling factor (R). It further injects the jitter-free clock into the delay line. The first element uses both channels, and others use the single channel to avoid delay variation between the first and last delay elements in the loop. The silent feature of this delay element is that it provides two outputs, such as d_0 and d_4, with different phases and identical time periods of $2 \times N \times T_d$, as shown in Figure 10.35(b). This reduces power consumption and doubles the effective length of the delay line.

The time interval measurement is divided into three parts: coarse count, coarse interpolation, and fine interpolation. The phase d_0, which interpolates the reference clock by "2," is applied to the coarse counter. It counts the number of cycles between "start" and "stop." In the coarse interpolator, "start" and "stop" latch the state of intermediate clocks $d_0 - d_7$ with resolution $T_d = T_{\text{ref}}/2 \times (2 \times N \times R)$.

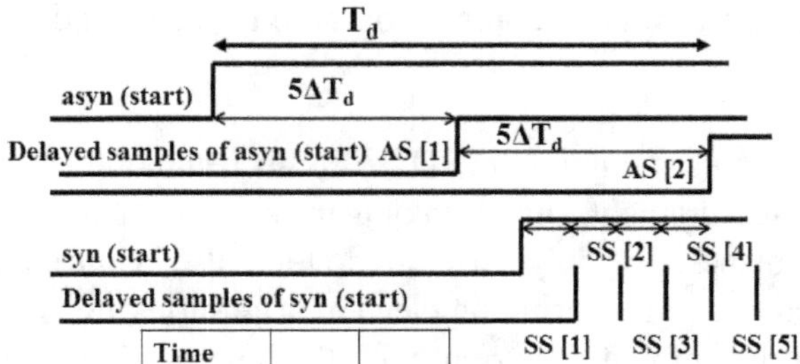

Time Samples	AS[1]	AS[2]
SS[1]	1	0
SS[2]	1	0
SS[3]	1	0
SS[4]	1	0
SS[5]	1	1

Figure 10.36. Timing Diagram of 2D Fine Interpolator in Three-Stage Interpolation

Table 10.1. Performance Comparison of Interpolation-Based Architecture

Parameters	Time Stamping	Nested DLL	Two Level Conversion	Re-cycling DLL
CMOS process	1 μm	0.8 μm	0.35 μm	0.35 μm
LSB	1.56 ns	92 ps	24–30 ps	12.2 ps
Clock frequency	40 MHz	85 MHz	130–160 MHz	5 MHz
Power	10 mW/ch	100 mW	<50 mW	40 mW
Area	25 mm^2	3.1 × 2.2 mm^2	—	2.5 × 3.0 mm^2
Delay elements	24	32	16	20
Interpolation ratio	16	128	256	16,384
Precision	100 ps	50 ps	—	8.1 ps

In the fine interpolator, the parallel delay line with capacitor load scaling is used. Here, the fine resolution ΔT_d is the *M*th part of T_d. Therefore, a minimum of *M* delay elements are needed. To further reduce the number of delay elements, a 2D arrangement of delay elements is used. As shown in Figure 10.36, one delay line contains two delay elements and another contains five delay elements. Thus, 10 interpolation slots with resolution ΔT_d can be obtained using seven delay elements. Therefore, the final achieved interpolation in the reference clock is $T_{ref}/4NRM$. Table 10.1

represents the performance comparison among interpolation-based TDC techniques.

- *Pulse shrinking delay line (PSDL)*: In this technique [82, 83] for time interval measurement, the shrinking in the pulse length is monitored until it vanishes. To apply shrinking to the pulse, a current starved buffer is used as a pulse-shrinking element. The shrinking is introduced by adjusting the amount of current with the help of bias voltage during a rising-edge input transition. So, the difference between rising edge delay (τr) falling edge delay (τf) is equal to the amount of shrinking and defines the LSB of time interval measurement. The pulse with a length equivalent to the time interval is applied to a cascaded delay line of pulse shrinking elements. A latch on each tap is set to logic 1 by the presence of the pulse. Figure 10.37 shows the timing diagram where, at the *n*th stage, the pulse vanishes so that the output of the corresponding latch is at logic 0. The first logic 1 to 0 transition count *n* in the output thermometer code multiplied by the LSB ΔT_d gives the time interval.

To reduce this nonlinearity, cyclic PSDL has been reported [83–85], as shown in Figure 10.38(a). Here, the number of rotations of the pulse

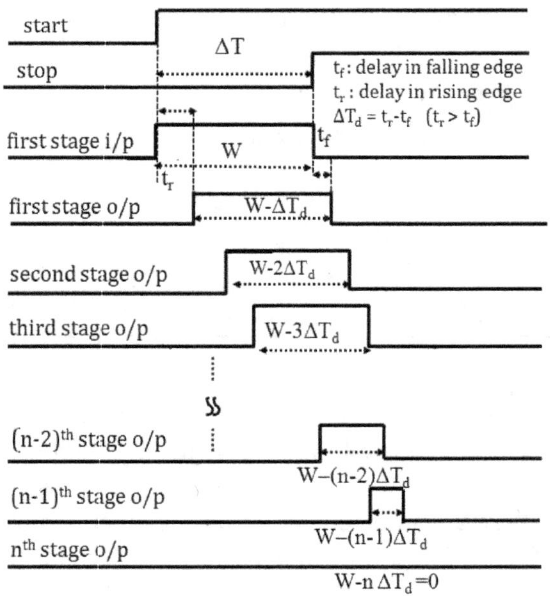

Figure 10.37. Timing Diagram of Pulse Shrinking Delay Line Technique

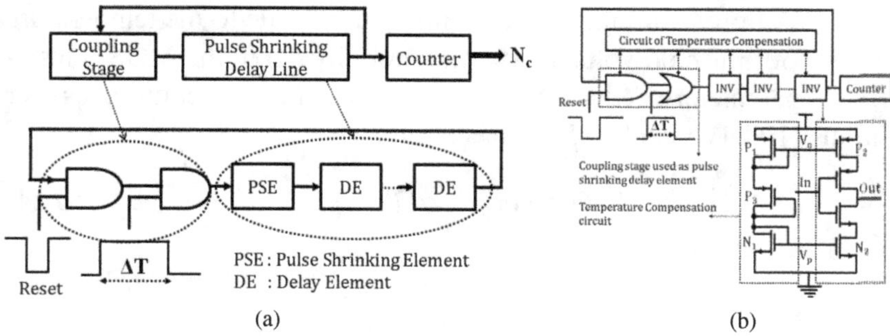

Figure 10.38. Cyclic Pulse Shrinking Delay Line: (a) Block Diagram, (b) Schematic Diagram with Temperature Compensation

(through the cyclic PSDL) until it vanishes is used for time interval measurement. The nonlinearity error reduces as only one element is used for shrinking. Even if more than one shrinking element is used in the delay line, nonlinearity error is still not encountered in cycle-to-cycle counting of pulse rotation.

The LSB of time interval measurement is calibrated by counting the rotations of the reference pulse of known duration through cyclic PSDL until it vanishes.

To set the high resolution, the amount of shrinking is adjusted by external reference voltage so that the pulse covers a large number of rotations before it vanishes. In the measurement phase, the calibrated LSB multiplied by the number of pulse rotations gives the time interval measurement.

To avoid bias adjustment for voltage-controlled shrinking, a standard cell-based pulse shrinking delay line has been reported [85]. Here, the shrinking is obtained by passing the pulse through two buffers, which have different driving strengths. This leads to different durations of pulses at the output of buffers. So, their logical AND gives a final pulse that is shrunken in length. The degree of shrinking is controlled by the adjustment of the aspect ratio of buffers. The other way is to utilize the inhomogeneous dimensions of the coupling stage for shrinking, as it differs from other inverters in the delay line. When a pulse passes through the interface of the delay line and coupling stage, it encounters different rising and falling edge delays. To ensure shrinking, the aspect ratio of the coupling stage and inverters is adjusted, so that rising edge delays are more than those of the falling edges. This digital pulse shrinking TDC achieves a 20 ps LSB and an 18 μs dynamic range. However, it has an LSB variation of ±25% over a temperature range of 0–100°C.

The variation in shrinking over temperature is due to the temperature sensitivity of threshold voltage (V_T) and mobility (μ) in the MOS transistor. Both parameters reduce with an increase in temperature, as per Equations (10.24) and (10.25), respectively:

$$V(T) = V(T_0) + \alpha(T - T_0), \tag{10.24}$$

$$\mu(T) = \mu(T_0)\left(\frac{T}{T_0}\right)^{K_m}, \tag{10.25}$$

where K_m ranges from −1.2 to −2.0 and T_0 is the reference temperature (e.g., 300 K), T is the environment's absolute temperature, and α ranges from –0.5 to –3.0 mV/K.

The variation in the above factors influences the drain current in opposite directions. Mobility dominates because of its exponential nature if the bias voltage for the transistor is much greater than the threshold voltage. Otherwise, the threshold voltage effect dominates. Using this feature, a temperature-compensated cyclic pulse shrinking-delay-line based TDC is reported [86], as shown in Figure 10.38(b). Here, the bias voltages (V_n and V_p) compensate the mobility loss of the logic inverter. These voltages are provided by a bias circuit, where transistor P_3 acts as the load of the two diode-connected (saturated) transistors P_1 and N_1. As temperature increases, the threshold voltages of P_1 and N_1 decrease, resulting in an increase in the conduction current of P_3. The current mirrors composed of (P_1, P_2) and (N_1, N_2) transfer the conduction current of P_3 with a positive temperature coefficient to the inverter, thus reducing the variation in delays.

In this temperature-compensated cyclic delay line, the variation in LSB reduces from 25% to 6% over a temperature variation from 0°C to 100°C. This scheme is also used to implement a TDC-based temperature sensor [35]. In Ref. [104], an offset error cancellation circuitry along with a pulse shrinking delay line is used to widen the dynamic range and to improve the accuracy of TDC. The cancellation circuitry for eliminating the offset error consists of a time adder with a delay line and a time subtractor with an identical delay line.

- *Interpolation of bin size of DLL by array of delay lock loop (ADLL)*: In time stamping, the bin size depends on the smallest attainable delay from a delay element. The current starved inverter-based delay element is frequently used in time stamping due to its salient features of low power and wide delay regulation range. However, the smallest delay is limited by the parasitic capacitance CMOS process node. As an alternative, a high-speed differential delay element with a large static current has been used in Ref. [63]. However, in portable and HEP experiments, power efficiency is desirable. Therefore, a scheme to interpolate the smallest delay of the current starved inverter (bin size of the DLL) has been reported in Ref. [87].

This scheme relies on using an array of several uniformly offset DLLs, as shown in Figure 10.39(a). To generate the precise offsets, a phase-shifting DLL (PSDLL) with an M number of delay elements is used. The unit delay (T_M) of PSDLL defines the required offset. It is chosen to be more than the process limited delay (T_N) by some fraction (F) to obtain the delay interpolation as $T_M = T_N + F^{-1}T_N$, where $F > 1$. The value of F defines the number of offset DLLs. The number of stages N (= T_{ref}/T_N) in each offset DLL and M (= T_{ref}/T_M) in PSDLL is chosen to lock them at one clock period.

The PSDLL provides the reference clocks (each having a initial phase difference of T_M from the previous one) to the offset DLLs, so that the interpolated bin size in offset DLLs is given by

$$\Delta T_d = T_M - T_N = \frac{T_N}{F}. \tag{10.26}$$

However, the sequencing of consecutive bins with a uniform time interval of ΔT_d is important before its utilization in time stamping. To find this sequence, the edge representation of delayed clocks on each tap of the offset delay lines is illustrated in Figure 10.39(b). The notations ($a, b, \ldots, f$) listed in

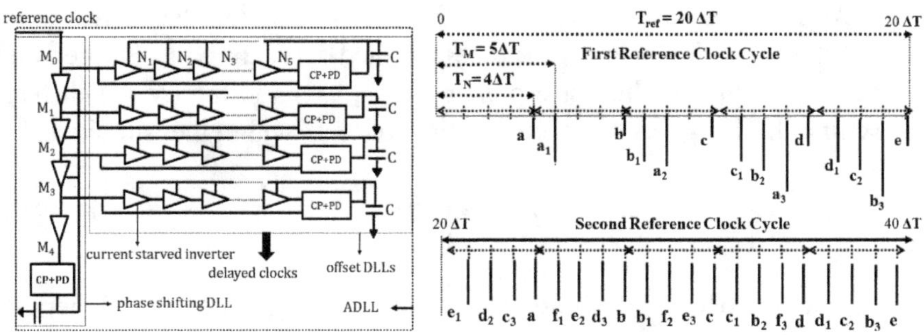

(a) (b)

Figure 10.39. (a) Block Diagram of ADLL-Based Time Interval Measurement. (b) Edge Representation of Delayed Clocks for 2D Matrix (4 × 5) of Delay Elements in ADLL

Table 10.2. Position of Delay Elements

Notation	[a to e]	[a_1 to f_1]	[a_2 to f_2]	[a_3 to f_3]
Position of DE	M_0 [N_1–N_5]	M_1 [N_1–N_5]	M_2 [N_1–N_5]	M_3 [N_1–N_5]

Table 10.2 represent the position of the delay elements. In the second reference clock cycle, a sequence of consecutive bins with a uniform time interval of ΔT_d can be achieved. The number of bins is given as $N \times F$.

The drawbacks of ADLL are power and area inefficiency as well as jitter accumulation in offset DLLs. By sharing ADLL among a number of channels of time stamping, its power and area efficiency can be improved. A single-hit TDC prototype in Ref. [88] has been reported using ADLL for sub-gate delay resolution.

- *Interpolation of bin size of DLL by RC-delay line*: The gate delay interpolation using ADLL technique requires a number of closed feedback loops in a complex topology. This invariably leads to high power and area consumption. Alternatively, a time-stamping scheme [89] to interpolate the process-limited bin size of DLL relies on acquiring M samples of the status of DLL on the arrival of the "hit" signal. This fractional sampling is obtained by delaying the "hit" signal using the tapped RC delay line. It provides M delayed versions of the "hit" signal with a time interval (ΔT_d) that is the Mth part of the bin size (T_d). The DLL bin size interpolation is therefore achieved by determining the number of delayed hit signals (m) that shift the DLL status to the next bin. As shown in Figure 10.40, the "hit" signal (S_0) occurs in between the Nth and ($N + 1$) th delayed clocks. From the samples of the status of delayed clocks, the transition in logic 0 to 1 is in the Nth bin. This provides the arrival time of hit $N \times T_d$ with bin size T_d (process limited delay) of DLL. Further, the delayed sample of hit corresponding to S_4 exists in the ($N + 1$)th bin, so $m = 4$. This gives the fractional time of "hit" within the Nth bin as

$$T_d - \frac{T_d \times m}{M}. \tag{10.27}$$

So, the occurrence time (T) of hit over a range of one clock period with sub-gate delay resolution of $\Delta T_d = T_{\text{ref}}/(M \times N)$ is given as

$$T = N * T_d + \frac{T_d(M - m)}{M} \; (0 \le m \le M). \tag{10.28}$$

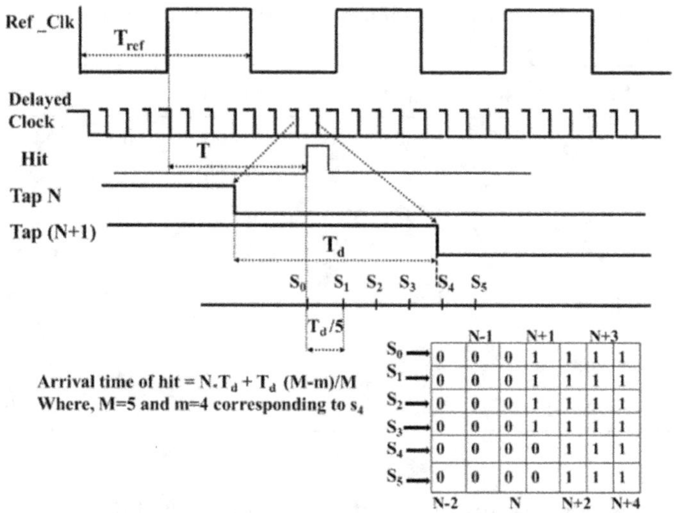

Figure 10.40. Timing Diagram of Interpolation of DLL Bin Size using RC Delay Line

Using this scheme, a TDC prototype with a single DLL and passive RC delay line with the same potential for interpolation as achieved from ADLL has been reported [88, 89].

- *Interpolation in bin size of DLL by multiple period locking*: The operating principle of conventional DLL architecture is based on one clock period locking [59]. If the delay of the voltage-controlled delay line with N stages is locked to multiple M clock periods as per Equation (10.29), the interpolation in the bin size of the DLL can be achieved with the condition that the numbers M and N are co-prime to each other:

$$M \times T_{\text{ref}} = N \times T_d, \tag{10.29}$$

$$\Delta T_d = \frac{T_d}{M}. \tag{10.30}$$

The interpolated delay is given by Equation (10.30). For instance, for $N = 32$, $M = 3$, $T_{\text{ref}} = 8$ ns, and $T_d = 750$ ps, the value of ΔT_d is 250 ps. The position of delay elements X to achieve the sequence of uniformly delayed clocks within a reference clock period (after three clock cycles) is listed in Table 10.3. This architecture has been used

Table 10.3. Sequence of Uniformly Delayed Clocks with respect to Position (X) of Delay Element

X	ΔT (ps)	X	ΔT (ps)	X	ΔT (ps)	X	ΔT (ps)
0	0	24	2,000	16	4,000	8	6,000
11	250	3	2,250	27	4,250	19	6,250
22	500	14	2,500	6	4,500	30	6,500
1	750	25	2,750	17	4,750	9	6,750
12	1,000	4	3,000	28	5,000	20	7,000
23	1,250	15	3,250	7	5,250	31	7,250
2	1,500	26	3,500	18	5,500	10	7,500
13	1,750	5	3,750	29	5,750	21	7,750

in the development of multi-hit TDC with a resolution of 50 ps [36] using the 0.8 μm CMOS process.

10.3.4.2.4. Miscellaneous techniques

(a) Re-triggerable Ring Oscillator Based Vernier Technique: The Vernier technique utilizes two oscillators, "slow" and "fast," with a slight difference in their time periods for time interval measurement [94–100]. The slow and fast oscillators start oscillating with time periods T_{oscst} and T_{oscsp} ($T_{\text{oscsp}} < T_{\text{oscst}}$) on the arrival of "start" and "stop" inputs, respectively, as shown in Figure 10.41. As T_{oscsp} is slightly less than T_{oscst}, in each cycle, the fast oscillator approaches the slow one by the step size of $\Delta T_d = T_{\text{oscst}} - T_{\text{oscsp}}$, which defines the resolution of the time interval measurement. Eventually, the fast oscillator coincides with or leads the slow oscillator. The phase coincidence or leading is detected by the phase detector, which toggles "eoc" and stops the ring oscillators [90–92].

Two counters count the elapsed cycles of both oscillators until phase coincidence. The coarse counter counts the number of slow oscillator cycles N_C until phase coincidence. It provides a long measurement range of $(2^b - 1) \times T_{\text{oscst}}$, where b is the number of bits of the coarse counter. The fine counter counts the cycles of the fast oscillator N_f corresponding to the number of ΔT_d within T_1 until phase coincidence.

The conversion equation of the time interval under measurement is given by

$$\Delta T = (N_c - 1)T_{\text{oscst}} - (N_f - 1)T_{\text{oscsp}}. \tag{10.31}$$

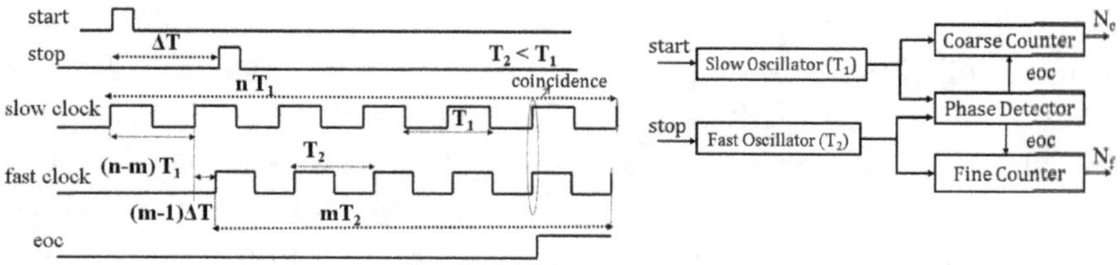

(a) (b)

Figure 10.41. Vernier Technique: (a) Block Diagram, (b) Timing Diagram

The above equation can be rearranged as

$$\Delta T = (N_c - N_f)\, T_{\text{oscst}} - (N_f - 1)\, \Delta T_d. \tag{10.32}$$

where $(N_c - N_f) \times T_{\text{oscst}}$ gives the coarse and $(N_f - 1) \times \Delta T_d$ gives the fine time measurements. The crucial design aspect is the number of bits in the fine counter. It is chosen in order to count at least the maximum number of ΔT_d within T_{oscst} given by the stretching factor S:

$$S = \frac{T_{\text{oscst}}}{T_{\text{oscst}} - T_{\text{oscsp}}} = \frac{T_{\text{oscst}}}{\Delta T_d}. \tag{10.33}$$

The key design challenges in the Vernier technique are the generation of two frequencies with a slight difference and their stabilization across PVT variations. In the TDC prototype [96], two-phase lock loops (PLLs) have been incorporated to generate the stabilized frequencies with a slight difference. To avoid the area- and power-inefficient DLL or PLL, in Refs. [98, 99], an on-chip time-period calibration method, which corrects the timing expression in PVT variation, has been used. The slight difference in frequencies of ring oscillators is obtained by using a fan-out difference of logic gates in Refs. [98, 99] and by different wire lengths in the automatic P&R of buffers in Ref. [100].

Theoretically, a high resolution ($T_{\text{oscst}} - T_{\text{oscsp}}$) can be achieved using this technique. However, the minimum detectable phase error by the phase coincidence detector limits the achievable resolution. In the reported phase detectors [98, 99], the phase detection uses sampling of the status of the slow clock by the fast clock using a *D* flip-flop. At the time of phase coincidence, the setup time violation of the flip-flop results in unpredictable (metastable) output. This situation can be avoided with the help of metastability-hardened flip-flops. However, its dead zone characteristic further limits the accurate phase coincidence detection and hence disables counters and oscillators. To mitigate the impact of dead zone and metastability on the accuracy of phase detection, in Ref. [101], a phase frequency detector (PFD) followed by a C^2MOS register has been reported. In the PFD shown in Figure 10.42(a), both oscillator clocks trigger two separate flip-flops with their inputs preset to logic 1. This avoids the violation of setup time and dead zone of flip-flops at the time of phase coincidence. The dynamic range of PFDs is limited by the resetting time of flip-flops. The resetting time is reduced by designing the PFD using dynamic logic with an in-built resetting using feedback of its outputs, as shown in Figure 10.42(b).

Further, to obtain "eoc," the C^2MOS register is used to sample the status of the "UP" signal by "DN," provided by PFD. When UP leads DN, the output of the register is at logic 1, which enables the counters. At the time of coincidence, the output is at logic 0, which disables the counter and oscillators.

The merits of Vernier technique are minimal logic resource consumption, area efficiency, process-independent resolution, and a large dynamic range. The limitation is its higher conversion time, $S \times T_{oscsp}$, which limits the measurement rate of TDC.

In the Vernier technique, the accuracy of the ring oscillator's time period cannot be maintained over a long time interval measurement due to the inherent accumulated jitter in the oscillators. A dual Vernier technique is reported in order to improve the accuracy of the Vernier technique.

(b) Dual Vernier Technique: In this scheme, two Vernier converters corresponding to "start" and "stop" are used. The difference between their counts in the conversion expression reduces the impact of jitter and systematic errors on the accuracy of time interval measurement.

This technique uses two event-triggerable oscillators with identical time periods. The identical time period (T_{osc}) of both oscillators is slightly

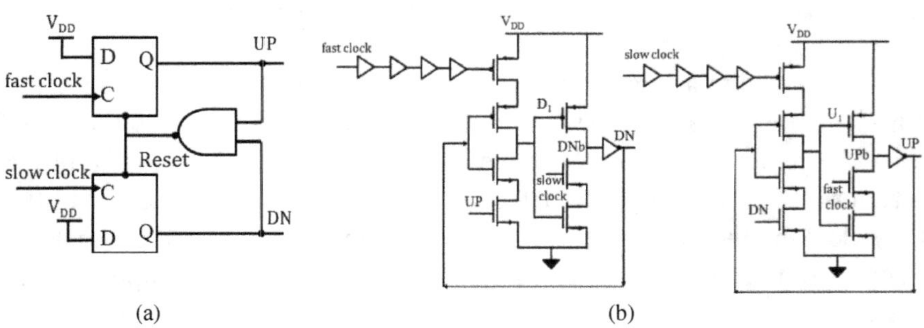

Figure 10.42. (a) PFD-Based Phase Detector. (b) Dynamic Logic-Based PFD

greater than the time period (T_{ref}) of the reference clock. Therefore, in each cycle, the reference clock independently approaches both oscillators with a step size of $\Delta T_d = T_{\text{osc}} - T_{\text{ref}}$, as shown in Figure 10.43(a). Eventually, they achieve phase coincidence with the reference clock.

In this technique, the cycles of the oscillators and reference clock are counted by three counters: one coarse and two fine. The fine counter counts the cycles of start N_{f1} and stop N_{f2} until their phase coincidence with the reference clock, as shown in Figure 10.43(b). These counts correspond to the number of ΔT_d within T_{ref}. The coarse counter counts the cycles of reference clock N_c within the time interval corresponding to individual phase coincidences of both oscillators with the reference clock. It provides a dynamic range of time intervals between "start" and "stop." The conversion expression is given by

$$\Delta T = N_{f1} \times T_{\text{osc}} + N_c \times T_{\text{ref}} - N_{f2} \times T_{\text{osc}}. \tag{10.34}$$

The number of bits in the fine counter is a crucial design parameter. It is chosen to count at least the maximum number of ΔT_d within T_{ref}. Here, ΔT_d defines the resolution of time measurement. This technique can measure negative time intervals, i.e., when "stop" comes before "start." Using this technique, a time interval counter named HP5370A has been reported with a 20 ps single shot resolution [93].

In Ref. [105], a variable-frequency-based time interval measurement is reported. This scheme uses two ring oscillators with their time periods T_{osc1} and T_{osc2}. The time interval (ΔT) is measured using Equation (10.35) by calibrating the oscillator frequencies:

$$T_{\text{osc2}} = T_{\text{osc1}} + \Delta T. \tag{10.35}$$

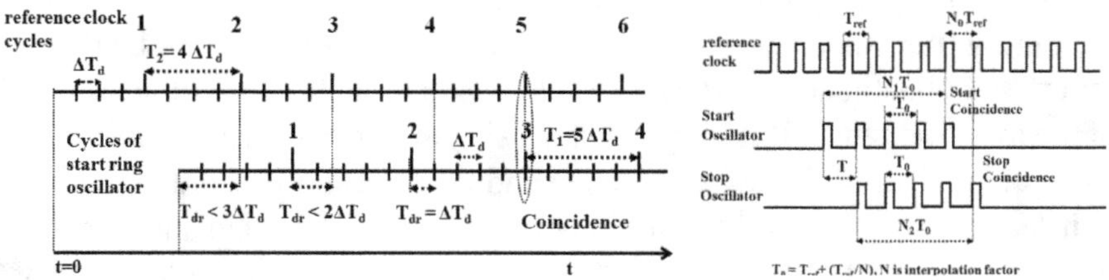

(a) (b)

Figure 10.43. (a) Edge Representation of Start Oscillator Coincidence with Reference Clock in Dual Vernier Technique. (b) Timing Diagram of Dual Vernier Technique

In order to calibrate the frequency $T_{osc1,}$ two counters A and B are used. Counter A runs at reference clock frequency (f_0), and counter B at f_{osc1}. A "disable" signal is asserted when the count of counter B matches the programmable preset number N. This signal also latches the count N_1 of counter A, which follows the equation

$$T_{osc1} \times N = T_0 \times N_1. \tag{10.36}$$

To calibrate the frequency f_{osc2}, it is applied to counter B. Similar to the previous operation, it again asserts a "disable" signal when its count matches the preset value N. It latches the count N_2 of A such that

$$T_{osc2} \times N = T_0 \times N_2. \tag{10.37}$$

Using Equations (10.35)–(10.37), the time interval can be evaluated as

$$\Delta T = \frac{(N_2 - N_1) \times T_0}{N}. \tag{10.38}$$

(c) Time Interval Using Successive Approximation Technique: Successive approximation is a well-known principle for ADC implementation that achieves high resolution but exhibits a long conversion time. In the time domain also, this technique can be implemented to measure time intervals using the binary search method. Here, the time interval between "start" and "stop" is compared in N iterations with delay adjustments in "start" or "stop," where N is the number of bits in the result. Each comparison converges one bit (starting from MSB) at a time. The algorithm advances in the time domain such that the leading signal (start or stop) is delayed every time after a comparison. Before each comparison, the amount of delay is successively reduced to half of the full-scale value.

The timing diagram of this technique is shown in Figure 10.44 for a measuring time interval of 10.4 LSB over a full scale (TFS) of 16 LSB. The "start" signal is first delayed by TFS/2 to compare with "stop" at the middle of the dynamic range. If it lags the "stop" signal, it means that the time interval is below half of the dynamic range and the MSB bit is set to be logic 0. If it leads from "stop," the time interval is greater than half the dynamic range, and so, the corresponding MSB bit is set to logic 1. Therefore, "start" is further delayed by TFS/4 and compared with the "stop" signal. As it lags "stop," it means the time interval is in between $t1$ + TFS/2 and $t1$ + TFS/2 + TFS/4. So, the next bit is set to logic 0. Further, the "stop" signal is delayed by TFS/8 and compared with the delayed "start" [102–105]. As "start" leads

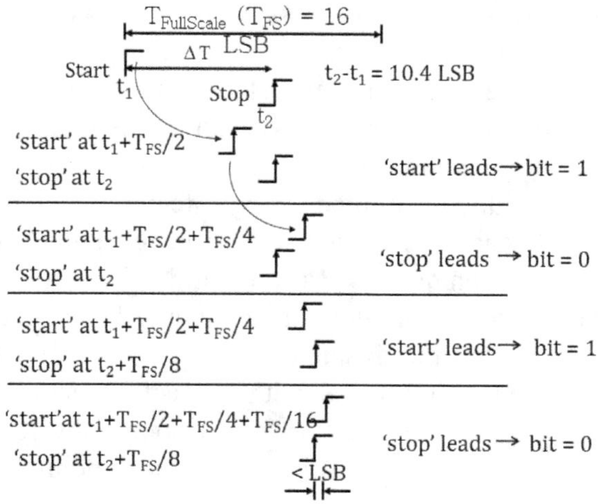

Figure 10.44. Time Interval Measurement Using Successive Approximation Technique

"stop," the corresponding bit is set to logic 1. Finally, "start" is again delayed by TFS/16 (LSB), and it approximately coincides with "stop." Therefore, the corresponding bit is set to logic 0. As the signals are approximately aligned with an error of less than one LSB, the condition refers to the end of the conversion. At the end of conversion, the bits "1010" correspond to the conversion result. Using this technique, a time interval measurement with 1 ps resolution over a 312.5 ps range has been reported [106–110].

10.4. Summary

This chapter reports a comprehensive review of various time domain signal processing blocks since their inception. The emphasis is placed on TDC implementation techniques along with their wide range of applications. Further, it discusses the key performance parameters of TDCs, which form the basis of comparison among various TDC implementation techniques. The implementation of each technique is elaborated upon with the support of design principles, block diagrams, timing diagrams, and conversion expressions.

The techniques are broadly classified into analog and digital techniques. Analog techniques (TAC-based ADC, time stretching, coarse and fine counting, and TAC-based TDL) provide high resolution and precision. However, they are limited by high power and area consumption as well as high costs. The integrated solution has provided a reduction in area and cost; however, it is not robust across PVT variations and thus degrades the precision of measurement.

In digital and mixed signal techniques (TDL and its various topologies, time stamping, and time memory chips), TDL is the main pillar that provides better resolutions compared to direct counting with small conversion times. Here, the resolution is limited by the gate delay, and the precision is limited by the nonlinearity error due to the mismatch among tapped delays. The sub- gate delay technique, such as the Vernier delay line (VDL), pulse shrinking, and time amplification, offer technology-independent high resolution but are limited by the dynamic range. Their corresponding cyclic structures offer good performance with respect to both precision and dynamic range. However, they suffer from a long conversion time. The hierarchical structures provided a solution to reduce nonlinearity error by reducing the length of delay lines at moderate conversion time.

The roadblocks to achieving high accuracy in CMOS TDC are the PVT variations and timing jitter. This makes LSB calibration a crucial step in ensuring the correctness of measurement. This can be carried out by using an inbuilt digital calibrator or automatic calibration circuits (DLL or PLL). A separate identical reference channel can be used for calibration, or the measurement channel itself can be used in calibration mode. The first option features independent time interval measurement and resolution calibration at the cost of calibration accuracy against local mismatches. The second option reduces the inaccuracy in calibration; however, it limits the measurement rate of TDC. Moreover, the local mismatches, which are dominating in modern CMOS processes, limit the precision of long delay-line-based TDCs, thus preventing the efficient use of modern processes in terms of high resolution. Further, the techniques, which use ring oscillators, are limited to achieve high accuracy due to accumulated jitter.

Thus, techniques can be preferred according to the specific requirements of time interval measurement for a given application after thoroughly weighing their advantages and limitations. Therefore, a performance comparison based on their merits (M), demerits (D), and remarks (R) is also included (Table 10.A.1) in Appendix 10.A. Finally, this chapter concludes with a time line flow diagram (Figure 10.A.1) and a rich bibliography. In this chapter, care has been taken to incorporate almost all the techniques, and so, it provides adequate material for the comprehensive study of various TDC techniques. Thus, under the set of given requirements and specifications of TDC, this chapter can assist in selecting a particular time interval measurement techniques after evaluating their advantages and limitations.

Appendix 10.A

Table 10.A.1. Performance comparison of TDC implementation techniques

Digital Counter	Tapped Delay Line (TDL)	Differential and Shrinking Delay Line
1. Small nonlinearity error (M) depends on stability of clock 2. Theoretically infinite DR (M) 3. Negligible conversion time (M) 4. Resolution and accuracy depends on the clock frequency (D) 5. Trade-off between resolution and power consumption (D)	1. Negligible conversion time (M) 2. Resolution dependent on absolute gate delay and sensitive to PVT variation (D), so requires calibration (R) 3. De-skewing required at stop input (D) 4. DR (tens of ns) defined by number of stages 5. Loop (cyclic) structure is possible for theoretically infinite DR (R) 6. Low precision limited by matching among tapped delays (D) 7. Trade-off in resolution and dynamic range (D)	1. Sub-gate delay resolution (tens of ps) (M) 2. Large DR is possible in loop (M) 3. Large conversion time(D) 4. Trade-off between resolution and dynamic range (D) 5. Low precision limited by matching among tapped delays or amount of shrinking (D) 6. Power hungry if continuous calibration is required (R)
TAC with ADC	**Parallel Delay Line**	**Loop Structures**
1. High resolution (M) 2. DR limited by nonlinearity of current source (D) 3. Conversion time depends on ADC (R) 4. Not efficient for low-width CMOS process (D) 5. Susceptible to process variation and environmental noise (D) 6. Accuracy much affected by noise coupling and nonlinearity (D) 7. Trade-off between resolution and DR (D)	1. Sub-gate delay resolution (M) 2. Strongly affected by PVT variations (D) 3. High linearity as DNL does not accumulate (M) 4. Requires $N(N+1)/2$ capacitors for N stage (D) 5. Careful hand design of each stage (R) 6. No loop feasible (R) 7. Efficient to be used as fine interpolator in hierarchical structures	1. Number of stages is reusable to expand dynamic range of TI measurement (M) 2. Less nonlinearity error as pulse covers the delay line once in each cycle (M) 3. Chip-area efficient (M) 4. High resolution (M) 5. Large conversion time (D)

(Continued)

Table 10.A.1. (*Continued*)

Dual Slope Technique	Pseudo Differential Delay Line	Single DLL with RC Delay Line
1. Robust and free from calibration as gain depends upon ratio, instead of absolute value (M) 2. Area and power efficient due to elimination of separate ADC as compared to TAC with ADC (M) 3. High resolution and high precision time measurement 4. Large conversion time (D)	5. Sensitive to PMOS and NMOS strength mismatch (M) 6. Resolution is twice as compared to buffer-based tapped delay line (M) 7. Output is in the form of thermometer code (R) 8. Careful layout is required due to twisting at taps (R)	1. Uses single DLL to provide same potential of interpolation as that in ADLL (M) 2. Power and area efficient as compared to ADLL (M) 3. Start-up calibration is required for passive delay element (R)
Array of DLL	**Time Stamping**	**Vernier Ring Technique**
1. Sub-gate delay resolution (M) defined by interpolation factor 2. Power and area inefficient due to number of DLLs (D) 3. Accumulation of jitter in number of DLLs (D) 4. Number of bins is not represented by power of "2" (R) 5. Dynamic range is equal to one reference clock period (R) 6. Better performance in multi-channel TDC (R)	1. Less conversion time, so more efficient. 2. For multi-hit time stamping (M) 3. Capable of multi-channel integration by sharing counter and TDL 4. Large DR limited by number of bits 5. of counter (M) 6. LSB depends on the interpolation 7. Ratio of reference clock (R) 8. Power dominated by DLL, so provides power and area efficiency in multi-channels (R)	1. Sub-gate delay resolution (M) 2. Sensitivity of resolution is small for process variation. 3. Theoretically infinite DR (M) 4. Accuracy limited by jitter in ring oscillators (R) 5. Area efficient (M) 6. Large conversion time (D) 7. Negligible nonlinearity due to local mismatch (R)

Dual Vernier Ring Technique	Nutt's Interpolation	Hierarchical Techniques
1. Sub-gate delay resolution (M) 2. Theoretically infinite DR (M) 3. High accuracy due to reduces effect of jitter obtained by difference in counts of both Vernier converters (M) 4. Large conversion time (D) 5. Requires reference clock (R) 6. Negligible nonlinearity due to local mismatch (R) 7. Negative time interval measurement is possible (M)	1. High resolution and large dynamic range simultaneously for TI measurement (M) 2. Needs two interpolators corresponding to "start" and "stop" (R) 3. Dynamic range of interpolator is defined by clock period (R) 4. Needs synchronizer to obtain TIs for interpolators and to synchronize them with counter	1. Efficient in reducing the length of delay line at low reference clock frequency (M) 2. High precision due to reduced non linearity error (M) 3. Area and power efficient due to less number of stages (M) 4. Moderate conversion time (D) 5. Synchronizers are required at each interpolation level (R)
Time Amplifiers	**Cyclic DLL**	**Successive Approximation**
a. Very small dynamic range (hundreds of ps) (D) b. Sub-gate delay resolution (ones of ps) c. Moderate conversion time d. Low power	1. Sub-gate delay resolution (tens of ps) (M) with small number (20) of delay elements and low-frequency clock (5 MHz) 2. Requires MUX-based delay element to match the feedback (R) 3. High precision due to less nonlinearity in short delay lines and stable clock (M) 4. Low power consumption 5. Conversion time depends on the required rotations to match one clock period	1. Sub-gate delay resolution (ones of ps) (M) with low-cost (0.35 μm) CMOS process 2. Requires digital-to-time converters in the implementation (R) 3. Dynamic range (hundreds of ns) limited by number of conversion bits 4. Low power consumption 5. Large conversion time

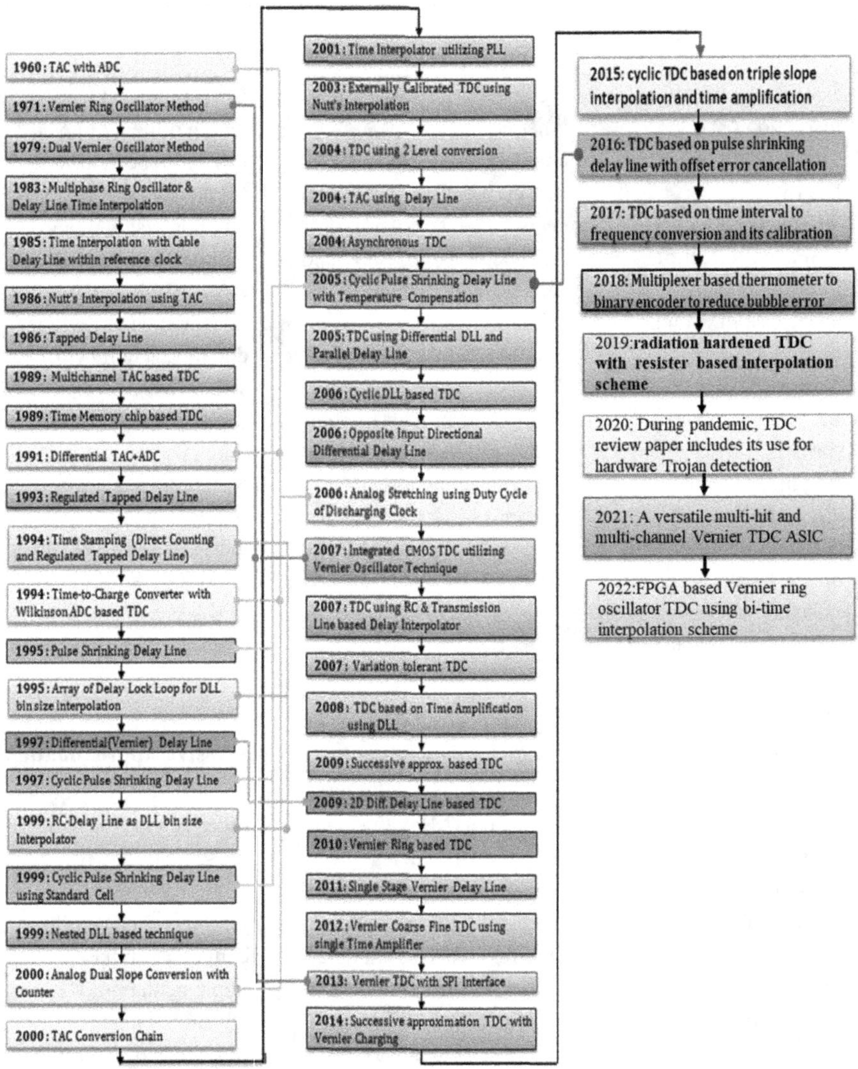

Figure 10.A.1. Flow Diagram Representing the Chronological Order of Reported Time Interval Measurement Techniques

References

1. Kang, J.-G., Kim, K., and Yoo, C. (2020). Time-domain analog signal processing techniques. *Journal of Semiconductor Engineering, 1*(2), 64–73.
2. Wagih, M. and Ismail, E. (2015). *Design of Time Based Analog to Digitalconverter (tb-adc): New Designmethodology Forvoltage-to-time-converter (vtc) Circuits*. Faculty of Engineering, Electronics, *11*, 902. Cairo Universitygiza: Egypt.
3. Pekau, H., Yousif, A., and Haslett, J. W. (2006). A CMOS integrated linear-voltage-to-pulse-delay-time converter for time-based analog-to-digital converters. In *IEEE International Symposium on Circuits and Systems,* 14–16 May 2006, Kos, Greece.
4. Elgreatly, A., Dessouki, A., Mostafa, H., Abdalla, R., *et al.* A novel highly linear voltage-to-time converter (VTC) circuit for time-based analog-to-digital converters (ADC) using body biasing. MDPI, Electronics 2020, *9*(12). www.mdpi.com/journals/electronics.
5. Mostafa, H., and Ismail, Y. I. (2013). Highly-linear voltage-to-time converter (VTC) circuit for time-based analog-to-digital converters (T-ADCs). In *IEEE 20th International Conference on Electronics, Circuits, and Systems (ICECS).*
6. Siddiqui, W. H. Voltage-to-time converter based ADC. https://mycourses.aalto.fi/pluginfile.php/1234138/mod_resource/content/1/Waqas_TB_ADC_ELEC_L3520.pdf.
7. Panetas-Felouris, O. (2022). A 3rd-order FIR filter implementation based on time-mode signal processing. https://www.researchgate.net/publication/359243318_A_3rdOrder_FIR_Filter_Implementation_Based_on_Time-Mode_Signal_Processing. Electronics, *11*(6), 902.
8. Stevens, A. E. (1989). A time-to-voltage converter and analog memory for colliding beam detectors. *IEEE, 24*, 1748–1752.
9. Koch, K., Hardel, H., Schulze, R., Badura, E., and Hoffmann, J. (2005). A new TAC based multi-channel front-end electronics for TOF experiments with very high resolution. *IEEE Transactions on Nuclear Science, 52*, 745–747.
10. Flemming, H. and Deppe, H. (2007). Development of high resolution TDC ASIC at GSI. In *IEEE Nuclear Science Symposium Conference Record*, Honolulu, HI, USA.
11. Deppe, H. and Flemming, H. (2009). The GSI event-driven TDC with 4 channels GET4. In *IEEE Nuclear Science Symposium Conference Record (NSS/MIC)*, Orlando, FL, USA, 24 October–01 November 2009.

12. Sasaki, O., Taniguchi, T., Ohska, T. K., Mori, H., Nonaka, T., Kaminishi, K., Tsukuda, A., Nishimura, H., and Takeda, M. (1989). 1.2 GaAs shift register IC for dead time less TDC. *IEEE Transactions on Nuclear Science, 36*, 512–516.
13. Ohsugi, T. and Arai, Y. (1989). TMC-A CMOS time-to-digital converter VLSI for high energy physics. *IEEE Transactions on Nuclear Science, 36*, 528–531.
14. Arai, Y. and Matsumura, T. (1992). A CMOS four channel X 1k time memory LSI with 1ns/b resolution converter. *IEEE Journal of Solid State Circuits, 27*, 359–364.
15. Arai, Y., IKeno, M., and Matsumura, T. (1992). A development of time-memory cell VLSI and a CAMAC module with 0.5 ns resolution. *IEEE Transaction on Nuclear Science, 39*(4), 784–788.
16. Arai, Y. and Ikeno, M. (2007). A time digitizer CMOS gate-array with a 250 ps time resolution. *IEEE Journal of Solid State Circuits, 31*(2), 212–220.
17. Maatta, K., Kostamovaara, J., and Myllylä, R. (1988). Time-to-digital conversion for fast and accurate laser range finding. In *Proceeding of International Congress on Optical Science and Engineering*, Vol. 1010, pp. 60–67.
18. Kostamovaara, J., and Myllylä, R. (1986). Time-to-digital Converter with analog interpolation circuit. *Review of Scientific Instrument*, Juha Kostamovaara, Risto Myllylä, *57*(11), 2880–2885.
19. Maatta, K. and Kostamovaara, J. (1993). Profiling of hot surfaces by pulsed TOF laser range finding application. *Applied Optics, 32*(27), 5334–5347.
20. W. Neagle, W., and Randell, D. R. Surface analysis techniques and applications. Special Publication No. 84. Royal Society of Chemistry, 1990.
21. Kawashima, S. (1995). A traffic condition monitoring by laser radar for advance safety driving. In *Proceeding Intelligent Vehicles Symposium*, Detroit, MI, USA, September 25–26, 1995, pp. 299–303.
22. Karadamoglou, K., Paschalidis, N., Stamatopouosl, N., and Kottarasl, G. (2004). A 32-bit high resolution asynchronous time-to-digital converter for space instruments, In *IEEE Aerospace Conference Proceeding*, Vol. 1010, pp. 2398–2403.
23. Deok-Woo Park, Shang-Yoon Hwang , Chungyong Kim and Gyu-Sik Kim. (2017). A study on an ultrasonic flow meter international journal of innovative science, *Engineering & Technology*, *4*(8), August 2017.
24. Swann, B. K., Blalock, B. J., Clonts, L. G., Binkley, D. M., Rochelle, J. M., Breeding, E., and Baldwin, K. M. (2004). A 100 ps time-resolution CMOS time-to-digital converter for positron emission tomography imaging applications. *IEEE Journal of Solid-State Circuits, 39*(11), 1839–1852.

25. Barbosa, A. F., Lima, H. P., deAndradeFilho, L. M., Lima, H. P., Barbosa, A. F. (2004). A TDC based system for X-Ray imaging detectors. *IEEE Transactions on Nuclear Science*, *52*(4), 932–937.
26. Pekau, H., Yousif, A., and Haslett, J. W. (2006). A CMOS integrated linear voltage-to-pulse-delay-time converter for Time based analog-to-digital converters. In *IEEE International Symposium on Circuits and Systems*.
27. Staszewki, R. B., Muhammad, K., Leipold, D., Hung, C.-M., Yo-Chuol Ho, Y.-C., and Wallberg, J. L. (2004). All digital TX frequency synthesizer and discrete time receiver for blue-tooth radio in 130 nm CMOS. *IEEE Journal of Solid State Circuits*, *39*(12), 2278–2291.
28. Ramakrishan, V., Balsara, P. T. (2006). A wide range, high resolution, compact CMOS time-to-digital converter. In *Proceeding of 19th International Conference on VLSI Design* (VLSI'06), Hyderabad, India, January 3–7, 2006, pp. 197–202.
29. Hsiao, M. J., Huang, J. R., and Chang, T. Y. (2004). A built-in parametric timing measurement unit. *IEEE Design and Test of Computers*, *21*(4), 332–330.
30. Park, K. and Park, J. (1998). 20 ps resolution time-to-digital converter for digital storage oscilloscope. In *IEEE Nuclear Science Symposium*, Toronto, ON, Canada, Vol. 2, pp. 876–881.
31. Park, K., Keunoh Park; Jaehong Park (1999). Time-to-digital converter of very high pulse stretching ratio for digital storage oscilloscope. *Review of Scientific Instruments*, *70*, 1568–1574.
32. Staszewski, R., Vemulapalli, S., Vallur, P., Wallberg, J., and Balsara, P. (2006). 1.3 V and 20 ps time-to-digital converter for frequency synthesis in 90-nm CMOS. *IEEE Transactions on Circuits and Systems II: Express Briefs*, *53*(3), 220–224.
33. Connor, V. O. and Phillips, D. (1984). *Time Correlated Single Photon Counting*. Academic Press: London.
34. Lata, K. and Manoj Kumar. (2013). All digital phase-locked loop (ADPLL): A survey 2013, *International Journal of Future Computer and Communication.*
35. Chen, P., Poki Chen, Chun-Chi Chen, Chin-Chung Tsai, Wen-Fu Lu. (2005). A TDC based CMOS smart temperature sensor. *IEEE Journal of Solid State Circuits*, *40*(8), 1642–1648.
36. Stellari, F., Zappa, F., Cova, S., and Vendrame, L. (1999). Tools for non-invasive optical characterization of CMOS circuits. In *Proceedings of IEDM*, Washington, DC, USA. pp. 487–490.
37. Legrele, C., and Chaminade, R. (1983). A one nano-second resolution time interval counter. *IEEE Transaction on Nuclear Science*, *NS-30*(1), 1–8.
38. Reverberi, R., Podini, P., and Lenzi, G. (1985). A versatile high quality time-to-digital converter. *Journal Physics-E: Science Instruments*, *18*(12).

39. Rahkonen, T., Kostamovaara, J., and Saynajakangas, S. (1988). ECL and CMOS ASIC devices for measurement of short time interval. *IEEE, 2*, 1593–1596.
40. Porat, D. I. (1973). Review of sub-nanosecond time interval measurement. *IEEE Transaction on Nuclear Science, 20*, 35–51.
41. Kalisz, J. (2004). Review of time interval measurement techniques with picoseconds resolution. *Metrologia, 41*(1), 1–6.
42. Razavi, B. (2001). *Design of Analog CMOS Integrated Circuits*. McGraw-Hill: NewYork.
43. Tanimori, T. (1989). Design and performance of semi-custom analog IC including two-TAC and two-current integrators for super-Kamiokande. *IEEE Transaction on Nuclear Science, NS36*, 497–501.
44. Ruotsalainen-Raisanen, E. (1999). A Bi-CMOS time-to-digital converter with 30ps resolution. In *Proceedings of the IEEE International Symposium on Circuits and Systems*, Orlando, FL, USA, 30 May–02 June 1999, Vol. 1, pp. 278–281.
45. Chen, P., Chen, C., Member, S., Zheng, J., and Shen, Y. (2006). A low cost low power CMOS time-to-digital converter based on pulse stretching. *IEEE transaction on Nuclear Science, 53*, 2215–2220.
46. Xin-Gang Wang, P., Yang, H.-G., *et al.* (2006). Successive approximation time-to-digital converter based on Vernier charging method. *IEICE Electronics Express, 11*, 1–6.
47. Gerds, E. J., Member, S., Spiegel, J. V., *et al.* (1994). A CMOS time-to-digital converter IC with two-level analog CAM. *IEEE Journal of Solid State Circuits, 26*, 1068–1076.
48. Sun, W., Chang, C., and Wang, N. (1994). A time-to-amplitude conversion chain for time interval measurement. *IOP Science Measurement and Technology, 11*(3), 1–8.
49. Farshidi, E. and Ebrahim FARSHIDI (2015). A novel cyclic time to digital converter based on triple-slope interpolation and time amplification. *Radioengineering, 24*(3), 800–807.
50. Gust, J. C. (2009). Stop watch and timer calibration. National Institute of Standards and Technology, Special Publication 960–12, U.S. Government Printing Office, Washington.
51. Sun, W., Chang, C., and Wang, N. (1994). Time interval averaging. Hewlett-Packard, Application Note: 162-1.
52. Raisanen-Ruosalainen, E., Rahkonen, T., and Kostamovaara, J. (2000). An integrated time-to-digital converter with 30 ps single shot precision. *IEEE Journal of Solid State Circuits, 35*, 1507–1510.
53. Kalisz, J., Szplet, R., and Pelka, R. (2000). Interpolating time counter with 200-ps resolution. *IEEE Transaction on Instrumentation and Measurement, 49*(4), 879–883.

54. Nissinen, I., Mantyiemi, A., and Kostamovaara, J. (2003). A CMOS time-to-digital converter based on a ring oscillator for A LASER RADAR. In *Proceedings of the 29th European on Solid-State Circuits Conference (ESSCIRC)*, Estoril, Portugal, September 16–18, 2003, pp. 469–472.
55. Martin, K. (2002). *Digital Integrated Circuit Design, Differential CMOS Circuits*, Chapter-5. Oxford University Press: New York.
56. Rahkonen, T., Kostamovaara, J., and Synjkangas, S. (1989). Time interval measurements using integrated tapped CMOS delay lines. In *Proceedings of 32nd Midwest Symposium on Circuits and Systems*, Urbana, IL, USA, Vol. 1, pp. 201–205.
57. Rehkonen, T. (1995). Circuit techniques and integrated CMOS implementation for measuring short time intervals. University of Tampere, Tampere, Finland, pp. 1297–1297.
58. Rahkonen, T. E. and Kostamovaara, J. T. (1993). The use of stabilize CMOS delay line for the digitization of short time-interval. *IEEE Journal on Solid State Circuits*, *28*, 3–10.
59. Pelgrom, M. J. M., Duinmaijer, A. C. J., and Welbers, A. P. G. (1989). Matching properties of MOS transistors. *IEEE Journal of Solid State Circuits*, *24*(5), 1433–1440.
60. Mota, M. (2000). Design and characterization of CMOS high-resolution time-to-digital converters. Ph.D. Thesis, Microelectronics Group, CERN, Geneva.
61. Mozsary, T. R., Andras, M., Chung, J.-F., and Rodriguez, A. (2006). Bio inspired 0.35 μm CMOS time-to-digital converter with 29.3ps lsb. In *32-European Solid State Circuit Conference*, Montreux, Switzerland, September 19–21, 2006, pp. 170–173.
62. Ljuslin, C. and Christianses, J. (1994). An integrated 16-channel CMOS time-to-digital converter. *IEEE Transaction on Nuclear Science*, *41*, 1104–1108.
63. Herve, C. and Torki. K. (2002). A 75ps RMS time resolution Bi-CMOS time-to-digital converter optimized for high rate imaging detectors. *Nuclear Instruments and Methods in Physics Research A*, *481*(1), 566–574.
64. Jansson, J. P., Mntyniemi, A., and Kostamovaara, J. (2009). Synchronization in a multilevel CMOS time-to-digital converter. *IEEE Transactions on Circuits and Systems I: Regular Papers*, *56*(8), 1622–1634.
65. Bigongiari, F. and Roncella, R. (1999). A 250-ps time-resolution CMOS multi-hit time-to-digital converter for nuclear physics experiments. *IEEE Transaction on Nuclear Science*, *46*, 73–77.
66. Arai, Y. (2001). A multi-hit time-to-digital converter VLSI for high-energy physics experiments. In *Proceedings of the Asia and South Pacific-Design Automation Conference*, 30 January 2001, pp. 5–6.

67. Rabey, J. M., Chandrakasan, A., and Nikolic, B. (2003). *Digital Integrated Circuits — A Design Perspective* (2nd edn.), Chapter 4. Pearson: Upper Saddle River, NJ, USA.
68. Yang, C. (2006). RC-delay chain-based RF time to digital converter. In *Proceeding of IEEE, International Conference on Mechatronics and Automation*, Luoyang, China, June 25–28, 2006, pp. 170–173.
69. Li, L., Zhou, W., Wang, F., Ou, X., Ning Ding, N., Chengzhi Zou, C., and Yu, M. (2007). A time-to-digitalconverter based on time-space relationship. In *Frequency Control Symposium, Joint with the 21st European Frequency and Time Forum IEEE International*, Geneva, Switzerland, 29 May–1 June 2007, pp. 815–819.
70. Kalisz, J. and Szplet, R. (1997). Field programmable gate array based time-to-digital converter with 200ps resolution. *IEEE Transaction on Instrumentation and Measurement*, *46*, 51–55.
71. Dudek, P., Szczepanski, S., and Hatfield, J. V. (1997). A high resolution CMOS TDC utilizing a Vernier delay line. In *Proceedings of Euro-sensors XI Conference*, Warsaw, Poland. Published in: *IEEE Journal of Solid-State Circuits*, *35*(2), 240–247.
72. Szplet, R., and Czuba, A. (2021). Two-stage clock-free time-to-digital converter based on Vernier and tapped delay lines in FPGA device, *Electronics 2021, 10*(18), 2190.
73. Vercesi, L., Liscidini, A., and Castello, R. (2009). Two dimensional Vernier time-to-digital converter. In *IEEE Conference on Custom Integrated Circuits*, San Jose, California, September 13–16, 2009.
74. Yu, J., Dai, F. F., and Jaeger, R. C. (2009–2010). A 12-bit Vernier ring time-to-digital converter in 0.13 μm CMOS technology. In *IEEE Symposium on VLSI Circuits, Digest of Technical Papers*, Vol. 3.
75. Abas, A. M., Russell, G., and Kinniment, D. J. (2004). Design of sub 10-picoseconds on-chip time measurement circuit. In *Proceeding of Design Automation Test Europe Conference*, Paris, France, February 16–20, 2004, Vol. 2, pp. 804–809.
76. Xie, D. K. and Zang, Q. C. (2005). Cascading delay Line TDC with 75ps resolution and reduced number of delay cells. *Review of Scientific Instruments*, *76*(1), 014701–014707.
77. Hwang, C., Member, S., Chen, P., and Tsao, H. (2004). A high-precision time-to-digital converter using two-level conversion scheme. *IEEE Transaction on Nuclear Science*, *51*(4), 1349–1352.
78. Mantyniemi, A., Rahkonen, T., and Kostamovaara, J. (1999). A high resolution digital CMOS time-to-digital converter based on nested delay lock loop. In *Proceedings of IEEE International Symposium on Circuits and Systems*, Vol. 2, pp. 537–540.

79. Waizman, A. (1994). A delay lock loop for frequency synthesis of de-skewed clock. In *IEEE International Solid-state Circuits Conference Digest of Technical Papers*, pp. 289–299.
80. Farjad-Rad, R., Dally, W., Hoik-Tiaq Ng., Poulton, J., Stone, T., Rathi, R., Lee, E., and Huang, D. (2002). A 0.2-2 GHz 12mW multiplying DLL for low-jitter clock synthesis in highly-integrated data communication chips. In *IEEE International Solid-state Circuits Conference Digest on Technical Papers*, San Francisco, CA, USA, February 7, 2002, Vol. 1, pp. 76–77.
81. Precision, T. S., Jansson, J., Mantyniemi, A., and Kostamovaara, J. (2006). A CMOS time-to-digital converter with single shot resolution better than 10 ps. *IEEE Journal of Solid-State Circuits*, *41*, 1286–1296.
82. Kostamovaara, J. (1995). Low-power CMOS time-to-digital converter. *IEEE Journal of Solid-State Circuit*, *30*, 984–990.
83. Chen, P. (1997). A low power high accuracy CMOS TDC. In *IEEE International Symposium on Circuits and Systems*, Hong Kong, China, June 12, 1997.
84. Sambles, R. (1997). Highly accurate cyclic CMOS time-to-digital converter with extremely low power consumption. *IEEE Electronics Letters*, *33*, 858–860.
85. Liu, S. and Chen, P. (1999). A cyclic CMOS time-to-digital converter with deep sub-nanosecond resolution. In *Proceedings of the IEEE on Custom Integrated Circuits*, San Diego, California, May 16–19, 1999, pp. 605–608.
86. Chen, C., Chang, W., and Chen, P. (2004). A precise cyclic CMOS time-to-digital converter with low thermal sensitivity. In *IEEE Nuclear Science Symposium Conference Record*, Rome, Italy, October 16–24, 2004, Vol. 3, pp. 1364–1367.
87. Christiansen, J. (1995). An integrated CMOS 0.15 ns digital timing generator for TDC's and clock distribution system. *IEEE Transaction on Nuclear Science*, *42*, 753–757.
88. Mota, M. (1998). A four channel self calibrating high resolution time-to-digital converter for TDC's and clock distribution system. In *IEEE International Conference on Electronics, Circuits and Systems*, Instituto Superior Tdcnico Lisboa, Portugal, September 7–10, 1998, pp. 409–412.
89. Mota, M. and Christiansen, J. (1998). A high-resolution time interpolator based on a delay locked loop and an RC delay line. *IEEE Journal of Solid State Circuits*, *34*(10), 1360–1366.
90. Mota, M. and Christiansen, J. (1999). A high-resolution time interpolator based on a delay locked loop and an RC delay line. *IEEE Journal of Solid State Circuits*, *34*(10), 1360–1366.

91. Christiansen, J., Marchioro, A., Moreira, P., and Mota, M. (2000). A data driven high performance time to digital converter. https://cdsweb.cern.ch/record/478865/files/p169.pdf.
92. Gorbics, M. S., Kelly, J., Roberts, K. M., and Sumner, R. L. (1998). A high resolution multi-hit time-to-digital converter integrated circuit. *IEEE Transactions on Nuclear Science*, *44*(3), 379–384.
93. Rashid, W. C. and Ahmadi, M. (2009). Short time interval measurement using CMOS time amplifier. In *Canadian Conference on Electrical and Computer Engineering*, St. John's, Newfoundland, Canada, May 3–6, 2009, pp. 000321–000324.
94. Rashid, W. C. and Ahmadi, M. (2009). A delay generation technique for narrow time interval measurement. *IEEE Transactions on Instrumentation and Measurement*, *58*(7), 2245–2252.
95. King, B. R. D. (1971). Two vernier time-interval-digitizer. *IEEE Transaction on Nuclear Instrumentation and Methods*, *97*, 359–370.
96. Chen, P., Chen, C., Member, S., and Shen, Y. (2007). PVT sensitive Vernier based time-to-digital converter with extended input range and high accuracy. *IEEE Transaction on Nuclear Science*, *54*, 294–302.
97. Hari Prasad, K., Chandratre, V. B., and Saxena, P. (2011). FPGA based time-to-digital converter. In *Proceedings of Nuclear Physics Symposium*, Andhra University, Visakhapatnam, India, December 26–30, 2011, Vol. 54, pp. 294–302.
98. Hari, K., Sukhwani, M., and Saxena, P. (2013). A CMOS standard cell based TDC. In *the International Conference on VLSI, Communication, Analog Designs, Signals & Systems and Networking (VCASAN-2013)*, Institute of Technology, Bangalore, India, July 17–19, 2013.
99. Hari, K., Sukhwani, M., Saxena, P., Chandratre, V. B, and Pithawa, C. K. (2014). A four channel time-to-digital converter ASIC with in-built calibration and SPI interface. *Nuclear Instrumentation and Measurement A*, *737*, 117–121.
100. Park, Y. and Wentzloff, D. D. A cyclic vernier time-to-digital converter synthesized from 65 nm standard celllibrary. In *Proceedings of 2010 IEEE International Symposium on Circuits and Systems*, Paris, France, 30 May–2 June 2010.
101. Lin, C.-H., and Syrzycki, M. (2011). Single stage Vernier time-to-digital converter with sub-gate delay time resolution. *Circuits and Systems, Scientific Research*, *2*, 365–371.
102. Mantyniemi, A., Rahkonen, T., and Kostamovaara, J. (2009). A CMOS TDC based on cyclic time domain successive approximation interpolation method. *IEEE Journal of Solid Statephysics*, *44*(11), 3067–3078.
103. Chen, C.-C., Hwang, C.-S., Lin, Y., and Chen, G.-H. All-digital pulse-shrinking time-to-digital converter with improved dynamic range. *Review of scientific Instruments*, *87*(4), 046104–046109.

104. Latha, P., *et al.* (2018). Implementation of mux based encoder for time to digital converters architecture. *International Journal of Engineering and Techniques*, *4*(3), 520–526.

105. Maamoun, M., Arami, S., Beguenane, R., Benbelkacem, A., and Meraghni, A. (2017). A 3ps resolution time-to-digital converter in low-cost FPGA for laser rangefinder. In *Proceedings of the World Congress on Engineering*, WCE, London, UK, July 5–7, 2017, Vol. 1.

106. Tewari, A. and Gupta, B. B. (2020). Secure timestamp-based mutual authentication protocol for IoT devices using RFID tags. *International Journal on Semantic Web and Information Systems (IJSWIS)*, *16*(3), 20–34.

107. Gupta, B. B., Yadav, K., Razzak, I., Psannis, K., Castiglione, A., and Chang, X. (2021). A novel approach for phishing URLs detection using lexical based machine learning in a real-time environment. *Computer Communications*, *175*, 47–57.

108. Jain, A. K. and Gupta, B. B. (2018). Two-level authentication approach to protect from phishing attacks in real time. *Journal of Ambient Intelligence and Humanized Computing*, *9*, 1783–1796.

109. Sahoo, S. R. and Gupta, B. B. (2021). Real-time detection of fake account in twitter using machine-learning approach. In *Advances in Computational Intelligence and Communication Technology: Proceedings of CICT 2019*, pp. 149–159. Springer: Singapore.

110. Agrawal, P. K., Gupta, B. B., Jain, S., and Pattanshetti, M. K. (2011). Estimating strength of a DDoS attack in real time using ANN based scheme. In *Proceedings of 5th International Conference on Information Processing, ICIP 2011, Computer Networks and Intelligent Computing*, Bangalore, India, August 5–7, 2011, pp. 301–310. Springer: Berlin, Heidelberg.

https://doi.org/10.1142/9789811279539_0011

Chapter 11

Design and Development of Multipurpose Agriculture Robot (Agrobot)

Shailendra Singh[†], Puspraj Singh Chauhan[*,‡], Chirag Sethi[§], Ankit Jain[¶], and Suneet Kumar Agnihotri[||]

Department of Electronics and Communication Engineering, Pranveer Singh Institute of Technology, Kanpur, India

[†]*shailendras.nitj@gmail.com*
[‡]*puspraj.chauhan@gmail.com*
[§]*sethichirag646@gmail.com*
[¶]*jaina21@gmail.com*
[||]*agnihotrisk@gmail.com*

Abstract

The cornerstone of the Indian economy is the agricultural sector. A large percentage of individuals around the world — more than 40% — have agriculture as their primary line of employment. The advancement of autonomous vehicles in agriculture has gained considerable spotlight in recent years. Soil quality is unaltered by robotization, which also significantly reduces manpower expenses and boosts overall productivity. The machinery described in this project is proficient at performing several farming operations at once. An environmentally safe source of energy is utilized by this equipment: Energy from the sun's radiation is

[*]Corresponding author.

accumulated and transformed into electrical energy, and a direct current motor then transforms this electrical power into mechanical power to rotate an anchor during a digging operation. For irrigation operations and seed sowing, respectively, we deploy seed hoppers and water reservoirs. The machine can change its course and maneuver over varied ground contours, excavate, sow seed, and water the ground. It also decreases the expense of planting seeds and the manpower required.

Keywords: Microcontroller, Bluetooth, motor driver L298, agriculture purpose

11.1. Introduction

The limited availability of skilled manpower in the agriculture sector is currently widespread, which hinders the rates of growth of developing nations that depend heavily on the sector, such as India. As India's population is growing, so does the demand for food, resulting in increased agricultural yields per acre. To tackle these issues, farmers should embrace the most recent technological advancements for various farming techniques, including digging, planting, and irrigation, that are more efficient and require less time. The main task of the sowing process is to sow seeds at the appropriate depth and with the proper spacing between two implanted seeds. This is achievable with the intervention of an agricultural robot (agrobot) capable of digging the trench and sowing the seeds. It should then cover the sown seeds with dirt and spray water after the seeds are set in the furrow field [1]. This agrobot minimizes labor and saves time, resulting in substantial savings and the assurance of accurate seed spreading.

Sixty years after their inception in industrial workplaces, robotic machinery are finding their way into our daily lives. The deployment of robots in agricultural activities, such as plowing, seed dispersal, and harvesting, will result in a shift in our conventional way of agriculture practice. In turn, this enhances the rate of productivity, yielding better outcomes. In India, human labor is traditionally utilized for hauling, lifting, weed control, and fruit harvesting. Due to the lack of utilization of robots in the agriculture sector, farmers are currently unable to get tangible benefits. So, these strategies must be implemented in agriculture. When spraying pesticides and herbicides in farming, farmers are confronted with hazardous environmental conditions. Fields are compacted by tractors because they are powerful, confined, and unable to maneuver challenging terrain [2, 3]. The agrobot has the potential to help farmers in many ways. Robots can function in any environment, irrespective of the weather, in compliance with an automated program that tells them to implement the appropriate exercise. The greatest advantage of using robots in agriculture is their light weight, which prevents them from compressing the soil in the field. Self-guided machinery, such as harvesters, has supported farmers for some time. More recently, autonomous multi-functional machinery have emerged and been deployed in orchards, groves, nurseries, and other agricultural contexts to automate or supplement tasks such as trimming, thinning, harvesting, mowing, and spraying. The robotics communities' capability to overcome technical, economical, and regulatory barriers to deployment and adoption will be crucial to the successful commercialization of such vehicles. The chairman of the Robotics and Automation Society invited both academic and industrial robotics experts as well as farming investigators, engineers, practitioners, and students to build a future in which robots empowered safe, productive, and reasonably priced food productivity. The presenters proposed an agricultural-robotic machine capable of conducting all agricultural tasks autonomously while requiring little manual labor and dramatically enhancing the output. Not only can this self-driving agricultural robot enhance output, but it can also monitor ecological variables, such as climatic conditions and the humidity of soil [4]. Using this method for modeling a robot would alter conventional agricultural techniques and bring in a modern age of agriculture. The presenters demonstrated the operation of the robot, which involves live harvesting of crops and fertilization using two applications. The current objective is to develop devices that can operate in a natural environment. They used the sun's radiation as the energy source, which is more practical than other renewable energy sources.

Bini *et al.* [5] presented a cohesive concept of precision farming using autonomous surface vehicles and drones. The system was segmented in accordance with the agricultural environment in order to govern the vehicles' itinerary planning, task execution, and detection. Researchers are attempting hybrid methods in precision farming using machine vision and machine learning to enhance the effectiveness of the automated systems. The greatest obstacle to this technique is the erroneous detection caused by leaves and branches in a field.

An innovative aspect of the studies by Kayacan *et al.* [6, 7] is the proposal of a learning algorithm, based on the theory of sliding mode control, for the tuning of a parameter whose impact is assessed on a real-time basis for the management of the yaw dynamics of a robotic tractor.

In lieu of requiring the designer to choose a set value for the aforementioned parameter, this study enables the system to modify it live, thus adding a level of flexibility to the overall controller. Real-time findings demonstrate that when T2FNN is used in accordance with a conventional controller design, the system in its entirety learns the system dynamics to achieve superior outcomes in terms of a shorter settling time and near-zero steady-state error.

An agrobot may be deployed to plant certain types of seeds into the soil, which increases crop production overall. Depending on the biological conditions of a particular area, many characteristics, such as climatic conditions including temperature and humidity, may be adjusted [8]. This intelligent agricultural IoT deployment may enhance crop yields in agricultural fields, and cloud-based agriculture can examine and store data for future use.

The objective of the studies by Durmuş *et al.* [9, 10] is to improve agronomic productivity by making a remote, autonomous machine capable of processing and tracking operations in the fields, such as sprinkling remedies for precision farming, fertilization, diagnostic testing, analyzing production, soil, and other farmland practices. The primary barriers are dependability and longevity under field conditions, as well as reducing robot unit costs while ensuring massive yields.

Thus, the proposed technique has the benefit of recognizing the afflicted crop and permitting robots to spray prescribed pesticides based on attributes and segments. It is considerably more challenging to control and diagnose plant diseases manually. Consequently, managing diseases in plants with harmful pesticides necessitates a significant amount of effort and the assistance of field specialists, and meeting fertilization requirements takes a great deal of time. Therefore, an image processing technique is suggested for identifying and classifying plant ailments [11].

11.2. Agrobot Design Model

The harvesting component of the agrobot is located at the front. The cutter consists of a solar panel and metal sheet and is operated by a direct current (DC) motor at high speed. We employed the digging mechanism in the end frame's rear section, where a tooth-like iron pin is utilized for cultivating and can be turned into a robotic arm. The cylindrical framework is used to sow or drop seeds in the field. The seed is placed in the fields via the cylindrical form construction, and a hole is drilled on the shaft to excavate earth [12, 13]. This is constructed of acrylic sheet and is powered by a DC motor. A PCB board was employed to build a circuit that connected all of the equipment and regulated the whole farming activity. The main component employed is an ATmega328p microcontroller, popularly known as Arduino. A motor driver IC L2938 is capable of driving two DC motors simultaneously is included in the circuit, as is a power-backed circuit model that contains a voltage lever from 5 to 12 V to control electricity and the sun's radiation to power the agriculture robot. An Arduino Uno board is utilized for logical programming and multitasking. We utilized a separate sprinkler system for irrigation. It is connected to a humidity sensor, which monitors the moisture level of the soil.

11.3. Agrobot Mechanics and Norms of Operation

The agrobot operates in three forms:

Step 1: This mechanism is concerned primarily with sowing seeds into the soil.
Step 2: This mechanism focuses on reaping yields.
Step 3: This mechanism is established for robotic digging.

Let's have a brief look at the three working modes as follows:

Step 1: We describe the process of dispersing seeds into the soil while keeping an eye outside for roadblocks in this mode, we must incorporate a DC motor with high rotations per minute (RPM) [14]. The motor is affixed to the robot's gear, where the container is mounted. An Arduino microcontroller code configures the container flasks to shake at an imminent moment. It controls the rate of seed dispersion, as depicted in Figures 11.1(a) and 11.1(b).

Step 2: We must keep a close eye on crop harvesting in this mode. Figures 11.1(c) and 11.1(d) depict the circuit model with the Arduino and the solar panel. When crops are harvested, cylindrical rods are employed to gather and collect them. The thresher is driven by the RPM of the DC motor, and the speed and other attributes are regulated by code.

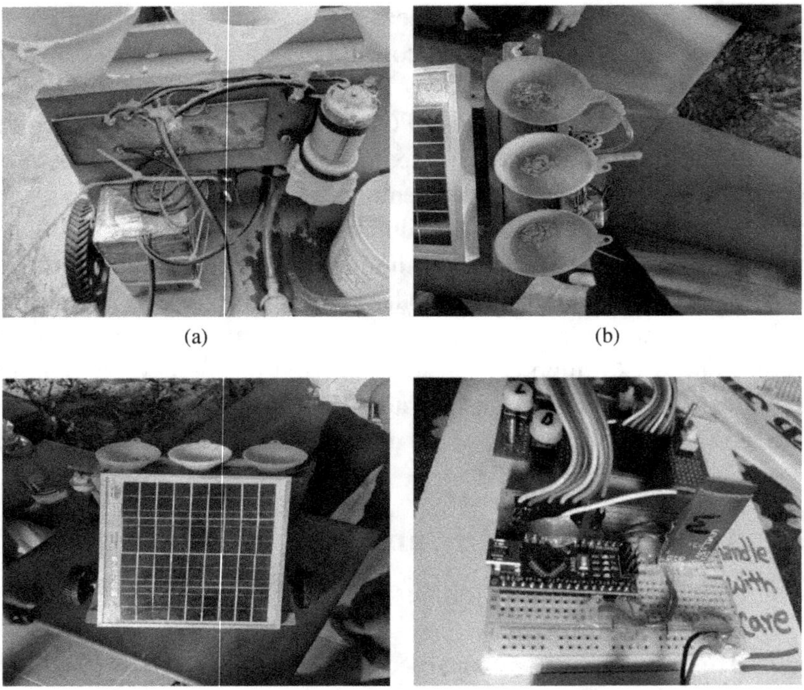

(a) (b) (c) (d)

Figure 11.1. (a) High-RPM DC Motor is Affixed, (b) Shaking of the Flasks Storing the Seeds, (c) Robot Equipped with Solar Panel, (d) Arduino Uno Mounted on the Surface to Control Operations

Step 3: In this mode, we observe the plowing activity of the robot. A plow is constructed of crisscrossed teeth that are linked to the rear of the robot, which digs the land, enabling the seeds to be sown in a straight line into the soil [15].

11.4. Circuit Interconnections

An additional circuit is necessary to connect the motors and water pump to the Arduino because these components require 12 V to operate while the Arduino only generates 5 V. Particularly, we employed a preboard and NPN power transistors designated as TIP122 [16–18]. The above-described circuit is sufficient for operating the robot. To alleviate the workload on the

Figure 11.2. Circuit Interconnections

Arduino's power pins, we also patched them with a male–female connector system in order to provide +5 V and ground rails for transmitting electricity. Figure 11.2 shows the two ends of the solder, one each for +5 V and ground.

Figures 11.3 and 11.4 shows transmitter and receiver circuit of the given module. The following are the connections for the circuit:

(1) The driver motors should be attached to the driver board, and the following pin connections should be made on the driver board:

- Signal — D6 on the Arduino board
- Signal — D9 on Atmega328p
- Signal — D10 on Atmega328p
- Signal — D11 on Atmega328p
- +5 V corresponds to +5 V on Atmega328p
- +5 V corresponds to +5 V on the Arduino board
- +12 V (motors will operate at this voltage) — batteries will be attached later on

(2) The Bluetooth connections comprise the following:

- Vcc — +5 V on Atmega328p
- Vcc — +5 V on Atmega328p
- Vcc — +5 V on Atmega328p
- Ground — Ground on Atmega328p
- Receiver — Transmitter on Atmega328p
- Transmitter — Receiver on Atmega328p
- +12 V is supplied straight from the batteries to the transistor circuit.

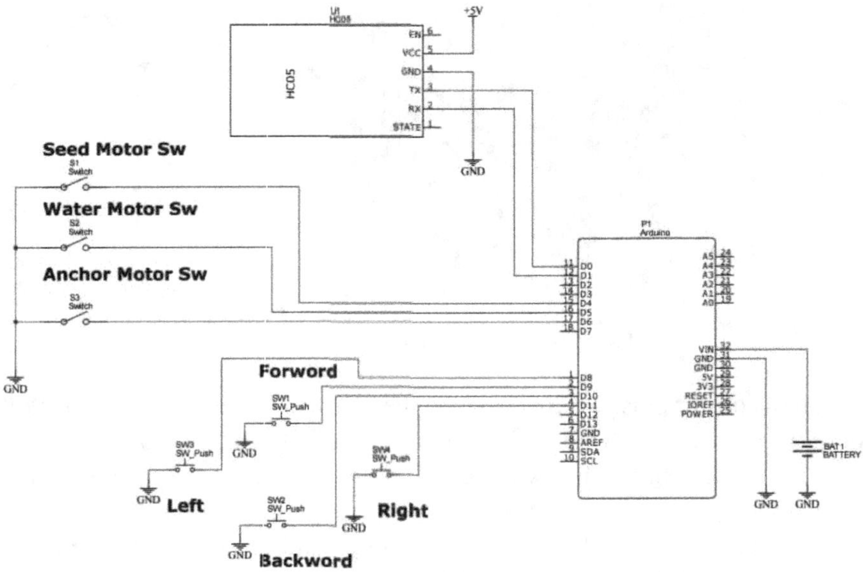

Figure 11.3. Transmitter Circuit

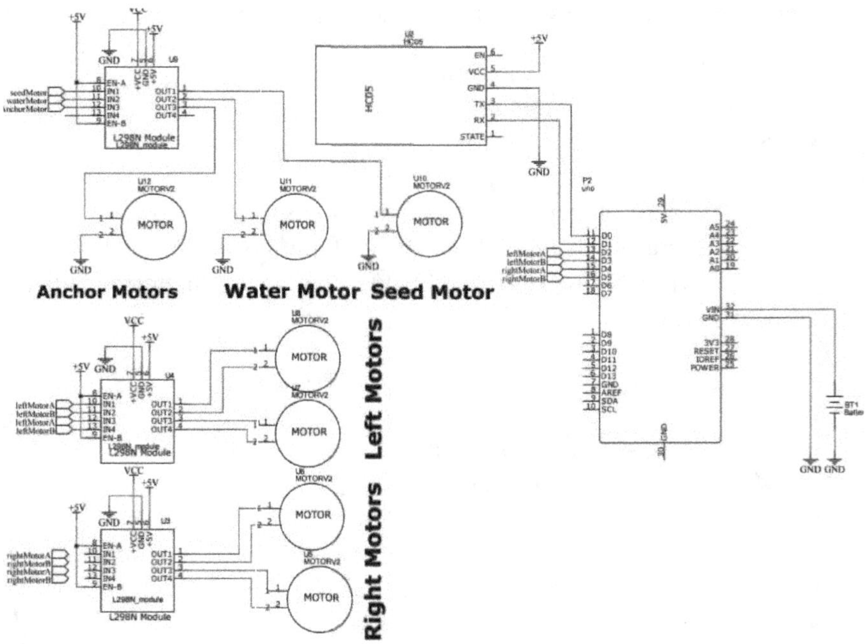

Figure 11.4. Receiver Circuit

(3) Then, 12 V will be supplied to the Arduino's Vcc and the driver circuit of the motor, and servo motors are also attached in the following manner [37–45].

(4) The connections of the servo motors are as follows:

- Vcc to +5 V on Atmega328p
- Ground to Ground on Atmega328p
- Signal to D3 on Atmega328p

(5) The reusable acid battery and the Arduino Uno are loaded with the code and are powered by a 12 V lead.

(6) The Arduino is configured and installed, and the controller application that utilizes Bluetooth for wireless communication along with the Arduino is configured and deployed [46–52].

11.5. Methodology

The entire robot system is powered by a battery and is managed by a remote control [19]:

- In this device, solar energy is collected by a solar panel and converted into electrical energy, which then supplies the required power to the DC motor [53–60].
- The robot's core structure is built with four interconnected wheels that are powered by a DC motor.
- An anchor is attached to the frame's front end, so it can dig into the ground.
- A plastic funnel is used to store seeds, which then pass through the funnel through a hole drilled in the shaft and into the ground [20].
- The water is sprayed from the hose for cultivation after the seed has already been sown.
- A 12 V battery is required to power the complete robot system.
- The robot's operations are controlled by the Arduino Uno board [61–68].
- A prototype drilling and seed-sowing system is implemented using the few resources that are available.
- Using cutting-edge methods, the system may be subjected to further development. Table 11.1 lists the components required.

Table 11.1. Circuit Components Used for Agrobot

S. No.	Name
1	12 V battery
2	9 V battery
3	Arduino Nano
4	Arduino UNO_R3
5	10 W solar panel
6	Toggle switch
7	Push switch
8	HC05 Bluetooth
9	L298N_Module (motor driver)
10	Motor for vehicle, seed motor, anchor motor
11	Motor clamp
12	Water pump
13	Hard board

11.6. Description of Hardware

11.6.1. *Microcontroller Arduino Mini*

The ATmega328 is used to make the Arduino Mini, as shown in Figure 11.5, which is a board with a microcontroller. It has 14 digital input/output pins, six analog inputs, a 16 MHz ceramic resonator, a USB connection, a power jack, an ICSP header, and a reset button. Six of the digital input/output pins can be used as PWM outputs [69–74]. It comes with everything needed to support the microcontroller [21, 22]. It is connected to a computer with a USB cable or powered with an AC-to-DC adapter or a battery to get started. The Mini is different from all the boards that came before it because it doesn't use the FTDI USB-to-serial driver chip. Instead, it has an Atmega16U2 (or an Atmega8U2 up to version R2) that has been programmed as a USB-to-serial converter. The name "Mini" comes from the Italian word for "one." It was chosen to mark the release of Arduino 1.0 [75–82].

Each of the Mini's six pin inputs, which are labeled A0–A5, has a resolution of 10 bits (i.e., 1024 different values). By default, they measure

Figure 11.5. Arduino Mini

Figure 11.6. ATmega328P

from ground to 5 V. However, the AREF pin and the analogue reference function can be used to change the upper end of their range. Also, some pins are made to perform specific functions [83–88].

11.6.2. *Atmega328p Microcontroller*

Based on the AVR enhanced RISC architecture, the ATmega328P is an 8-bit CMOS microcontroller which utilizes little power.

The ATmega328P, shown in Figure 11.6, achieves maximum throughput nearer to 1 MIPS per MHz by running powerful instructions in a single clock cycle [23, 24]. This allows the system to be designed to make the best use of both power and processing speed [89–95].

11.6.3. *Servo Motor*

A servo rotatory actuator is a device controlled by electricity that has the capability of pushing or rotating an object with great accuracy. Servo

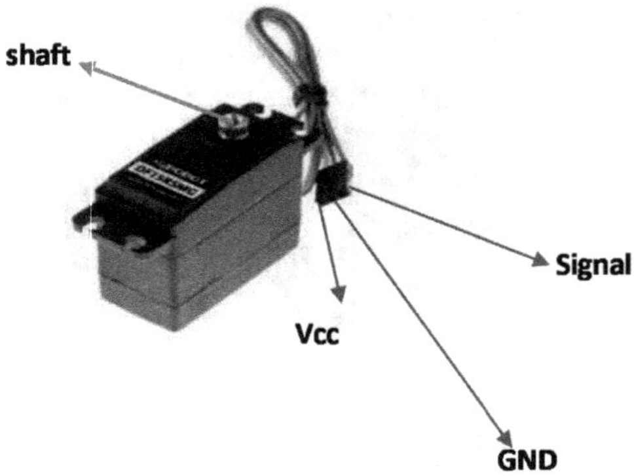

Figure 11.7. Servo Motor

motors, as depicted in Figure 11.7, are used when something needs to be rotated at a certain angle or moved a certain distance [96–102]. It's a simple motor that keeps running on a servo mechanism. If the motor is powered by DC current, it is called a DC servo motor, and if it is powered by alternating current (AC), it is called an AC servo motor. We can get a servo motor that has an exponential torque in a tiny, lightweight package [25]. Owing to these qualities, they are utilized in numerous products, from remote-controlled toys, such as cars, helicopters, and planes, to robotic machinery.

Servo motors are rated in kg/cm, with most rated at 3, 6, or 12 kg/cm. These numbers indicate what the maximum load capacity of your rotary actuator is at a certain separation [103–108]. For example, a 6 kg/cm servo motor is capable of lifting 6 kg if the load is slung 1 cm from the axis of the motor [26]. The wider the gap, the less the load it can carry. A servo motor's position is monitored by an electrical pulse, and its circuitry is installed next to the motor [109–115].

11.6.4. *HC-05 Bluetooth Module*

The HC-05 is a Bluetooth module, shown in Figure 11.8, designed for wireless communication. This module can be employed as either a master or a slave. It is used in a lot of consumer goods, such as wireless headsets, game controllers, and wireless mice and keyboards. It can reach up to about 100 m, depending on the transmitter and receiver, the weather, the terrain, and the locality. It follows a standard protocol called IEEE

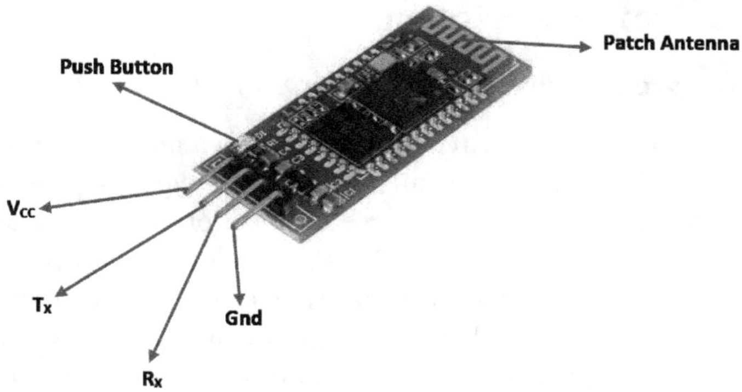

Figure 11.8. HC-05 Bluetooth Module

Figure 11.9. Motor Driver Module L293D

802.15.1 [27]. With it, you can build a wireless personal area network (PAN). It transmits messages over the air with frequency hopping spread spectrum (FHSS) radio technology. It communicates serially with devices, e.g., it talks to the microcontroller through the serial port (USART) [116–124].

11.6.5. *Motor Driver Module L293D*

The motor driver module is an IC with a 16-pin structure. This L293D motor driver IC has four input pins, four output pins, two enable pins, and a 7 V motor power supply, as shown in Figure 11.9 [125–132]:

$$Vcc = 5\text{ V},$$

$$Vss = 7\text{ V}.$$

11.7. Software Description

11.7.1. *Integrated Development Environment*

An integrated development environment (IDE) is a piece of software that provides computer programmers all the tools they require to develop software. Most IDEs have at least a source code editor, tools for automating the build process, and a debugger [28, 29]. There isn't a clear line separating an IDE from the rest of the software development environment. Sometimes, a version control system or tools that make it easier to build a graphic user interface are integrated. For object-based software development, many modern IDEs also have a class browser, an object browser, and a class hierarchal structure diagram [133–138].

11.8. Arduino IDE

The Arduino IDE is open-source software that is mostly used to write the code for the Arduino module and compile it, as illustrated in Figure 11.10. It is an official Arduino software that makes putting together code so easy that even a person who has no prior technical experience can learn it quickly. It can be easily installed on MAC, Windows, and Linux OS, and it operates through a programming language based on Java, which has

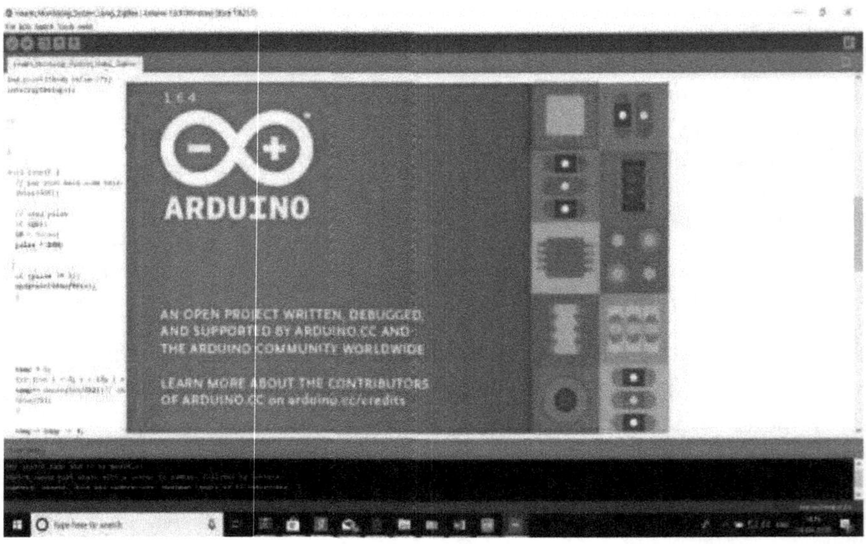

Figure 11.10. Arduino IDE

embedded functionalities and instructions that are very important for troubleshooting, editing, and curating code in the framework [30].

The primary code, which is also called a "sketch," is made on the IDE platform. This creates a hex file, which is then sent to the controller on the board and uploaded. The main parts of the IDE environment are the editor and compiler. The editor is utilized to encode the program, and the compiler is used to compile the code and upload it to the Arduino microcontroller module. Both C and C++ can be used in this environment [139–141].

The Arduino Mini's ATmega328 has been reconfigured with a bootloader that lets you add new code without requiring an external hardware program developer. It utilizes the Genuine STK500 protocol for communication (reference, C header files). You can also skip the bootloader and program the microcontroller through the in-circuit serial programming (ICSP) header. See the instructions in Ref. [31] for more information. The firmware source code for the ATmega16U2 (or 8U2 on the rev1 and rev2 boards) is available. The ATmega16U2/8U2 has a DFU bootloader already on it.

When you open the IDE software after downloading it, it will look as shown in Figure 11.11.

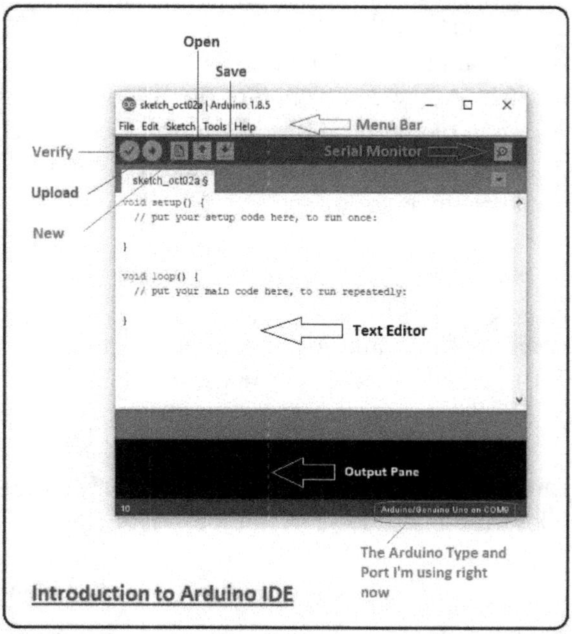

Figure 11.11. Arduino IDE Editor Page

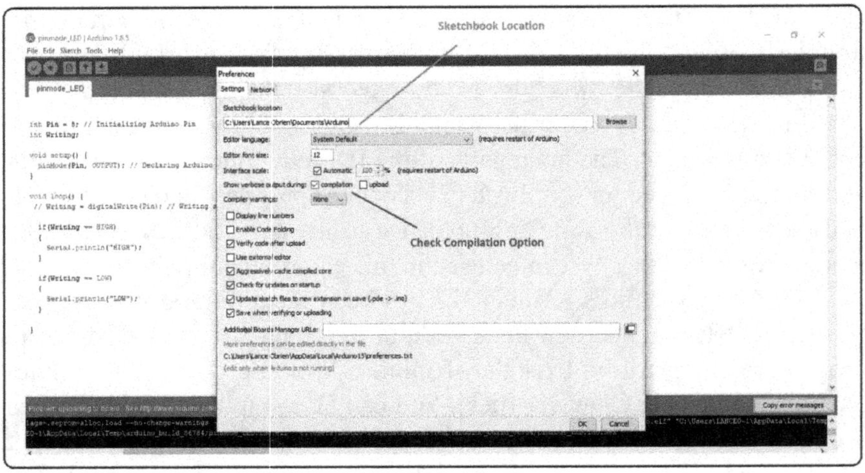

Figure 11.12. Arduino IDE New Window

The vertical strip at the top of your screen is called the Menu bar, and it features the five selections mentioned as follows. Figure 11.12 shows the File selection options: You may either use an existing window or create a new one to enter the code by selecting the File New window. The number of categories within which the file choice fits is depicted in the table that follows.

The code compilation will be shown in the Output pane as soon as you click the upload button in the preferences panel once you have checked the compilation box [32]. At the completion of the compilation process, you'll be able to see the hex file produced for the most recent sketch, which you can then upload to the Arduino board in order to carry out the desired action.

The other menus are as follows:

- The Edit menu is used for copying and pasting the code with further font changes.
- Sketch is used for programming and compilation.
- Tools menu is mostly used for testing projects. The new microcontroller's bootloader may be burned using the programmer part of this panel.
- Help: If you are unsure about using the software, thorough assistance is accessible, from installation to troubleshooting.

Figure 11.13 shows how the six options under the Menu tab fit into the currently active window. The code may be checked by clicking the circular button with the checkmark on it. When you are finished with your

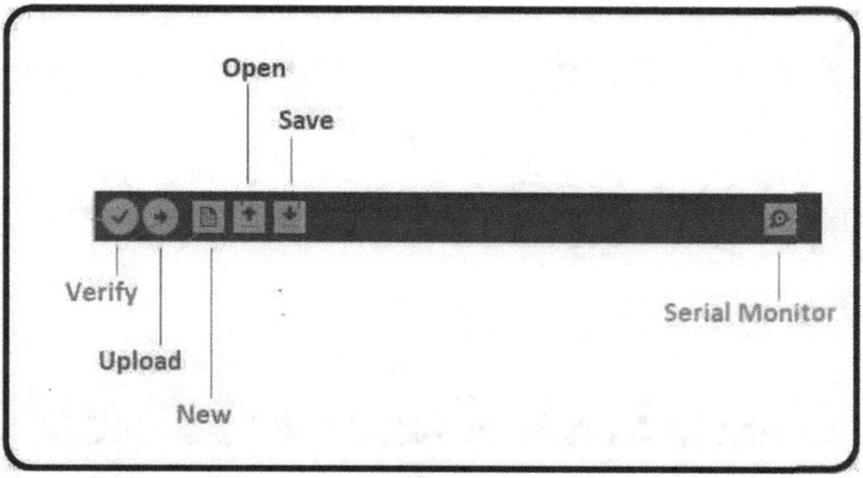

Figure 11.13. Arduino IDE Open and Save Tab

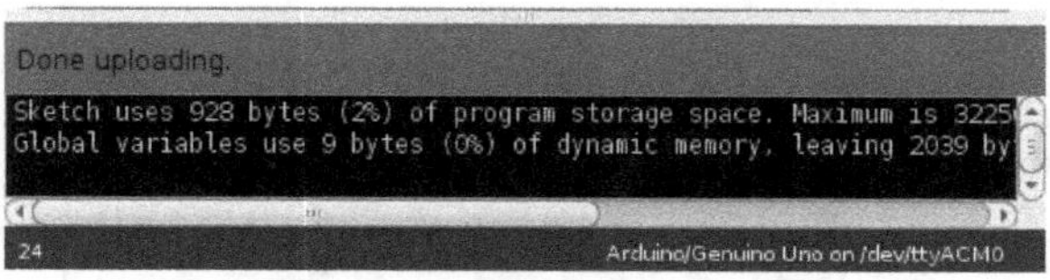

Figure 11.14. Generation of Hex File

code, click here. A program may be transferred to the Arduino board by pressing the Upload option. To create a new file, press the dotted paper icon.

In order to access a previously saved Arduino project, press the up arrow icon. Pressing the down arrow icon will save the code being executed. If you click the button in the upper right-hand corner, a window will show up, which will serve as a terminal for transmitting and receiving serial data; this is the serial monitor [33–35]. Alternatively, you may open the serial monitor by pressing Ctrl+Shift+M simultaneously or by selecting it from the Tools panel. You can get a handle on how your program is doing by using the serial monitor and debugging the sketches you've developed. The serial monitor cannot be opened unless an Arduino module is physically linked to a computer via a USB interface.

You must choose the baud rate of the Arduino board you are currently using. For the Arduino Mini used in this project, the baud rate is 9600; thus, when you enter the following code and click the serial monitor, the result will appear as seen in Figure 11.14 [36].

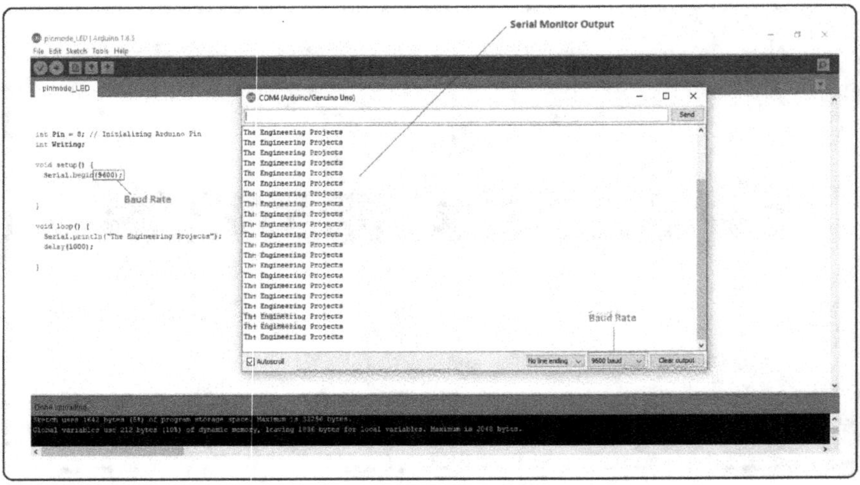

Figure 11.15. Serial Outputs

The bottom portion of the main screen is referred to as the Output pane, which focuses mostly on the compilation state of the running code: the memory utilized by the code and any mistakes that have occurred. Before you plan to upload the hex file to your Arduino module, you must solve these problems, as shown in Figure 11.15.

There are certain specialized libraries used for calling and executing particular functions on the board when using the Arduino C programming language.

11.9. Conclusion

There are a plethora of ways that robots could end up making farming more productive, and more and more robots are showing up on farms in different forms. Most likely, technology can solve the other problems that come with self-driving farm equipment. This kind of equipment might be deployed in the future, but there are important reasons to think that a computer might not just replace the driver. This could mean that the way crops are produced needs to be rethought. It could be easier and cheaper to grow crops with a lot of small machines than it is with a few big ones.

One benefit of smaller machines is that they might be simpler for individuals who do not work on farmlands to become accustomed to. Agricultural production jobs are hard, dangerous, and require intellectual ability and quick, but very monotonous, decisions. This is why robots can be a good replacement for human operators. Machines can pick up on the color, firmness, weight, density, ripeness, size, and shape of the better products. Robots can make our lives better, but they can also hurt us.

Both the software and hardware specifications for Arduino are open source, which means that enthusiasts can put together the simplest Arduino modules by hand. You can buy more complex and challenging pre-built Arduino modules for a reasonable cost. The hardware offers a variety of sizes and shapes, from small devices that can be worn to larger modules that are mounted on the surface. USB is the most common way for computers to connect with each other, but there are also Bluetooth, serial, and Ethernet connections. The software for Arduino is free and can be utilized by anyone. The popular language Wiring is used as the basis for the programming platform. The IDE is based on Processing, a well-known language for designers. Arduino is distinct from most microcontroller integrations in that it can be run on Windows, Linux, and macOS [142–145].

Open-source platforms, such as Arduino makes it simpler for developers to experiment with interactive objectives. Using the Arduino platform, these inventors will be able to quickly make prototypes and test interactive devices before making a product that is ready for mass production. One day, the next Mark Zuckerberg or Steve Jobs might come up with new ways for computers to talk to the real world. Arduino is a great way to try out different ways that smart devices can be utilized.

11.10. Scope for Further Improvements in the Robot

The following are some possible ways our project can be improved:

- The robot can be completely automated by using ultrasonic sensors and developing path-finding mechanisms and algorithms to allow for a fully automated solution.
- The framework can be made stronger so that it can be used in any place and on any scale.

11.11. Scope For Future Work

The purpose of this study was to outline potential future applications and illustrate the present situation of agricultural and autonomous system implementation and trends. Three main categories of field operations — crop establishment, plant care, and selective harvesting — were recognized as the first possible practical applications for autonomous vehicles in agriculture. These applications were evaluated and contrasted with traditional systems.

Our goal was to create a multifunctional farm robot prototype that can carry out the following tasks:

- To operate as a robot truck that could plunge into the soil, plant seeds, and spray water. All of the robot's equipment will be powered by batteries and solar power.
- To deploy a tiny robot to reduce human labor in the agriculture sector.
- To conduct all three actions at once, increasing output and saving time.
- To perform a significant amount of work in a shorter amount of time.
- To allow the farmer to control this robot remotely while observing from the side.

The potential for labor price reductions, the consequences of farm design and size for operations, routine working hours, potential environmental effects, expenditures on energy, and concerns regarding security were all considered. However, our current project is capable of numerous improvements and enhancements. Some of them are listed as follows:

- Feedback sensor nodes, such as optical encoders, can be added to enhance the robot's ability to navigate its way around.
- A display can be added to show what steps have been taken.
- Lithium polymer batteries can be employed to make the robot lightweight, which can decrease the amount of energy it requires.

References

1. Sharma, S., Singh, P., Singh, V., and Srivastava, P. K. (2020). Design and implementation of Agrobot by using IoT. *International Journal of Engineering Applied Sciences and Technology, 5*(1), 319325.
2. Mutharasu, S., Divya, V., Dhivya Bharathi, K., Ellakiya, M. V., and Janani, E. (2019). Design and implementation of agrobot by using IoT. *International Journal Advanced Research, Ideas and Innovations in Technology, 5*(2).
3. Kauser, N., Banu, S., and Yuvaraja, T. (2018). IoT based agricultural application using robota — Grobot. *Journal of Computational and Theoretical Nanoscience, 15*(6), 2137–2139.
4. Palacin, J., Salse, J. A., Valganon, I., and Clua, X. (2004). Building a mobile robot for a floor-cleaning operation in domestic environments. In IEEE Transactions on Instrumentation and Measurement, *53*(5), pp. 1418–1424.
5. Bini, D., Pamela, D., and Prince, S. (2020). Machine Vision and Machine Learning for Intelligent Agrobots: A review. In *2020 5th International Conference on Devices, Circuits and Systems (ICDCS)*, Coimbatore, India, pp. 12–16.
6. Schofield, M. (1999). "Neither master nor slave..." A practical case study in the development and employment of cleaning robots. In *Proceedings of 1999 7th IEEE International Conference on Emerging Technologies and Factory Automation, ETFA'99,* (Cat. No.99TH8467), Barcelona, Spain, Vol. 2, pp. 1427–1434.
7. Kayacan, E., Kayacan, E., Ramon, H., Kaynak, O., and Saeys, W. (2014). Towards agrobots: Trajectory control of an autonomous tractor using type-2 fuzzy logic controllers. *IEEE/ASME Transactions on Mechatronics, 20*(1), 287–298.
8. Ragavi, B., Pavithra, L., Sandhiyadevi, P., Mohanapriya, G. K., and Harikirubha, S. (2020). Smart Agriculture with AI Sensor by Using Agrobot. *2020 Fourth International Conference on Computing Methodologies and Communication (ICCMC)*, Erode, India, pp. 1–4. IEEE.
9. Durmuş, H., Güneş, E. O., Kırcı, M., and Üstündağ, B. B. (2015). The design of general purpose autonomous agricultural mobile-robot: "AGROBOT". *2015 Fourth International Conference on Agro-Geoinformatics (Agro-geoinformatics)*, Istanbul, Turkey, pp. 49–53. IEEE.
10. Dharanika, T., Karthik, S. R., Vel, S. S., Vyaas, S., and Yogeshwaran, S. (2021). Automatic Leaf Disease Identification and Fertilizer Agrobot, *2021 7th International Conference on Advanced Computing and Communication*

Systems (ICACCS), IEEE, Coimbatore, India, March 19–20, 2021, pp. 1341–1348.

11. Vrochidou, E., Bazinas, C., Mavridou, E., Pachidis, T., Mamalis, S., Koundouras, S., Gkrimpizis, T., and Kaburlasos, V. G. (2022). Considerations for a multi-purpose Agrobot design toward automating skillful viticultural tasks: A study in Northern Greece Vineyards, *Proceedings of HAICTA 2022*, Athens, Greece, September 22–25, 2022.
12. Kitić, G., Krklješ, D., Panić, M., Petes, C., Birgermajer, S., and Crnojević, V. (2022). Agrobot Lala — An autonomous robotic system for real-time, in-field soil sampling, and analysis of nitrates. *Sensors, 22*(11), 4207.
13. Gupta, M. S., Babu Vemuri, V. P., Sai Kumar, A., Venkata Rambabu, G., and Ganesh, M. (2022). Field study of remote controlled Agrobot. *Materials Today: Proceedings, 64*, pp. 865–869.
14. Krishnan, R. S., Lakshmi Narayanan, K., Golden Julie, E., Boopesh Prashad, V. A., Marimuthu, K., and Sundararajan, S. (2022). Solar powered mobile controlled Agrobot. In *2022 Second International Conference on Artificial Intelligence and Smart Energy (ICAIS)*, Coimbatore, India, February 23–25, 2022, pp. 787–792. IEEE.
15. Sobh, A. S., Hussien, H., and El-Aty, A. (2022). Designing and developing of a renewable Agrobot to enhance harvest productivity and reduce water consumption. In *Recent Advances in Manufacturing Processes and Systems*, pp. 415–426. Springer: Singapore.
16. Bini, D., Pamela, D., Shamia, D., and Prince, S. (2022). Intelligent Agrobots for crop yield estimation using computer vision. *Computer Assisted Methods in Engineering and Science, 29*(1–2), 161–175.
17. Praseena, S., Sanjena, S., Thejaswini, S. M., and SenthamilSelvi. M. (2019). Sensor based Agrobot for sowing seeds. *International Research Journal of Engineering and Technology (IRJET)*, *6*, pp. 7729–7732.
18. Prajith, A. S., Nowfiya, B. S., Noushad, N., Subi, S., and Lal, D. (2020). Automatic agricultural robot — Agrobot. In *2020 IEEE Bangalore Humanitarian Technology Conference (B-HTC)*, Vijiyapur, India, pp. 1–5.
19. Peter, A. M., Priya Saju, P., and Anila, P. V. (2019). Agrobot: Farming machine for sowing and irrigation. *International Journal for Research in Engineering Application & Management (IJREAM)*, *5*(3), 15–129.
20. Jadhav, P. K., Deshmukh, S. S., and Khairnar, P. N. (2019). Survey paper on AgRo-bot autonomous robot. *International Research Journal of Engineering and Technology (IRJET), 6*(12), 434–441.
21. Apat, S. K., Mishra, J., Srujan Raju, K., and Padhy, N. (2022). A study on smart agriculture using various sensors and Agrobot: A case study. In *Smart Intelligent Computing and Applications*, Vol. 1, pp. 531–540. Springer: Singapore.

22. Moysiadis, V., Tsolakis, N., Katikaridis, D., Sørensen, C. G., Pearson, S., and Bochtis, D. (2020). Mobile robotics in agricultural operations: A narrative review on planning aspects. *Applied Sciences, 10*(10), 3453.
23. Shamshiri, R., Redmond, C. W., Hameed, I. A., Yule, I. J., Grift, T. E., Balasundram, S. K., Pitonakova, L., Ahmad, D., and Chowdhary, G. (2018). Research and development in agricultural robotics: A perspective of digital farming. *International Journal of Agricultural and Biological Engineering, 11*, pp. 1–14.
24. Vasconez, J. P., Kantor, G. A., and Auat Cheein, F. A. (2019). Human–robot interaction in agriculture: A survey and current challenges. *Biosystems Engineering, 179*, 35–48.
25. Fue, K. G., Porter, W. M., Barnes, E. M., and Rains, G. C. (2020). An extensive review of mobile agricultural robotics for field operations: Focus on cotton harvesting. *AgriEngineering, 2*, pp. 150–174.
26. Steward, B., Gai, J., and Tang, L. (2019). The use of agricultural robots in weed management and control. In *Robotics and Automation for Improving Agriculture*, pp. 161–186. Burleigh Dodds Science Publishing.
27. Gerhards, R., Sanchez, D. A., Hamouz, P., Peteinatos, G. G., Christensen, S., and Fernandez-Quintanilla, C. (2022). Advances in site-specific weed management in agriculture — A review. *Weed Research, 62*(2), 123–133.
28. Zhang, W., Miao, Z., Li, N., He, C., and Sun, T. (2022). Review of current robotic approaches for precision weed management. *Current Robotics Reports, 3*, 139–151.
29. Ghobadpour, A., Monsalve, G., Cardenas, A., and Mousazadeh, H. (2022). Off-road electric vehicles and autonomous robots in agricultural sector: Trends, challenges, and opportunities. *Vehicles, 4*(3), 843–864.
30. Chebrolu, N. (2021). Spatio-temporal registration techniques for agricultural robots. PhD dissertation, Universitäts-und Landesbibliothek, Bonn.
31. Dutta, A., Roy, S., Kreidl, O. P., and Bölöni, L. (2021). Multi-robot information gathering for precision agriculture: Current state, scope, and challenges. *IEEE Access, 9*, 161416–161430.
32. Oliveira, L. F. P., Moreira, A. P., and Silva, M. F. (2021). Advances in agriculture robotics: A state-of-the-art review and challenges ahead. *Robotics, 10*(2), 52.
33. Yan, Y., Zhang, B., Zhou, J., Zhang, Y., and Liu, X. (2022). Real-time localization and mapping utilizing multi-sensor fusion and visual–IMU–wheel odometry for agricultural robots in unstructured, dynamic and GPS-denied greenhouse environments. *Agronomy, 12*(8), 1740.
34. Chandra, R. and Collis, S. (2021). Digital agriculture for small-scale producers: Challenges and opportunities. *Communications of the ACM, 64*(12), 75–84.

35. Pedersen, S. M., Fountas, S., Sørensen, C. G., Van Evert, F. K., and Blackmore, B. S. (2017). Robotic seeding: Economic perspectives. In *Precision Agriculture: Technology and Economic Perspectives*, pp. 167–179. Springer: Cham.
36. Pedersen, S. M., Fountas, S., and Blackmore, S. (2008). Agricultural robots — Applications and economic perspectives. In *Service Robot Applications*, IntechOpen, pp. 369–382.
37. Bhardwaj, A., Kumar, P., Raj, B., and Anand, S. (2022). Design and performance optimization of doping-less vertical nanowire TFET using gate stack technique. *Journal of Electronic Materials (JEMS), 41*(7), 4005–4013.
38. Singh, J. and Raj, B. (2019). Tunnel current model of asymmetric MIM structure levying various image forces to analyze the characteristics of filamentary memristor. *Applied Physics A, 125*(3), 203.1 to 203.11.
39. Goyal, C., Ubhi, J. S., and Raj, B. (2019). Low leakage zero ground noise nanoscale full adder using source biasing technique. *Journal of Nanoelectronics and Optoelectronics, 14*, 360–370.
40. Singh, G., Sarin, R. K., and Raj, B. (2016). A novel robust exclusive-OR function implementation in QCA nanotechnology with energy dissipation analysis. *Journal of Computational Electronics, 15*(2), 455–465.
41. Wadhera, T., Kakkar, D., Wadhwa, G., and Raj, B. (2019). Recent advances and progress in development of the field effect transistor biosensor: A review. *Journal of Electronic Materials, 48*(12), 7635–7646.
42. Wadhwa, G., Kamboj, P., Raj, B. (2019). Design optimisation of junctionless TFET biosensor for high sensitivity. *Advances in Natural Sciences: Nanoscience and Nanotechnology, 10*(7), 045001.
43. Bansal, P. and Raj, B. (2016). Memristor modeling and analysis for linear dopant drift kinetics. *Journal of Nanoengineering and Nanomanufacturing, 6*, 1–7.
44. Singh, A., Khosla, M., and Raj, B. (2016). Circuit compatible model for electrostatic doped Schottky barrier CNTFET. *Journal of Electronic Materials, 45*(12), 4825–4835.
45. Ashima, D. V. and Raj, B. (2021). Performance analysis of charge plasma induced graded channel Si nanotube. *Journal of Engg. Research EMSME Special Issue*, 146–154.
46. Tomar, A. S., Magraiya, V. K., and Raj, B. (2013). Scaling of access and data transistor for high performance DRAM cell design. *Quantum Matter, 2*, 412–416.
47. Jain, N. and Raj, B. (2018). Parasitic capacitance and resistance model development and optimization of raised source/drain SOI FinFET structure for analog circuit applications. *Journal of Nanoelectronics and Optoelectronins, 13*, 531–539.

48. Singh, S., Vishvakarma, S. K., and Raj, B. (2018). Analytical modeling of split-gate junction-less transistor for a biosensor application. *Sensing and Bio-Sensing, 18*, 31–36.
49. Gopal, M. and Raj, B. (2013). Low power 8T SRAM cell design for high stability video applications. *ITSI Transaction on Electrical and Electronics Engineering, 1*(5), 91–97.
50. Raj, B., Mitra, J., Bihani, D. K., Rangharajan, V., Saxena, A. K., and Dasgupta, S. (2010). Analysis of noise margin, power and process variation for 32 nm FinFET based 6T SRAM cell. *Journal of Computer (JCP), 5*(6), 1–8.
51. Sharma, D., Mehra, R., and Raj, B. (2021). Comparative analysis of photovoltaic technologies for high efficiency solar cell design. *Superlattices and Microstructures, 153*, 106861.
52. Kaur, P., Buttar, A. S., and Raj, B. (2021). A comprehensive analysis of nanoscale transistor based biosensor: A review. *Indian Journal of Pure and Applied Physics, 59*, 304–318.
53. Yadav, D., Raj, B., and Raj, B. (2020). Design and simulation of low power microcontroller for IoT applications. *Journal of Sensor Letters, 18*, 401–409.
54. Singh, S. and Raj, B. (2020). A 2-D analytical surface potential and drain current modeling of double-gate vertical t-shaped tunnel FET. *Journal of Computational Electronics, 19*, 1154–1163.
55. Singh, J. and Raj, B. (2019). An accurate and generic window function for non-linear memristor model. *Journal of Computational Electronics, 18*(2), 640–647.
56. Kaur, M., Gupta, N., Kumar, S., Raj, B., and Singh, A. K. (2021). Comparative RF and crosstalk analysis of carbon based nano interconnects. *IET Circuits, Devices & Systems, 15*(6), 493–503.
57. Kandasamy, N., Ahmad, F., Ajitha, D., Raj, B., and Telagam, N. (2020). Quantum dot cellular automata based scan flip flop and boundary scan register. *IETE Journal of Research, 66*, 535–548.
58. Sharma, S. K., Raj, B., and Khosla, M. (2019). Enhanced photosensivity of highly spectrum selective cylindrical gate $In_{1-x}Ga_xAs$ nanowire MOSFET photodetector. *Modern Physics Letter-B, 33*(12), 1950144.
59. Singh, J. and Raj, B. (2019). Design and investigation of 7T2M NVSARM with enhanced stability and temperature impact on store/restore energy. *IEEE Transactions on Very Large Scale Integration Systems, 27*(6), 1322–1328.
60. Bhardwaj, A. K., Gupta, S., Raj, B., and Singh, A. (2019). Impact of double gate geometry on the performance of carbon nanotube field effect transistor structures for low power digital design. *Computational and Theoretical Nanoscience, 16*, 1813–1820.

61. Jain, N. and Raj, B. (2019). Thermal stability analysis and performance exploration of asymmetrical dual-k underlap spacer (ADKUS) SOI FinFET for security and privacy applications. *Indian Journal of Pure & Applied Physics (IJPAP)*, *57*, 352–360.
62. Singh, A., Khosla, M., and Raj, B. (2019). Design and analysis of dynamically configurable electrostatic doped carbon nanotube tunnel FET. *Microelectronics Journal*, *85*, 17–24.
63. Jain, N. and Raj, B. (2019). Dual-k spacer region variation at the drain side of asymmetric SOI FinFET structure: Performance analysis towards the analog/RF design applications. *Journal of Nanoelectronics and Optoelectronics*, *14*, 349–359.
64. Singh, J., Sharma, S., Raj, B., and Khosla, M. (2018). Analysis of barrier layer thickness on performance of $In_{1-x}Ga_xAs$ based gate stack cylindrical gate nanowire MOSFET. *JNO*, *13*, 1473–1477.
65. Jain, N. and Raj, B. (2018). Analysis and performance exploration of high-k SOI FinFETs over the conventional low-k SOI FinFET toward analog/RF design. *Journal of Semiconductors (JoS)*, *39*(12), 124002-1-7.
66. Goyal, C., Ubhi, J. S., and Raj, B. (2018). A reliable leakage reduction technique for approximate full adder with reduced ground bounce noise. *Journal of Mathematical Problems in Engineering*, *2018*, 16. Article ID 3501041.
67. Anuradha, J. S., Raj, B., and Khosla, M. (2018). Design and performance analysis of nano-scale memristor-based nonvolatile SRAM. *Journal of Sensor Letter*, *16*, 798–805.
68. Wadhwa, G. and Raj, B. (2018). Parametric variation analysis of charge-plasma-based dielectric modulated JLTFET for biosensor application. *IEEE Sensor Journal*, *18*(15), 6070–6077.
69. Singh, J. and Raj, B. (2018). Comparative analysis of memristor models for memories design. *JoS*, *39*(7), 074006-1-12.
70. Yadav, D., Chouhan, S. S., Vishvakarma, S. K., and Raj, B. (2018). Application specific microcontroller design for IoT based WSN. *Sensor Letter*, *16*, 374–385.
71. Singh, G., Sarin, R. K., and Raj, B. (2018). Fault-tolerant design and analysis of quantum-dot cellular automata based circuits. *IEEE/IET Circuits, Devices & Systems*, *12*, 638–64.
72. Singh, J. and Raj, B. (2018). Modeling of mean barrier height levying various image forces of metal insulator metal structure to enhance the performance of conductive filament based memristor model. *IEEE Nanotechnology*, *17*(2), 268–267.
73. Jain, A., Sharma, S., and Raj, B. (2018). Analysis of triple metal surrounding gate (TM-SG) III–V nanowire MOSFET for photosensing application. *Opto-electronics Journal*, *26*(2), 141–148.

74. Jain, A., Sharma, S., and Raj, B. (2016). Design and analysis of high sensitivity photosensor using cylindrical surrounding gate MOSFET for low power sensor applications. *Engineering Science and Technology*, *19*(4), 1864–1870.
75. Singh, A., Khosla, M., and Raj, B. (2017). Analysis of electrostatic doped Schottky barrier carbon nanotube FET for low power applications. *Journal of Materials Science: Materials in Electronics*, *28*, 1762–1768.
76. Kumar, G. S., Singh, A., and Raj, B. (2018). Design and analysis of gate all around CNTFET based SRAM cell design. *Journal of Computational Electronics*, *17*(1), 138–145.
77. Singh, G. P., Sohi, B. S., and Raj, B. (2017). Material properties analysis of graphene base transistor (GBT) for VLSI analog circuits. *Indian Journal of Pure & Applied Physics (IJPAP)*, *55*, 896–902.
78. Singh, A., Khosla, M., and Raj, B. (2016). Comparative analysis of carbon nanotube field effect transistor and nanowire transistor for low power circuit design. *Journal of Nanoelectronics and Optoelectronics*, *11*, 388–393.
79. Kumar, S. and Raj, B. (2017). Estimation of stability and performance metric for inward access transistor based 6T SRAM cell design using n-type/p-type DMDG-GDOV TFET. *IEEE VLSI Circuits and Systems Letter*, *3*(2), 25–39.
80. Sharma, S., Kumar, A., Pattanaik, M., and Raj, B. (2013). Forward body biased multimode multi-threshold CMOS technique for ground bounce noise reduction in static CMOS adders. *International Journal of Information and Electronics Engineering*, *3*(3), 567–572.
81. Singh, H., Kumar, P., and Raj, B. (2013). Performance analysis of majority gate SET based 1-bit full adder. *International Journal of Computer and Communication Engineering (IJCCE)*, *2*(4).
82. Bhardwaj, A. K., Gupta, S., and Raj, B. (2020). Investigation of parameters for Schottky barrier (SB) height for Schottky barrier based carbon nanotube field effect transistor device. *Journal of Nanoelectronics and Optoelectronics*, *15*, 783–791.
83. Bansal, P. and Raj, B. (2017). Memristor: A versatile nonlinear model for dopant drift and boundary issues. *JCTN*, *14*(5), 2319–2325.
84. Jain, N. and Raj, B. (2016). An analog and digital design perspective comprehensive approach on fin-FET (fin-field effect transistor) technology — A review. *Reviews in Advanced Sciences and Engineering (RASE)*, *5*, 1–14.
85. Sharma, S., Raj, B., and Khosla, M. (2017). Subthreshold performance of $In_{1-x}Ga_xAs$ based dual metal with gate stack cylindrical/surrounding gate nanowire MOSFET for low power analog applications. *Journal of Nanoelectronics and Optoelectronics*, *12*, 171–176.

86. Raj, B., Saxena, A. K., and Dasgupta, S. (2009). Analytical modeling for the estimation of leakage current and subthreshold swing factor of nanoscale double gate FinFET device. *Microelectronics International*, *26*, 53–63.
87. Soniya, S. S., Wadhwa, G., and Raj, B. (2019). An analytical modeling for dual source vertical tunnel field effect transistor. *International Journal of Recent Technology and Engineering (IJRTE)*, *8*(2), 603–608.
88. Singh, S. and Raj, B. (2019). Design and analysis of hetrojunction vertical T-shaped Tunnel field effect transistor. *Journal of Electronics Material*, *48*(10), 6253–6260.
89. Goyal, C., Ubhi, J. S., and Raj, B. (2019). A low leakage CNTFET based inexact full adder for low power image processing applications. *International Journal of Circuit Theory and Applications*, *47*(9), 1446–1458.
90. Raj, B., Saxena, A. K., and Dasgupta, S. (2008). A compact drain current and threshold voltage quantum mechanical analytical modeling for FinFETs. *Journal of Nanoelectronics and Optoelectronics (JNO)*, *3*(2), 163–170.
91. Wadhwa, G. and Raj, B. (2020). An analytical modeling of charge plasma based tunnel field effect transistor with impacts of gate underlap region. *Superlattices and Microstructures*, *142*, 106512.
92. Singh, S. and Raj, B. (2020). Modeling and simulation analysis of SiGe hetrojunction double gatevertical t-shaped tunnel FET. *Superlattices and Microstructures*, *142*, 106496.
93. Singh, A., Saini, D. K., Agarwal, D., Aggarwal, S., Khosla, M., and Raj, B. (2016). Modeling and simulation of carbon nanotube field effect transistor and its circuit application. *Journal of Semiconductors (JoS)*, *37*, 074001-6.
94. Jain, N. and Raj, B. (2016). Device and circuit co-design perspective comprehensive approach on FinFET technology — A review. *Journal of Electron Devices*, *23*(1), 1890–1901.
95. Kumar, S. and Raj, B. (2016). Analysis of I_{ON} and ambipolar current for dual-material gate-drain overlapped DG-TFET. *Journal of Nanoelectronics and Optoelectronics*, *11*, 323–333.
96. Anjum, N., Bali, T., and Raj, B. (2013). Design and simulation of handwritten multiscript character recognition. *International Journal of Advanced Research in Computer and Communication Engineering*, *2*(7), 2544–2549.
97. Sharma, S., Raj, B., and Khosla, M. (2016). A Gaussian approach for analytical subthreshold current model of cylindrical nanowire FET with quantum mechanical effects. *Microelectronics Journal*, *53*, 65–72.

98. Singh, K. and Raj, B. (2015). Performance and analysis of temperature dependent multi-walled carbon nanotubes as global interconnects at different technology nodes. *Journal of Computational Electronics*, *14*(2), 469–476.
99. Kumar, S. and Raj, B. (2015). Compact channel potential analytical Modeling of DG-TFET based on evanescent–mode approach. *Journal of Computational Electronics*, *14*(2), 820–827.
100. Singh, K. and Raj, B. (2015). Temperature dependent modeling and performance evaluation of multi-walled CNT and single-walled CNT as global interconnects. *Journal of Electronic Materials*, *44*(12), 4825–4835.
101. Sharma, V. K., Pattanaik, M., and Raj, B. (2014). INDEP approach for leakage reduction in nanoscale CMOS circuits. *International Journal of Electronics*, *102*(2), 200–215.
102. Singh, K. and Raj, B. (2015). Influence of temperature on MWCNT bundle, SWCNT bundle and copper interconnects for nanoscaled technology nodes. *Journal of Materials Science: Materials in Electronics*, *26*(8), 6134–6142.
103. Anjum, N., Bali, T., and Raj, B. (2013). Design and simulation of handwritten Gurumukhi and Devanagri numerical recognition. *International Journal of Computer Applications*, *73*(12), 16–21.
104. Khandelwal, S., Gupta, V., Raj, B., and Gupta, R. D. (2015). Process variability aware low leakage reliable nano scale DG-FinFET SRAM cell design technique. *Journal of Nanoelectronics and Optoelectronics*, *10*(6), 810–817.
105. Sharma, V. K., Pattanaik, M., and Raj, B. (2014). ONOFIC approach: Low power high speed nanoscale VLSI circuits design. *International Journal of Electronics*, *101*(1), 61–73.
106. Khandelwal, S., Raj, B., and Gupta, R. D. (2015). FinFET based 6T SRAM cell design: analysis of performance metric, process variation and temperature effect. *Journal of Computational and Theoretical Nanoscience*, *12*, 2500–2506.
107. Singh, S., Yadav, S., Rahul, J., Srivastava, A., and Raj, B. (2015). Impact of HfO_2 in graded channel dual insulator double gate MOSFET. *Journal of Computational and Theoretical Nanoscience*, *12*(6), 950–953.
108. Sharma, V. K., Pattanaik, M., and Raj, B. (2014). PVT variations aware low leakage INDEP approach for nanoscale CMOS circuits. *Microelectronics Reliability*, *54*, 90–99.
109. Raj, B., Saxena, A. K., and Dasgupta, S. (2013). Quantum mechanical analytical modeling of nanoscale DG FinFET: Evaluation of potential, threshold voltage and source/drain resistance. *Journal of Material Science in Semiconductor Processing*, *16*(4), 1131–1137.

110. Gopal, M., Prasad, S. S. D., and Raj, B. (2013). 8T SRAM cell design for dynamic and leakage power reduction. *International Journal of Computer Applications*, *71*(9), 43–48.
111. Pattanaik, M., Raj, B., Sharma, S., and Kumar, A. (2012). Diode based trimode multi-threshold CMOS technique for ground bounce noise reduction in static CMOS adders. *Advanced Materials Research*, *548*, 885–889.
112. Raj, B., Saxena, A. K., and Dasgupta, S. (2011). Nanoscale FinFET based SRAM cell design: Analysis of performance metric, process variation, underlapped FinFET and temperature effect. *IEEE Circuits and System Magazine*, *11*(2), 38–50.
113. Sharma, V. K., Pattanaik, M., and Raj, B. (2013). Leakage current ONOFIC approach for deep submicron VLSI circuit design. *International Journal of Electrical, Computer, Electronics and Communication Engineering*, *7*(4), 239–244.
114. Chawla, T., Khosla, M., and Raj, B. (2021). Design and simulation of triple metal double-gate germanium on insulator vertical tunnel field effect transistor. *Microelectronics Journal*, *114*, 105125.
115. Kaur, P., Gill, S. S., and Raj, B. (2022). Comparative analysis of OFETs materials and devices for sensor applications. *Journal of Silicon*, 14, 4463–4471.
116. Sharma, S. K., Kumar, P., Raj, B., and Raj, B. (2022). $In_{1-x}Ga_x$ as double metal gate-stacking cylindrical nanowire MOSFET for highly sensitive photo detector. *Journal of Silicon*, *14*, 3535–3541.
117. Raj, B., Saxena, A. K., and Dasgupta, S. (2009). Analytical modeling of quasi planar nanoscale double gate FinFET with source/drain resistance and field dependent carrier mobility: A quantum mechanical study. *Journal of Computer (JCP)*, *4*(9), 1–8.
118. Bhushan, S., Khandelwal, S., and Raj, B. (2012). Analyzing different mode FinFET based memory cell at different power supply for leakage reduction. In *Seventh International Conference on Bio-Inspired Computing: Theories and Application, (BIC-TA 2012) Advances in Intelligent Systems and Computing*, Vol. 202, pp. 89–100.
119. Singh, J. and Raj, B. (2018). Temperature dependent analytical modeling and simulations of nanoscale memristor. *Engineering Science and Technology*, *21*, 862–868.
120. Singh, S., Bala, S., Raj, B., and Raj, B. (2020). Improved sensitivity of dielectric modulated junctionless transistor for nanoscale biosensor design. *Sensor Letter*, *18*, 328–333.
121. Kumar, V., Vishvakarma, S. K., and Raj, B. (2020). Design and performance analysis of ASIC for IoT applications. *Sensor Letter*, *18*, 31–38.
122. Jaiswal, A., Sarin, R. K., Raj, B., and Sukhija, S. (2018). A novel circular slotted microstrip-fed patch antenna with three triangle shape defected ground structure for multiband applications. *Advanced Electromagnetic (AEM)*, *7*(3), 56–63.

123. Wadhwa, G. and Raj, B. (2018). Label free detection of biomolecules using charge-plasma-based gate underlap dielectric modulated junctionless TFET. *Journal of Electronic Materials (JEMS)*, *47*(8), 4683–4693.
124. Singh, G., Sarin, R. K., and Raj, B. (2017). Design and performance analysis of a new efficient coplanar quantum-dot cellular automata adder. *Indian Journal of Pure & Applied Physics (IJPAP)*, *55*, 97–103.
125. Singh, A., Khosla, M., and Raj, B. (2017). Design and analysis of electrostatic doped Schottky barrier CNTFET based low power SRAM. *International Journal of Electronics and Communications, (AEÜ)*, *80*, 67–72.
126. Kaur, P., Pandey, V., and Raj, B. (2020). Comparative study of efficient design, control and monitoring of solar power using IoT. *Sensor Letter*, *18*, 419–426.
127. Bhardwaj, A. K., Gupta, S., and Raj, B. (2020). Development & analysis of compact model for double gate Schottky barrier CNTFET. *Journal of Nanoelectronics and Optoelectronics*, *15*, 1199–1208.
128. Wadhwa, G., Kamboj, P., Singh, J., and Raj, B. (2020). Design and investigation of junctionless DGTFET for biological molecule recognition. *Transactions on Electrical and Electronic Materials*, *22*, 281–289.
129. Chawla, T., Khosla, M., and Raj, B. (2020). Optimization of double-gate dual material GeOI-vertical TFET for VLSI circuit design. *IEEE VLSI Circuits and Systems Letter*, *6*(2), 13–25.
130. Verma, S. K., Singh, S., Wadhwa, G., and Raj, B. (2020). Detection of biomolecules using charge-plasma based gate underlap dielectric modulated dopingless TFET. *Transactions on Electrical and Electronic Materials (TEEM)*, *21*, 528–535.
131. Jain, N. and Raj, B. (2017). Impact of underlap spacer region variation on Electrostatic and analog/RF performance of symmetrical high-k SOI FinFET at 20 nm channel length. *Journal of Semiconductors (JoS)*, *38*(12), 122002.
132. Singh, S. and Raj, B. (2020). Analytical modeling and simulation analysis of T-shaped III–V heterojunction vertical T-FET. *Superlattices and Microstructures*, *147*, 106717.
133. Singh, G., Sarin, R. K., and Raj, B. (2017). Design and analysis of area efficient QCA based reversible logic gates. *Journal of Microprocessors and Microsystems*, *52*, 59–68.
134. Singh, A., Khosla, M., and Raj, B. (2016). Compact model for ballistic single wall CNTFET under quantum capacitance limit. *Journal of Semiconductors (JoS)*, *37*, 104001–8.
135. Singh, S., Khosla, M., Wadhwa, G., and Raj, B. (2021). Design and analysis of double-gate junctionless vertical TFET for gas sensing applications. *Applied Physics A*, *127*(16), 1–8.
136. Singh, I., Raj, B., Khosla, M., and Kaushik, B. K. (2020). Potential MRAM. *Technologies for Low Power SoCs*, *10*(4), 2050027.

137. Singh, S. and Raj, B. (2020). Parametric variation analysis on hetero-junction Vertical t-shape TFET for Supressing ambipolar conduction. *Indian Journal of Pure and Applied Physics*, *58*, 478–485.
138. Soniya, S. S., Wadhwa, G., and Raj, B. (2019). Design and analysis of dual source vertical tunnel field effect transistor for high performance. *Transactions on Electrical and Electronics Materials*, *21*, 74–82.
139. Kaur, M., Gupta, N., Kumar, S., Raj, B., and Singh, A. K. (2020). RF performance analysis of intercalated graphene nanoribbon based global level interconnects. *Journal of Computational Electronics*, *19*, 1002–1013.
140. Wadhwa, G. and Raj, B. (2019). Design and performance analysis of junctionless TFET biosensor for high sensitivity. *IEEE Nanotechnology*, *18*, 567–574.
141. Singh, J. and Raj, B. (2019). Enhanced nonlinear memristor model encapsulating stochastic dopant drift. *JNO*, *14*, 958–963.
142. Nedjah, N. (2022). *Robotics and AI for Cybersecurity and Critical Infrastructure in Smart Cities.* A. A. Abd El-Latif, B. B. Gupta, and L. de Macedo Mourelle (eds.). Springer, Singapore.
143. Gupta, B. B. and Nedjah, N. (eds.). (2020). *Safety, Security, and Reliability of Robotic Systems: Algorithms, Applications, and Technologies*. CRC Press, Boca Raton.
144. AlZu'bi, S., Hawashin, B., Mujahed, M., Jararweh, Y., and Gupta, B. B. (2019). An efficient employment of internet of multimedia things in smart and future agriculture. *Multimedia Tools and Applications*, *78*, 29581–29605.
145. Tewari, A. and Gupta, B. B. (2020). Secure timestamp-based mutual authentication protocol for IoT devices using RFID tags. *International Journal on Semantic Web and Information Systems (IJSWIS)*, *16*(3), 20–34.

https://doi.org/10.1142/9789811279539_bmatter

Index

www.ingramcontent.com/pod-product-compliance
Lightning Source LLC
LaVergne TN
LVHW010328220926
848484LV00004BB/856